Economic Geography

Critical Introductions to Geography

Critical Introductions to Geography is a series of textbooks for undergraduate courses covering the key geographical sub-disciplines and providing broad and introductory treatment with a critical edge. They are designed for the North American and international market and take a lively and engaging approach with a distinct geographical voice that distinguishes them from more traditional and outdated texts.

Prospective authors interested in the series should contact the series editor:

John Paul Jones III
School of Geography and Development
University of Arizona
jpjones@email.arizona.edu

Published

Cultural Geography
Don Mitchell

Geographies of Globalization
Andrew Herod

Geographies of Media and Communication
Paul C. Adams

Social Geography
Vincent J. Del Casino Jr

Mapping
Jeremy W. Crampton

Research Methods in Geography
Basil Gomez and John Paul Jones III

Political Ecology, Second Edition
Paul Robbins

Geographic Thought
Tim Cresswell

Environment and Society, Second Edition
Paul Robbins, Sarah Moore, and John Hintz

Urban Geography
Andrew E.G. Jonas, Eugene McCann, and Mary Thomas

Health Geographies: A Critical Introduction
By the right of Tim Brown, Gavin J. Andrews, Steven Cummins, Beth Greenhough, Daniel Lewis, and Andrew Power

Economic Geography

A Critical Introduction

Trevor J. Barnes

Brett Christophers

WILEY Blackwell

This edition first published 2018
© 2018 John Wiley & Sons Ltd

The right of Trevor J. Barnes and Brett Christophers to be identified as the authors of this work has been asserted in accordance with law.

Registered Office(s)
John Wiley & Sons, Inc., 111 River Street, Hoboken, NJ 07030, USA
John Wiley & Sons Ltd, The Atrium, Southern Gate, Chichester, West Sussex, PO19 8SQ, UK

Editorial Office
9600 Garsington Road, Oxford, OX4 2DQ, UK

For details of our global editorial offices, customer services, and more information about Wiley products visit us at www.wiley.com.

Wiley also publishes its books in a variety of electronic formats and by print-on-demand. Some content that appears in standard print versions of this book may not be available in other formats.

Library of Congress Cataloging-in-Publication Data

Names: Barnes, Trevor J., author. | Christophers, Brett, 1971– author.
Title: Economic geography : a critical introduction / by Trevor Barnes and Brett Christophers.
Description: Hoboken, NJ : John Wiley & Sons, 2018. | Series: Critical Introductions
    to Geography | Includes bibliographical references and index. |
Identifiers: LCCN 2017030688 (print) | LCCN 2017042733 (ebook) | ISBN 9781118874301 (pdf) |
    ISBN 9781118874288 (epub) | ISBN 9781118874332 (cloth) | ISBN 9781118874325 (pbk.)
Subjects: LCSH: Economic geography.
Classification: LCC HF1025 (ebook) | LCC HF1025 .B3328 2018 (print) | DDC 330.9–dc23
LC record available at https://lccn.loc.gov/2017030688

Cover image: © World History Archive / Alamy Stock Photo
Cover design by Wiley

Set in 10/12.5pt Minion by SPi Global, Pondicherry, India
Printed and bound in Malaysia by Vivar Printing Sdn Bhd

1   2018

Contents

Acknowledgments

At Wiley-Blackwell, we would like to thank Justin Vaughan for his support for this project from the start and for his patience during the extended writing process. We also thank Wiley-Blackwell's production team that included Sarah Keegan, Liz Wingett, and Joe White for seeing the book through to fruition. We are enormously appreciative of the skills of the picture editor, Kitty Bocking, as well as the copy-editor, Giles Flitney, who often understood better what we meant than we did. We are further grateful to those who have read and commented on draft chapters for us – all the chapters, in the case of series editor J.P. Jones, and individual chapters in the case of Leigh Johnson, Jamie Peck, and Marion Werner. Needless to say, all errors of fact or interpretation are our responsibility. In addition, Brett Christophers would like to thank his family (Agneta, Elliot, Oliver, and Emilia) for allowing the book to encroach on two successive summers in Dalstuga. Trevor J. Barnes would like to thank Joan Seidl for dealing with the vexing issues around copyright permission for figures and photos, and for much, much more. He would also like to dedicate the book to the now thousands of undergraduate students who have taken his economic geography course at the University of British Columbia over the past 35 years.

Vancouver and Uppsala, July 2017

List of Figures

Chapter 1

Why Economic Geography Is Good For You

1.1 Introduction

One of the great, albeit controversial, episodes of the American National Public Radio show *This American Life* was "Mr Daisey and the Apple factory" (first aired January 6, 2012). It was at the time the most downloaded single episode of the show ever with close to 900,000 downloads.[1] Hosted by Ira Glass and produced by WBEZ Chicago, *This American Life* presents each week one or more stories based on some particular theme. Glass did not make explicit the theme for the week of "Mr Daisey and the Apple factory," but there was absolutely no doubt it was economic geography. The episode demonstrated both the fundamental importance of the subject, and, given the download figures, the potential public interest in the subject. It showed why economic geography was good for you, the theme of this introductory chapter.

Mr Daisey is a self-described technology geek. As he said on the show: "of all the kinds of technology that I love in the world, I love the technology that comes from Apple the most, because I am an Apple aficionado. I'm an Apple partisan. I'm an Apple fanboy. I'm a worshiper in the cult of Mac" (*This American Life* 2012). For fun and relaxation Mr Daisey will strip his MacBook Pro into its 43 component pieces, clean them with compressed air, and put them back together again. One day he was reading a website dedicated to Apple products, and came across a post from someone who, when they turned on their just-bought iPhone, found four pictures on it that had been taken at the factory where it was made. The pictures had obviously been taken to test the device but had not been erased.

Those pictures became for Mr Daisey his economic geographic "aha" moment. As he put it, "until I saw those pictures, it was only then I realized, I had never thought, ever, in a dedicated way, about how [Apple products] were made" (*This American Life* 2012). Mr Daisey did some preliminary inquiries and discovered that the pictures were taken in the Foxconn plant in Shenzhen in Southern China, just outside of Hong Kong. He continues:

> the most amazing thing is, almost no one in America knows its name. Isn't that remarkable? That there's a city where almost all of our crap comes from, and no one knows its name? I mean, we think we do know where our crap comes from. We're not ignorant. We think our crap comes from China. Right? Kind of a generalized way. China.
>   But it doesn't come from China. It comes from Shenzhen. It's a city. It's a place.
>
> (*This American Life* 2012)

In raising these questions, Mr Daisey was beginning to do economic geography. He was asking *where* something was produced, why it was produced *there and not somewhere else*, and how and why it *geographically moved* from that place to Mr Daisey's place, America. All of these issues are fundamentally economic geographic, requiring Mr Daisey to begin to acquire economic geographic knowledge. And like any aspiring economic geographer, after gathering preliminary facts at home, Mr Daisey went on to do fieldwork, to visit China, to go to Shenzhen, to the city, to the place. He began to interview people, visiting the Foxconn factory (or at least standing outside of it).

As he found out more about the lives of the people who made Apple products (which he had originally figured were "made by robots"), Mr Daisey was increasingly unsettled.[2] Economic geography turned out not to be as soothing as stripping down the 43 component parts of his MacBook Pro and cleaning them with compressed air. Still, Mr Daisey came to believe that knowing where all one's "crap" comes from, assembled from across the four corners of the world, and the conditions under which it is produced, was something that as a citizen of the world one ought to know. He began to appreciate why economic geography was good for him. We believe too that the discipline is good for you. Although we are less smitten with our computerized devices than Mr Daisey, we are smitten with economic geography. We are economic geography aficionados, economic geography partisans, and economic geography fanboys. There is no economic geography cult (as far as we know), but if there were, we would worship at its altar.

The purpose of this book is to try to persuade you to join us in our admiration of economic geography. We want to convince you that knowing something about the subject is essential to living in the twenty-first century. Virtually everyone now on the planet, together numbering a total of more than 7 billion people, is a global citizen whether they like it or not. The changes to the economy over the past 40 years, as it has become globalized, affecting so many people in so many different ways, are economic and also fundamentally geographical. Accordingly, economic geography is relevant not just as a background setting, but to understand why economic change occurs at all. It enters into the very frame of that change, its skeletal structure. And what is bred in the bone comes out in the marrow. We don't mean to exaggerate our claim. There are large swathes of life with which economic geography does not deal. But those with which it does are nut and bolt issues; they hold together the scaffolding on which much contemporary social and cultural life is constructed: from the clothes you wear, to the food you eat, to the music you hear, to the videos you watch, to the schools you attend, and even to the university classes you take. It is our belief that if you fail to understand that scaffolding, you fail to understand one of the bases of contemporary social and cultural life itself. You fail to understand your own life and the lives of others around you both far and near.

The rest of this chapter pursues these larger themes and in doing so introduces our book. The chapter is divided into three unequally sized sections. The first and longest, continuing from Mr Daisey, argues that there is something about the present juncture – right now – that makes the study of economic geography especially important and pressing. We present the argument in two parts. First, we set out the leading features of the present moment, and contend that they uncannily fit with the interests and conceptual framework of economic geography. It is the right discipline for right now and right here. Second, we suggest that suitability is a result of economic geography having remade itself as an intellectual project over the last 40 years. During that period, it has become increasingly open-minded, with an ability to roll with change, yet it still retains a memory of the past, possessing a broad, catholic base of knowledge, as well as an ability to synthesize and link across seemingly yawning subject divides. These qualities, along with its ingrained spatial awareness, have given economic geography enormous purchase in understanding the present.

The second section discusses the meaning and implications of the subtitle of our book, *A Critical Introduction*. We clarify what it means to think critically, and in particular, what it means to think critically about the discipline of economic geography. We adopt two strategies. The first is to try to get behind or beneath the discipline, and not to accept it at face value. It involves asking why things are done as they are, as well as what decisions were taken in the past that now determine the economic geographic knowledge we listen to in lecture halls and read in journals and books like this one. This also frequently necessitates examining internal disciplinary processes and external contextual historical factors. The second is to provide critical evaluations of different types of economic geography. This means *not* sitting on the fence. Instead, we try, where appropriate, to take a position, to say what we think is working well in the discipline (and why) and what is not.

The last section of this chapter sets out the structure and argument of the rest of the book, which is divided into two main parts. The first is concerned with economic geography as a discipline, the second with economic geography's examination of a changing economic geographic world. In both cases, we aspire to live up to our book's subtitle and to be critical in both senses just described.

1.2   "May You Live in Interesting Times": Economic Geography's World

1.2.1   Interesting times

"May you live in interesting times" is often termed "the Chinese curse." There is no evidence that this expression was Chinese, or that it was a curse, however. Contemporary economic geographers certainly live in interesting times, though. Further, those times have been propitious for the discipline rather than a curse, raising its stature and profile. Here are six features that have made the times interesting for economic geography, bearing on the subject, lending themselves to its disciplinary analysis.

- First, and going directly to Mr Daisey and the Apple factory, is pervasive globalization, a fundamentally economically geographic phenomenon. Economic geography inheres in globalization's very definition; it is part of the term's conceptual furniture. By globalization we mean the ever-increasing economic geographic integration of the world as measured by the movements across national borders of: (i) goods, services, and capital; (ii) labor (people); (iii) knowledge and information (communication); and (iv) cultural goods and activities – sports, cuisines, electronic games, films, music, TV shows, and so on.

  Globalization goes back a long way, of course. The Ancient Greek poet Homer (that is, if he ever existed!) was writing about globalization as early as around 800 BCE in his epic poem *The Odyssey*. It featured the soldier and sailor Odysseus and his spectacular global travels (at least global for 800 BCE), involving murderous Cyclops, enchanting and beautiful but deadly temptress Sirens, and evil sea monsters Scylla and Charybdis (Figure 1.1). Globalization is enormously long-standing, but over the last 30 years or so there has been a quantitative shift in its pace and range, and a qualitative shift in its form. Since 1980 there has been a tenfold growth in the value of world trade in goods and services (US$2.4 trillion in 1980 versus US$23.3 trillion in 2013 in constant dollars); a more than twenty fold increase in the value of the stock of global foreign direct investment (a little over US$1 trillion in 1980 and more than US$22 trillion in 2013); a more than doubling in the

Figure 1.1 Globalization ancient style? Odysseus is strapped to the mast to prevent him from doing something he shouldn't when hearing the Sirens' bewitching but mortally dangerous songs. By plugging their ears with melted beeswax his crew were protected (from a Greek vase, 500–480 BCE, British Museum).

number of foreign workers now employed across the world (currently about 3% of the entire world's population); a fiftyfold increase between 1980 and 2008 in the number of minutes that US telephone subscribers spend annually making international calls; and a substantial increase in overseas compared to domestic box office receipts for Hollywood films – in 1980 the bulk of box office receipts for Hollywood films came from the US domestic market, whereas in 2014 that had reversed with two thirds of receipts coming from abroad. Things are clearly not the same as they once were. It is not "déjà vu all over again," and this is what makes times so interesting for economic geographers. In view of its significance, and its inherently economic geographic nature, we devote a whole chapter (Chapter 8) to globalization and the economic geographic study of it.

- A second notable feature of our times is a communications revolution that began in the 1960s. It has been a key factor in remaking contemporary economic geographic relations by permitting almost costless, instantaneous communication, aural and visual, between parties scattered across the four corners of the world. That revolution was inaugurated by combining computerization with existing electronic forms of communication. The origins of the latter went back to the early nineteenth century with the beginning of the telegraph. The computer was much more recent, originally developed to meet military needs during World War II (to decipher Nazi coded messages). Further refined during the Cold War, the computer became a backbone of the US nuclear defense system. By the mid-1950s, it started to move beyond its initial military setting, increasingly used by corporate business, becoming the basis for new forms of communication by voice, text, and image. At first new telecommunication technologies were very expensive, with older forms of communication continuing to exist alongside them. Even in the late 1970s, telegrams – small squares of yellow paper on which were glued ribbons of typed words – were still used to transmit urgent information. In the scheme of things, they were fast and cheap. It wasn't until the mid-1990s, after email became generally available, that the telegram began to become redundant, finally being confined to history a decade or so later with the establishment of text messaging, Skype, and a myriad of social media platforms.

- A third key feature turns on technological improvements in physical transportation, and is bound up with larger, faster, and more efficient vehicles (planes, trucks, boats, and trains). In combination they colossally reduced freight costs, facilitating the current cornucopia of cheap commodities from around the world. Transportation costs have fallen so much that the economists Edward Glaeser and Janet Kohlhase (2004, p.200) write that it is now "better to assume that moving goods is essentially costless [rather] than to assume that moving goods is an important component of the production process." The lower costs are primarily the result of economies of scale, that is, a declining cost per unit brought about by large-scale transportation utilizing ever-more capacious vehicles. Planes have gotten larger (the triple decker Airbus A380 accommodates 853 passengers); trains have gotten longer (one in Australia stretches 7.3km, consisting of 683 separate rail cars); trucks have gotten bigger (a North American double trailer articulated truck can carry 80,000kg); and cargo boats have gotten lengthier, taller, and vaster (the largest cargo vessel in the world, *CSCL Globe*, is 400m long, 59m high, and carries 19,000 containers, that is, enough room to transport 108 million pairs of sneakers). The rise of containerization has been especially important. Container shipping has become "the great hidden wonder of the world, a vastly underrated business…. It has shrunk the

Figure 1.2   The *Ideal-X* loading the first ever containers ("the box") at the Newark, NJ, port for their maiden voyage to Houston, Texas, April 26, 1956. Source: The Port Authority of New York and New Jersey.

planet and brought about a revolution because the cost of shipping boxes is [now] so cheap" (Shah 2000). That revolution can be precisely dated. On April 26, 1956, 58 metal boxes were loaded by crane on to a converted oil tanker, the *Ideal-X*, moored in Newark, New Jersey (Figure 1.2). It arrived in Houston, Texas, five days later. The boxes were taken off the vessel by crane, hitched to waiting truck cabs, and taken to their destinations. This was the beginning of the container revolution, which, as Mark Levinson (2006) writes, "made the world smaller, and the world economy bigger." By 2002, for example, as a result in large part of "the box," American consumers had four times the variety of goods to select from compared to 30 years before.

- A fourth feature goes to the break-up sometime during the 1970s and 1980s of the previous socioeconomic system that held in the Global North, called Fordism (Box 1.1: Fordism and the Keynesian Welfare State). Fordism was the manufacturing system of mass production initially pioneered by Henry Ford at his Dearborn, Michigan, automobile plant from the first decade of the twentieth century. The technique of mass production that Ford forged was later transferred to other industrial sectors, and adopted also by other industrial countries such as the United Kingdom, Germany, France, and Italy (it was in fact an Italian Marxist, Antonio Gramsci, who in 1934 coined the term "Fordism"; Forgacs 2000, pp.275–300).

    After World War II, Fordist manufacturing also became linked to a particular form of government in Global Northern manufacturing countries, known as the Keynesian Welfare State (KWS). The KWS ensured that Fordism was provided with everything that it needed for continued production, such as energy, roads, and even labor ("produced" in this case by state-owned schools and colleges). At the same time, the government guaranteed that people had sufficient income to buy the manufactured goods that Fordism was so good at producing (which meant among other things providing income to people who would otherwise not receive it, such as retirees, the unemployed, and those unable to work). The important point for our purposes is that

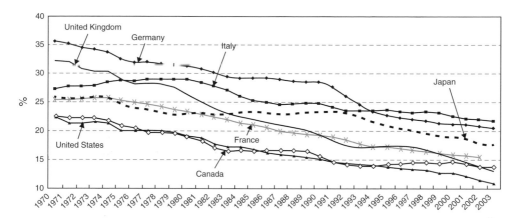

Figure 1.3 Deindustrialization in the Global North. The decline in the percentage of employment in manufacturing in G7 countries over the period 1970–2003. Source: Pilat, D. et al. 2006, p.6.

this system of Fordism and the KWS was organized nationally. Each country had its own particular combination of a KWS and a system of Fordist manufacturing, and these combinations produced different levels of success. While Scandinavian countries were generally high-fliers, the United Kingdom was a perennial laggard.

From the early 1970s, however, Fordism and the associated KWS ran into massive difficulties. Industrial productivity fell, labor costs rose precipitously, and consequently profits were squeezed. Something had to give. In this case, a number of manufacturers in the Global North began moving their operations to the Global South, drawn in large part by low wages and the two technological processes of the communications and transportation revolutions already discussed. As this process gained momentum, the former national-based system of Fordism and the KWS was increasingly undone in the Global North. Tens of millions of manufacturing jobs were lost in a process called "deindustrialization," turning former manufacturing regions into rust belts. The share of manufacturing employment in seven major industrialized countries in the Global North fell sharply over the period 1970–2003 (Figure 1.3). In the United Kingdom, manufacturing employment fell from a third of all jobs to a little over a tenth. Certain regions that had specialized in manufacturing, like the Upper Mid-West and North East of the United States, and the northern part of Britain, were decimated, hollowed out, becoming shadows of their former selves. At the same time, a new global system emerged, involving massive manufacturing investment within the Global South. Newly industrialized countries (NICs) emerged initially in Hong Kong, Singapore, Taiwan, and Korea, and then later in Brazil, Mexico, Thailand, Malaysia, and, of course, China. None of these places historically had undertaken large-scale industrial manufacturing, but under the new regime of globalization they would.

Germane here was that all of these processes – the establishment of different national Fordisms along with associated KWSs, their later decline in the form of industrial restructuring and deindustrialization, and the appearance of their mirror image in the rise of the NICs – were all to their core economic geographic. They were made for economic geography, perfect grist for the disciplinary mill.

Box 1.1   Fordism and the Keynesian Welfare State

Henry Ford with his Model T. Source: Getty Images.

Henry Ford (1863–1947) got the ideas for production line mass manufacturing from visiting a meat packing plant in Michigan at the beginning of the 20th century. There, live animals would enter the factory at one end and, by means of conveyor belts and hooks moving overhead, they would become progressively "disassembled" as they journeyed through the plant. By the end, they left the factory gates as T-bone steaks, hamburger patties, and breakfast sausages. Ford's genius was to take that process but reverse it. Instead of disassembling, he assembled. Raw inputs would arrive at one end of his factory – coal, iron ore, rubber, cloth – and over the course of their journey through the plant they would be "assembled," in this case, put together most famously as a Model T car that would roll off at the other end of the production line. The iconic plant where this happened was the River Rouge Complex on the Detroit Red River (Figure 1.4). The main production plant (NE quadrant in the photo, by the billowing smoke) was over 3km long, 1km wide, and at its height employed 80,000 workers. Production line labor worked robotically, standing in line along a conveyor belt, undertaking the same repetitive task over and over and over again (see Box 12.1: Taylorism).

Ford was notoriously politically conservative. He once said, "people are never so likely to be wrong than as when they are organized." He wouldn't have been pleased with the rise of the KWS, which was definitely left wing in its politics. Nonetheless, it joined with Fordism after World War II to produce hitherto unprecedented levels of general economic wealth, "the affluent society" as the American economist John Kenneth Galbraith (1958) called it, or, as the British Prime Minister Harold Macmillan told his electorate, "most of our people have never had it so good."

Figure 1.4 The Ford Rouge Plant, Dearborn, Michigan, 1927.

John Maynard Keynes. Source: Getty Images.

The Keynesian part of the title KWS came from John Maynard Keynes (1883–1946) (pronounced Canes), a Cambridge (UK) economist. Keynes was appalled by the devastation of the 1930s Great Depression. His book, *The General Theory of Employment, Interest and Money* (1936), offered a way out. The volume functioned as a how-to manual, identifying various economic levers to be pulled and buttons to be pushed to prevent the Great Depression occurring again. Keynes's great insight was that for the economy to be sustained, for it not to drop down into the dark pit of recession, general (macro) consumption levels must be maintained. That meant everyone must consume whether they had a job or not. Hence, the rise of the Welfare State. It provided income to people who otherwise would have none, ensuring there were always eager-beaver consumers able to buy goods from Fordism's bounty. Contra Henry Ford, though, Keynes also showed the importance of the role of state intervention in the economy.

- A fifth feature was a critical change in the nature of labor markets and work. Deindustrialization cut a grim swathe across Global Northern manufacturing regions from the 1970s, eliminating an enormous number of relatively high-paying, often unionized manufacturing jobs held typically by men. Such changed work practices became fodder even for popular culture. For example, the 1997 movie *The Full Monty*, later a Broadway and West End musical, featured a group of male Sheffield steel workers who, when they were made redundant, took off their boiler suits and steel-capped boots and entered what was traditionally a female service-sector occupation, exotic dancing. While that film may have been a fantasy, it pointed to a fundamental change around work and its culture. The Global North was moving toward a postindustrial, service-based economy, turning upside down the older Fordist nostrum of mass manufacturing. Employment now came in two main forms. There were McJobs – these were low-paid and required little by way of skills or experience, although there could be variation even in this general category. At the higher end was employment in establishments of conspicuous consumption such as boutique cafes or designer clothing stores. At the lower end was flipping burgers in fast-food restaurants or being an "associate" working the floor of a big-box discount retailer. The other type of service-sector employment was in the creative economy, or new economy, or cultural economy. Workers here were often highly paid, and the work frequently required skills that were formally credentialed. Work could be in existing sectors that had been reconstituted, such as advertising or higher education, or in entirely new sectors, such as video game or web design. Both kinds of service work – the low-end and the high-end – required embodied interactions with customers and with other workers, calling for cultural, social, and even emotional (soft) skills. Unlike in manufacturing, where one worker's body was much like another, simply the repository of brawn to be dispensed in the industrial process, in the service sector the body was often a vital part of the product sold. The kind of body made a difference, and how it was dressed, comported, spoke, and looked could be crucial.

   Again, in all of this, economic geography mattered: from the global redistribution of employment that produced a new international division of labor (see Chapter 8), to the emergence of specialized high-tech, creative industry regions such as, most famously, Silicon Valley in California (see Box 12.3: Silicon Valley), to the micro-spaces as varied as a derivatives trading room or a Starbucks café where work was actually performed.

- Finally, there is neoliberalism, which from 1980 increasingly replaced the old KWS. Neoliberalism is a form of government that relies on market-based incentives, that is, price signals, to achieve its ends. Its introduction was especially associated with deregulation, that is, dispensing with government rules and laws around the economy, and denationalization and privatization, that is, selling off government-owned economic assets and sectors to private firms. In both cases, the neoliberal justification was that the free market served the economy better than the central state did. The ground should therefore be cleared of state interference for the market's effective operation. The Chilean military dictator (and embarrassingly a geographer) Augusto Pinochet first systematically tried out neoliberalism in his country in 1973. Neoliberalism as a practice, though, became most associated with the administrations of US President Ronald Reagan (1980–88), and perhaps even more so with the UK Prime Minister Margaret Thatcher (1979–1990). Thatcher was the more zealous. She oversaw an enormous privatization of UK state-owned business (e.g., railways, telecommunications, water, coal, iron and steel, power). And she made it her personal quest to dismantle the KWS, which she called "the nanny state," and which she believed pampered and coddled rather than exposed people to the necessary tough love of enterprise and the market. For neoliberals like Reagan and Thatcher economic regulations imposed by countries only gummed up and retarded the market. Consequently, not only did individual countries need to be freed up for business, but so did the whole world. To that end, from the late 1980s international institutions like the World Bank and the International Monetary Fund (joined together in the so-called Washington Consensus) imposed neoliberalism particularly on countries of the Global South to ensure they toed the neoliberal line. Such institutions believed that an uncluttered globalization would be the fullest expression of the open market. When the market was found everywhere in the world, when there was a perfect economic geography of globalization, it would be "the end of history," as the neoliberal apologist Francis Fukuyama (1992) put it. Historical change would stop. Utopia was at hand. Consequently, it also implied the end of geography. Utopia, from the Greek, literally means no place.

It is likely the conceit of every new generation of economic geographers to believe that the period in which they entered the discipline is the most interesting. But it really seems so for economic geographers who have come of age since the 1980s. They – we – have witnessed a series of wrenching and profound geographical transformations in the economy. The results have been variable, from despair and trauma to energetic purposefulness and optimism. But however the changes are described, undeniably they signaled interesting times. Everything the discipline had previously claimed about the importance of geography came home to roost. The present moment is impossible to understand without an economic geographic sensibility. That's why we believe economic geography is good for you.

1.2.2 Interesting discipline

While the times may have favored economic geography, that did not necessarily mean the discipline would seize the opportunity. There were also other social sciences keen to conceptualize the present economic geographic moment (see Chapter 4). Would economic geography, always a bit shy and lacking confidence, up its game and assert its special contribution? We suggest it has. It has become an interesting discipline. Elements making up its structure as an academic subject have allowed economic geography to take on and say something important about the present moment.

The first such element is a recognition of, and stress on, geographical difference. That emphasis was there from the discipline's beginning. Doing geography meant understanding how and why one place, or region, or country, was different from another. On first blush, it might appear that recognizing geographical difference would not be so helpful when examining perhaps the most important force now shaping economic geography: globalization. Globalization is often presented as an inexorable, relentless power, leveling, crushing, and pulverizing geographical difference. Thomas Friedman (2005), the well-known *New York Times* columnist, is in that camp. To use his terms, "the world is flat." Globalization for Friedman is like a steamroller creating a smooth, homogeneous surface over any space that it passes. It produces a world like the one depicted in the cartoon in Figure 1.5, where everywhere is the same. Against that interpretation, however, economic geographers have kept to their intellectual roots, and emphasized instead continued geographical difference and variation. Accordingly, they've suggested that as globalization unfolds over space, it rubs up against, abrades, and is in friction with specific places and regions, ones defined by their unique combination of material and institutional forms. Globalization may be a powerful force, but it is never powerful enough to plane down and erase geographical difference altogether. The form that geographical difference takes may be changed in the interaction with globalization, but it is not eradicated. In turn, continuing difference makes a difference, transforming globalization into a patchwork of variegated local forms. The geographical world, rather than becoming one vast monolithic plain – a flatscape – remains spatially dappled and mottled (Chapter 8). As always, geography matters.

The second element is economic geography's open-mindedness toward theory and method, and hence its willingness to remake itself in light of changing circumstances. In contrast, other disciplines, like orthodox economics, for example, appear unchanging, set in aspic jelly, more or less pursuing the same theory and method since their institutionalized beginning. In the case of economics, the constant is the maximizing method based on the mathematical technique of constrained maximization that first entered the discipline nearly 150 years ago (Mirowski 1989). That's not economic geography, which has re-created itself several times over during its history (Chapter 3). In the current version of the last 30 years or so, increasingly anything goes. In particular, the discipline has been unflinching in its inclination to experiment. The nineteenth-century American essayist and poet Ralph Waldo Emerson (1911, p.302) once said, "All life is an experiment. The more experiments you make the better." That's economic geography. Only rarely over the course of its history has it believed that there were any holy cows to violate. You choose the theoretical framework that seems most appropriate for the (shifting) geographical process at hand rather than choosing one because it was always used. As a result, economic geographers might take their theoretical framework from science and technology studies, or from cultural theory, or from feminism, or from Marxism, or from institutionalism, or from orthodox economics (see Chapter 5). It might come in a pure form, drawing on a single tradition. Or it might be a mix and match, an eclectic style of theorization, a bricolage, a DIY approach to theory construction. Likewise, and discussed further in Chapter 6, the methods used to assemble the data to be explained are also increasingly varied and assorted. The data might be a set of statistics requiring numerical presentation and analysis; or they might be a set of interview transcripts to be textually interpreted; or they might be a set of ethnographic field notes literally taken on the job. One never knows until one determines the specific object of inquiry. That is the rub.

Figure 1.5 The death of geography under the steamroller of globalization? Or a death that is greatly exaggerated? Source: Louis Hellman.

Economic geography, as we show in the next chapter, has become increasingly open about the object it studies; its boundaries are ever-more porous and blurred, and thus requiring receptivity to an expanding range of different theories and methods. Again, there is a stark contrast with orthodox economics, which has a fixed definition of its object of study, circumscribed by a sharp border between what it is and what it isn't. That couldn't be further from contemporary economic geography. Its more open-ended approach, we believe, is better suited to keeping up with the evolving and multifaceted character of economic geography on the ground. The geography of the economy is inconstant, always on the move, morphing, bleeding together new combinations in new spaces: bodily, gendered, institutional, textual, performative, and virtual. But precisely because contemporary economic geography has no hard and fast confines, it can keep up. It is not fazed by the fact that "we're not in Kansas anymore." It is so much more exciting. We're in the whole world.

The third element, and again another longstanding feature of economic geography, concerns its catholic and synthetic inclinations. By catholic we mean economic geography's broad inclusiveness of subject matter (again, see Chapter 2). Almost anything can be an object of disciplinary investigation: the tangible and the intangible, the human and the non-human, the living and the dead. Of course, the case for inclusion must be argued, not simply asserted. But the threshold for inclusion is low. It is low in part because historically the disciplinary objective was geographical synthesis. There was an expectation that all the many different things found in a place or region could be brought together, related, linked, synthesized to form a more complex, integrated geographical entity. This synthetic impulse goes back to the very beginnings of geography as a project about describing places and regions.[3] To represent a place or a region meant trying to connect, to synthesize, the many different things, human and non-human, found in that place or region. A place or a region was not just an inventory of everything that was there. They were to be allied, associated, joined, that is, a synthesis made. The catholic and synthetic nature of economic geography is ideal for representing the present in which there is such a collision of so many different things in so many different places; contemporary economic geography is just that mash-up. That is what synthetic economic theory strives to accomplish. Synthesis is in the very marrow of economic geography, perfectly positioning it as an intellectual project to represent the present.

Finally, there is an overwhelming attention to spatial process, that is, an assertion that discussions of spatial process are not mere background atmospherics, simply to add color, but are fundamental, integral to the objects described. Take again the example of globalization. Spatial process, as we argued, inheres in the very term, joined at the hip from the beginning. For example, Erica Schoenberger (2008) argues that economic geographic processes were necessarily involved as first Ancient Athens then later Ancient Rome attempted to globalize their empires through imperial military conquest. Up until about the fifth century BCE, military campaigns were highly restricted, involving sometimes only a single pitched battle. But that changed during the second half of the fifth century BCE when for the first time there were standing armies and navies. However, maintaining a standing military required economic geography. Soldiers and sailors needed to be paid, and they, along with slave rowers (the Athenian navy required as many as 40,000 slave rowers), also needed to be fed and provisioned. When the Athenian or Roman military wanted to expand to new spaces, to globalize, that necessitated transportation, local production, money, exchange, and the organization of markets where none existed before. It meant creating an economic geography. The two went hand in hand. And they have continued to go hand in hand through to the present. Of course, there have been enormous changes in the character of globalization and associated economic geography. For example, the main instruments of globalization are now large multinational corporations (MNCs) and transnational corporations (TNCs) rather than imperial armies and navies. But the link to economic geography is no less strong. MNCs and TNCs are fundamentally economic geographic institutions. Their very definition and rationale is based on the idea of spatially differentiated process, that different kinds of firm operations are located in different places. Their very institutional nucleus is economic geography, inseparable in this case from globalization.

Economic geography has been a discipline in waiting. The various features that characterize it as an intellectual project now appear made for registering, recording, situating, analyzing, and explaining the present. It is an interesting discipline for interesting times.

1.3 Being Critical: In What Sense a "Critical" Introduction to Economic Geography?

The word critical derives from the Greek word *kritikos*, meaning judgment or discernment. That makes it different from being simple negation, a knee-jerk contrary view. One of Monty Python's famous comedy sketches involved Michael Palin paying £1 to have a five-minute argument with a professional arguer played by John Cleese. The argument took the form of alternating contradicting claims: "No, it isn't!" "Yes, it is!" Michael Palin quickly tired of this game, however. In an exasperated voice he said to John Cleese, sitting across the desk from him, "An argument isn't just contradiction." "Well! it CAN be!" said Cleese, and the alternating contradicting statements began all over again until £1's worth of arguing time ran out.[4] Palin was clearly right. An argument is more than the "automatic gainsaying of anything that the other person says." It also involves "an intellectual process." Similarly, being critical means more than saying, "No, it isn't." It connotes, as its Greek root suggests, considered assessment and discrimination, that is, hard mental labor.

The purpose of our book is to provide an explicitly critical introduction to the discipline. That is what makes it different from other introductions to the discipline, of which there are several. Other introductions essentially describe the discipline. They tell you what it is and what it does. We do that, too, but we do something more. We critique it, using two separate lines of questioning.

1.3.1 Questioning the status quo

The first entails questioning why things are as they are; that is, not taking them for granted – "that's just the way it is" – or at face value. *Why* does economic geography today look like it does? Why do economic geographers do things in the particular ways they do? This manner of being critical is partly about exercising a historical imagination (see Chapter 3): What historical developments led to things being the way they are, what intellectual battles were fought, and why did the winners win? But it is not only about history. The present needs to be questioned in and of itself. What assumptions underlie this or that economic geographic study? Why does the author use this method rather than another one? Why is the study of phenomenon X popular and validated, but not the study of phenomenon Y? Who decides what is worthy of economic geographic analysis? Departmental professors? Disciplinary leaders? Funding bodies? Being critical means asking all such questions.

Such critical questioning can be thought of as a form of denaturalization. In the case of a particular text, it is about contesting the text's seemingly evident ("natural") meaning, the one that lies on the surface. The text may look straightforward: a set of sturdy words bolted together as an impeccable sentence, or a crisp photograph, or an uncluttered graph. Once subject to critical interpretation, however, each text can reveal a set of hitherto unrecognized social and cultural judgments made in their construction, that seep deeply, coloring meaning. As a result, the text begins to look less straightforward and sturdy, its meaning no longer so obvious and natural.

Practicing critique in a discipline like economic geography necessarily involves understanding the internal and external processes that bear on it, and which are often bound up with differential power relationships. It is those processes that often shape the real, underlying meanings of the texts that a discipline produces. It is they that need to be uncovered and laid bare.

Take for example the very first text in academic economic geography written in English, George Chisholm's *Handbook of Commercial Geography*, published in 1889 (discussed further in Chapter 3). On the surface, Chisholm's looks like an innocent text, although admittedly a little dull. It contains table after table of numerical statistics relating to the production, price, and trade volumes of innumerable commodities from all over the world, from tons of iron and steel to pounds of tea. There are also several pull-out regional maps of various continents and subcontinents showing areas of production, internal railroad lines, shipping routes, and population densities. The overriding message of the book seems innocuous too, if not a bit naïve. Economic activity is determined by "natural advantages" found in any given place (Chisholm 1889, p.1). Once we begin burrowing into the text, critically interpreting it, it is not so innocent and naïve, however.

As we will suggest in Chapter 3, one of the purposes of Chisholm's book is to establish a new discipline within British universities, economic geography. The intent of the *Handbook* is to mark out academic turf, to claim scholarly space for a new university subject. It is a forceful act, an assertion of power. The *Handbook* as artefact is one of the means to make that assertion. Its very materiality demonstrates that the new field possesses substance. Economic geography must exist because there is now a bulky volume printed by an established publisher, Longman, Green, and Co., printed in its name. Apart from its materiality, there is the text itself that lies between the book's covers as words, as numbers, as maps. They lay out literally what this new discipline will look like, the kinds of practices that must be undertaken to conform to it, and what an economic geographer must do to be an economic geographer. And it pays off. At age 58, Chisholm is offered a job at the University of Edinburgh as an economic geographer. This internal context is not somehow ancillary to understanding the text of the *Handbook*. It forms part of its very meaning, and it is necessary to excavate it when undertaking critical interpretation.

It is the same with the external context, which in this case is the peak of British imperialism (discussed further in Chapter 3). Chisholm, writing the *Handbook* in London, was at the very epicenter of the British Empire, where decisions about the colonies were made, and more broadly, Britain's militant role within the world was determined. That context clearly seeps into the pores of Chisholm's text, determining the statistics he tabulates, the maps he provides (including information written on them as well as the map's legend), the particular commodities described, and the countries represented. There is nothing innocent about colonialism, and there is nothing innocent about the text of Chisholm's *Handbook* that represents it. It is a *Handbook* of the colonial project written at the center of empire, and requiring critical interpretation.

A critical introduction to economic geography necessitates worrying away at the status quo of the discipline, as materialized in institutions, theories, texts, and so on. To worry means dwelling on something, thinking about it one way, then another, not accepting it without providing further thought, scrutiny, and assessment. Our book aims to do exactly that.

1.3.2   Critical evaluation

The second of this book's lines of critical questioning involves evaluation. Rather than merely describing economic geography in its various guises, the traditional textbook approach, we endeavor to evaluate them. We put different disciplinary forms under a critical spotlight and ask pointed questions about each. Does this type of economic geography actually serve any useful purpose? Does that economic geographic approach live up to its billing? Are those economic geographers who practice this methodology blazing a trail we should all be following, or are they in fact heading off down a blind alley?

If the rationale for questioning economic geography's status quo is obvious, the rationale for providing a specifically evaluative introduction may be less so. Why take a position? Why offer an opinion, when it is arguably safer merely to describe? The reason is that no textbook is merely descriptive, however hard it tries to be. To the extent that it is read and absorbed and thought about by practitioners and students, any introductory economic geography text necessarily shapes how economic geography develops in the future; it never simply "reflects." And if our book is liable to shape economic geography's future – however minimally or marginally – then we want it to do so positively and productively. For we are not only economic geography "fanboys," we *care*, passionately, about the discipline. If we think false steps are being taken, we feel obliged to say so. If we think other steps look more promising, on the other hand, we say that too.

Recalling the Monty Python sketch, our evaluation is never simply knee-jerk negation. We do not flatly contradict; we endeavor rather to discern and carefully judge. We follow the political theorist Wendy Brown (2005, p.16) in seeing critique as "a process of sifting and distinguishing" and we strive, like her, and however imperfectly, to exercise it as "a practice of affirming the text it contests." As Brown goes on to say, critique "does not, it cannot, reject or demean its object. Rather, as an act of reclamation, critique takes over the object for a different project than that to which it is currently tethered."

In the case of this book that "project" is, quite simply, to provide a lively, relevant, and discriminating introduction to economic geography *from our authorial perspective*. For the critique we offer is, of course, very much *ours*. It comes from somewhere and is necessarily mediated by our particular views and values. Just as our critique is not negation, then, nor is it free-floating, unmoored from its authors. As Richard Bernstein asks, "critique in the name of what? What is it that we are implicitly or explicitly affirming when we engage in critique?" (Bernstein 1992, p.317). So we try to be honest and transparent in this regard: a critical introduction to economic geography should, we think, be clear about the basis for its criticisms. Thus in evaluating different strands of economic geography we explain *how* we are evaluating them and *why* we think this is a fair or useful basis for evaluation.

1.4   Outline of the Book

The book is divided into two main parts. The first is about economic geography as a discipline, the second is about some of the pressing substantive subjects that economic geography studies. We bring to both parts a critical sensibility.

Textbooks generally do not focus on the internal workings of a discipline, or reflect systematically on its larger intellectual character. There may be fleeting reference to the subject's history, possibly something on its methods, but likely that's it. The implicit rationale for ignoring these internal features is that students are not interested. As the British say, the topic attracts only trainspotters: geeks and nerds (as North Americans say) who are obsessed by arcane minutiae. We disagree. Apart from the fact that discussions about the make-up and character of a discipline involve compelling as well as politically and morally charged discussions, knowing about the discipline is absolutely necessary to understand the discipline's knowledge of substantive issues. The discipline does not gain its knowledge by simply holding up a mirror that passively reflects the economic geographic world outside. Rather, the discipline actively produces economic geographic knowledge. Consequently, to be critical one must know something of interior disciplinary imperatives that vigorously shape economic geographic knowledge, and requires knowledge of the subject's local rules, history, internal sociology, even its geography. It is this that the first part of this book provides.

The reference to economic geography's own geography warrants pause. We do not pretend to introduce economic geography in its totality, because we do not know it in anything like its totality. We are painfully aware of our limitations, and especially linguistic ones. This book introduces and assesses economic geography *written in (or translated into) English*. That does not mean it is *only* about Anglo-American economic geography; scholars working in many non-English-speaking parts of the world have in recent decades increasingly chosen – or been institutionally compelled – to publish more of their research in English. But the reality, for better or worse, is that the book is *mainly* about Anglo-North American economic geography.

Chapter 2, the first of Part I, begins by reviewing definitions of the discipline, and critically discussing the intellectual and sociological work that such definitions perform. We provide our own provisional definition of economic geography, although, following Immanuel Wallerstein (2003, p.453), we remain "dubious" about it, believing that such a definition should always remain up for grabs, subject to contestation and criticism. We argue that economic geography is best understood less as a set of disciplinary institutions, or even as the study of tightly specified subject matter, than as something like an intellectual "sensibility" or perspective. Nevertheless, economic geography's institutions *are* important. By formalizing and institutionalizing economic geography's "sensibility," and giving it a home and intellectual visibility and credibility in the way Chisholm's book did, those institutions nourish it and help to keep it alive. Chapter 3 provides a history of the discipline from its first institutionalization as a university subject during the late nineteenth century. We emphasize both internal and external forces, bound up often with asymmetries of power, which shape and direct that history, producing what we call different invented versions of the discipline, with each requiring critical interpretation. Chapter 4 is about the border country of economic geography, that is, the space that it interacts in and shares with adjoining disciplines in the social sciences. We argue in the chapter that this space is potentially productive for critical inquiry, both calling into question taken-for-granted disciplinary truths, and at the same time relaxing disciplinary constraints, allowing people to think thoughts not usually thought. Chapter 5 is a critical review of the changing meaning, role, and use of theory in economic geography. We are concerned especially to check the vital signs of theoretical practice in economic geography, which we conclude for the most part show robust good health. Chapter 6 is about a topic that is often hidden in one of the discipline's dark corners: methodology. We are concerned to shine a bright light on it in this chapter to reveal the range, diversity, and critical potential of the various methods available. Finally, Chapter 7 takes the lid off the black box of economic geography as a lived, performed discipline to reveal what actually goes on. It is not always a pretty sight, involving messy entanglements of material things, immaterial ideas, human bodies, and social institutions, but it usually gets the job done, except of course when it doesn't.

The second part of the book is more like the standard fare of economic geographic textbooks. It is a review of a set of core substantive topics to which economic geographers have contributed both theoretically and empirically. The five topics we have selected all meet two key criteria. First, they are important topics, ones central to the economy of the twenty-first century; and second, they are fundamentally economic geographic topics, which is to say topics that require an economic geographic sensibility being brought to bear to adequately unravel them. But what is not standard fare, we believe, is the consistent critical perspective that we apply to each of those topics and the analysis of them.

Chapter 8 is about globalization and uneven economic development more generally. We examine conventional representations of globalization – including the Friedmanite flat-earth geographies mentioned earlier – and contrast these with the altogether more imbalanced and uneven geography of the global economy as depicted by economic geographers. Chapter 9 considers a crucial infrastructure of the global economy: modern money and finance. We live not only in interesting but also in financial (or "financialized") times; the chapter explains why and how, showing the indispensability of a geographical sensibility to any understanding of monetary and financial phenomena. In Chapter 10 we examine in turn the phenomena viewed by many as the very motors of today's economy: cities. Given the importance of cities, individually and collectively, to the economy, we argue that economic geography provides unique insights into contemporary economic processes and outcomes. Chapter 11 shifts focus to what has traditionally been seen as the city's "other" – nature – but continues with the same theme, namely the centrality of geography (here, the environment) to the economy. Just as capitalism today would look markedly different if it were not a predominantly and increasingly urban capitalism, it would also look markedly different if it adopted an alternative relation to nature from the processes of privatization and commodification that presently prevail. The last chapter of Part II, Chapter 12, assesses technological and industrial change, and posits these processes of transformation as fundamentally geographical, demanding an economic geographic analysis.

Chapter 13, the concluding chapter, emphasizes two themes that run throughout the book. The first is economic geography's ability to accommodate, and to find germane relationships among, an inordinately wide range of different phenomena. Rather than a discipline that has become narrower and narrower, it has become wider and wider, recording an ever-expanding web of synthetic connections. The second is the hope of economic geography and is bound to the disciplinary critical practices we highlight throughout the volume. Those critical practices are not undertaken for their own sake, but in the audacious hope of making a better world.

1.5   Conclusion

We have tried to write a different kind of textbook for economic geography. Not that we are casting aspersions on other existing textbooks in the field. They are generally excellent. The discipline absolutely needs them. Without them, there would be no discipline. But it doesn't mean that they all need to be the same.

The distinctiveness of our book is given by the subtitle of our volume. We provide a critical introduction to the field. Critical does not mean just being negative. Being critical is in fact a necessary step to learning, to becoming edified. By contesting, by excavating, by prodding and questioning, by exposing one's values, by trying things out, by experimenting, by registering reactions and push-back, one learns, one becomes educated. That is what Mr Daisey was trying to do when he began his work in economic geography. And you can do so too.

Notes

1 The transcript is available at: http://www.thisamericanlife.org/radio-archives/episode/454/transcript (accessed July 7, 2017).
2 Exactly what Daisey found out when he visited Shenzhen is disputed. *This American Life* later retracted the Daisey story because of concerns he had fabricated some of his experiences in China, concluding that his account comprised "a mix of things that actually happened when he visited China and things that he just heard about or researched, which he then pretends that he witnessed first hand" (see http://podcast.thisamericanlife.org/special/TAL_460_Retraction_Transcript.pdf; accessed July 7, 2017). Daisey's reporting of his economic geographic fieldwork, in short, was flawed.
3 The word geography comes from two Greek words *geo*, meaning earth, and *graphein*, meaning to write. Geography therefore means literally "earth writing," that is, using words to describe the world.
4 www.montypython.net/scripts/argument.php (accessed July 7, 2017).

References

This American Life. 2012. Mr Daisey and the Apple factory: transcript. http://www.thisamericanlife.org/radio-archives/episode/454/transcript (accessed June 14, 2017).

Bernstein, R.J. 1992. *The New Constellation: The Ethical-Political Horizons of Modernity and Postmodernity*. Cambridge, MA: MIT Press.

Brown, W. 2005. *Edgework: Critical Essays on Knowledge and Politics*. Princeton, NJ: Princeton University Press.

Chisholm, G.G. 1889. *A Handbook of Commercial Geography*. London: Longman, Green, and Co.

Emerson, R.W. 1911. *Journals of Ralph Waldo Emerson, with Annotations – 1841–1844*. Boston: Houghton Mifflin.

Forgacs, D. (Ed.) 2000. *The Gramsci Reader: Selected Writings 1916–35*. New York: New York University Press.

Friedman, J. 2005. *The World Is Flat: A Brief History of the Twenty-First Century*. New York: Farrar, Straus & Giroux.

Fukuyama, F. 1992. *The End of History and the Last Man*. New York: Free Press.

Galbraith, J.K. 1958. *The Affluent Society*. New York: Houghton Mifflin Harcourt.

Glaeser, E., and Kohlhase, J. 2004. Cities, Regions and the Decline of Transport Costs. *Papers in Regional Science* 83: 197–228.

Keynes, J.M. 1936. *The General Theory of Employment, Interest and Money.* London: Macmillan.

Levinson, M. 2006. *The Box: How the Shipping Container Made the World Smaller and the World Economy Bigger.* Princeton, NJ: Princeton University Press.

Mirowski, P. 1989. *More Heat Than Light: Economics as Social Physics, Physics as Nature's Economics.* Cambridge: Cambridge University Press.

Pilat, D., Cimper, A., Olsen, K., and Webb, C. 2006. *The Changing Nature of Manufacturing in OECD Countries.* STI Working Paper 2006/9. Science, Technology and Industry, OECD.

Schoenberger, E. 2008. The Origins of the Market Economy: State Power, Territorial Control and Modes of War Fighting. *Comparative Studies in Society and History* 50: 663–91.

Shah, S. 2000. A Simple Box that Changed the World. *The Independent*, 30 August.

Wallerstein, I. 2003. Anthropology, Sociology, and Other Dubious Disciplines. *Current Anthropology* 44: 453–60.

Part I

Thinking Critically about Economic Geography

Chapter 2

What Is Economic Geography?

2.1　Introduction

Defining the nature of a scholarly domain or discipline is invariably a hazardous business, fraught with potential pitfalls. This first sentence already signals as much: How *should* we label the thing we are trying to define? Is economic geography a "domain," or a "discipline," or perhaps both of those things? Or is it something else entirely? This chapter attempts to navigate a workable and useful path through such thorny questions, offering in the process an original definition of economic geography that then serves to frame and guide the rest of the book. We think providing a definition is important, albeit difficult; defining economic geography is a necessary first step to offering a critical introduction to it.

Our starting point is Immanuel Wallerstein, an American historical sociologist. (Already, a disciplinary label; they are hard to avoid.) In 2002, Wallerstein delivered a lecture entitled "Anthropology, Sociology, and Other Dubious Disciplines" (published as Wallerstein 2003). We recommend reading it, because Wallerstein displays a healthy but informed skepticism about academic fields and the boundaries that separate them. His perspective is critical. For him, fields like anthropology and sociology are "dubious" not because they lack rigor, substance, or value but because their definition is, and should always be, contested and unstable – "less solid," as he puts it, "than most participants imagine" (p.453). They are dubious in a positive, not negative, sense; they always contain the potential to be changed and improved.

We think of economic geography as just such a "dubious" discipline or field. Its definition is contingent; it is up for grabs. So, while we will give a definition of economic geography that we think is fit for purpose, we are not wedded to it, and we do not believe it should be taken for granted. The least that can be asked of readers of a critical introduction to a field is that they approach all definitions of that field, including ours, critically.

Motivated by Wallerstein's healthy skepticism, the chapter proceeds in four sections. The first digs deeper into the question of definition. How can we define an academic field? How has economic geography previously been defined? What key considerations have guided us in developing our own working definition? And last but not least, what is our definition?

The second section considers the nature of the "economic world" to which economic geography represents a geographical approach. The provisional nature of this world is laid bare. There never has been consensus about the boundaries separating the economic from the non-economic, the economy from other spheres of social life. Rather, what counts as the "economy" is an ever-movable feast, and hence so also is economic geography. Tracing the changing nature of understandings of the "economy" in society in general, and in economic geographic scholarship in particular, we ask questions inspired directly by our critical approach – questions such as who in society defines what is "economic," and what is not, and what are the implications?

The third section shifts from the economic to economic geography. The aim of this section is to render our definition of economic geography meaningful and tangible. It identifies and examines studies that exemplify economic geography as we have defined it in practice. In this way, we put some much-needed flesh on the definitional bones we provided. By the end of this section, readers should be well positioned to recognize economic geographic scholarship when they come across it.

The chapter's fourth and final section shifts from the general to the specific: from our (general) answer to the question "What is economic geography?" to the (specific) answers posited or presumed by the main "schools" of contemporary economic geographic inquiry. What distinctive slants do today's principal narratives and practices of economic geography throw on our wide definition? We identify four particularly prominent approaches, without claiming to be comprehensive. In doing so, we explicitly set the stage for the rest of the book inasmuch as each principal approach can be usefully conceived as a unique "assemblage" of crucial component parts: key themes/topics (Part II), conjoined in historically specific ways (Chapter 3) with different theories (Chapter 5), methods (Chapter 6), and insights (from cognate disciplines) (Chapter 4) by distinctive scholarly communities (Chapter 7).

2.2 Defining Economic Geography

Part of the reason why academic disciplines are inherently "dubious," says Wallerstein (2003), is that they are always "actually three things simultaneously." That's definitely true of economic geography. To be sure, economic geography differs superficially from anthropology or sociology – Wallerstein's primary examples – insofar as it is "only" a sub-field (of geography and, more narrowly still, of human geography). But what it shares with those other disciplines is much more fundamental (see also Chapter 4).

Like them, economic geography is first an *intellectual category*: a terminological construction used to assert "that there exists a defined field of study with some kind of boundaries, however disputed or fuzzy, and some agreed-upon modes of legitimate research" (Wallerstein 2003, p.453). We will turn to those modes of legitimate research in Chapter 6, and academic boundaries in Chapter 4. Second, academic disciplines are *institutional structures*: "There are departments in universities with disciplinary names. Students pursue degrees in specific disciplines, and professors have disciplinary titles. There are scholarly journals with disciplinary names" (p.453). And third, disciplines are *cultures*. Here, Wallerstein merits citing at length:

> The scholars who claim membership in a disciplinary grouping share for the most part certain experiences and exposures. They have often read the same "classic" books. They participate in well-known traditional debates that are often different from those of neighboring disciplines. The disciplines seem to favor certain styles of scholarship over others, and members are rewarded for using the appropriate style. And while the culture can and does change over time, at any given time there are modes of presentation that are more likely to be appreciated by those in one discipline than by those in another. (Wallerstein 2003, p.453)

Chapter 7 tries to grapple with both the institutional structures *and* the cultures of economic geography. In fact, the issue of economic geography's cultures permeates several of the chapters.

Given that academic disciplines are simultaneously (at least) these three things, it is no wonder that attempts at defining them take such varied forms. There are arguably three main approaches to disciplinary definition, mapping roughly, though not exactly, onto Wallerstein's tripartite classification: "what," "how" and "who." In the first case, definition derives from *what* is studied: particular topics, themes, issues, and so on. History is the study of the past, say. The second is "how" a discipline studies: for instance, *how* does physics, or philosophy, or social anthropology go about formulating the knowledge that we recognize *as* physics, philosophy, or social anthropology? This approach to definition will encompass at a minimum epistemology – baldly, how we come to "know" the world – and methodology – laboratory experimentation, or ethnography, or archival work, and so on. Lastly and most expediently, the definition stems from *who* studies. Political science, for example, would be defined as what political scientists do.

Where economic geography is concerned, most attempts at definition take the first of these three approaches and emphasize its subject matter, what it studies. Consider, for example, entries in the various dictionaries of human geography. One (Gregory et al. 2009, p.178) defines economic geography as a "sub-field of human geography concerned with describing and explaining the varied places and spaces where economic activities are carried out and circulate."[1] According to another (Castree, Kitchin, and Rogers 2013, p.118), it is a "subdiscipline of geography that seeks to describe and explain the absolute and relative location of economic activities, and the flows of information, raw materials, goods, and people that connect otherwise separate local, regional, and national economies."

Meanwhile, existing economic geography textbooks (e.g., Coe, Kelly, and Yeung 2007; Mackinnon and Cumbers 2007), perhaps wary of the dangers previously alluded to, tend to shy away from definitional exercises altogether, although Daniel Mackinnon and Andrew Cumbers come close(st) to providing one in emphasizing that economic geography is "concerned with concrete questions about the location and distribution of economic activity, the role of uneven geographical development and processes of local and regional economic development" (p.12). They also say that as "economic geographers, we are particularly concerned with the location of different types of [economic] activity, the economies of particular regions and the economic relationships between different places" (p.1). Here, too, subject matter is demonstrably to the fore.

In this book, we approach the question of definition differently. In doing so, however, our argument is not that the kinds of definitions of economic geography quoted above are somehow wrong. It is that they are limited and thus limiting. Though it is clearly not their intention, such definitions, in asserting that economic geography is concerned with the study of, say, X, inevitably run the risk of implying that economic geography is not – cannot be – the study of Y. For all sorts of reasons that will become apparent in what follows, we believe this is not a risk worth running. Rather than take economic geography's subject matter as given, we believe it is imperative to question it. We should ask, following Doreen Massey (1997, p.27; original emphasis), "*what histories lie behind the constitution of this object of study of 'economic' geography*" and also recognize that such histories often "involve conflict and struggle."

In pursuing an alternative approach to definition that relies on none of the aforementioned axes of what, how, or who, we strove to meet several ends. Perhaps most importantly, we aimed for a definition that is plural, broadly based, and open-ended. Our definition must not tie economic geography – and economic geographers – down to a fixed set of concerns or approaches. The future should be open, not closed. After all, economic geography as it is and has been is just one among the myriad possibilities for what economic geography, in practice, *could* be.

Furthermore, economic geography as we understand it is not, and should not be conceived as, the sole preserve of self-styled economic *geographers*. It can be, and is, practiced much more widely. Some of the most compelling and significant economic geographic scholarship, as this book will show, has been produced by those who would almost certainly not see themselves as "doing" economic geography.

At the same time, however, disciplinary affiliation and adherence is not something to be sneered at. While we do not define economic geography in terms of its institutional structures, we recognize that the institutionalization of economic geographic scholarship in and through economic geography *as an academic discipline* is important. If economic geography today is thriving, it is due in no small part to the strength of institutions that nourish, promote, and reproduce it. "Ways of thinking, of apprehending the world," Massey (1997, p.27) observes, "have to be produced and maintained." So they do, and this production and maintenance is clearly more important for "niche" fields like economic geography than for established behemoths like mainstream economics.

How then have we chosen to define economic geography for the purposes of this book? We propose that it can be productively defined as *an analytical attentiveness to, and an explanatory emphasis upon, the substantive implication of space, place, scale, landscape, and environment in (what are deemed to be) "economic" processes*. This definition being multidimensional and not necessarily self-evident, we now try to unpack it.

2.3   The "Economic" World

2.3.1   Producing the economy

If economic geography is centrally concerned with geography's multifaceted role in *economic* processes, it behooves us to begin our inquiry by closely considering this qualifying term: the "economic." What is meant by "economic," and therefore what types of processes – their conditions, their character, their outcomes – does their geographical study presume within economic geography? Our take-away message here is that there is no definitive answer. The "economic" and the "economy" are socially constructed, which is to say that ultimately they are what society deems them to be. As Massey (1997, p.35) says, "What we think of as 'the economic' is itself expressive of other aspects of our culture/society."

But there are some cautionary points. To say that the economy is socially constructed is not to say that the processes described as economic are somehow unreal. The production of commodities using wage labor is about as "real" as one can get; what *is* always arbitrary, though, is the categorization of such activity as economic and, therefore, as part of something called the economy. Nor is our observation intended to suggest that we should forsake definition – definitions of the economy can be very helpful, intellectually as well as socially, so long as their subjective nature remains in focus. And it is also important to distinguish between social constructionism and relativism. Insisting that what belongs within "the economy," and what sits outside of it, is a social construction is not the same as saying that all such social constructions are equally (in)accurate. Some representations of the "economic" are clearly more illuminating, meaningful, and useful than others, even if none are the absolute Truth.

That said, let us turn to the modern dictionary definition of the economy. At the risk of simplifying, the standard definition usually contains some combination of five main elements. The first three are all captured in the statement – taken here from the online version of the Oxford English Dictionary (OED) – that the economy comprises "the production and consumption of goods and services and the supply of money."[2] To clarify, the three discrete elements here are: "economic" processes (production and consumption), outputs (goods and services), and means of circulation (money). To these three, two further elements are frequently added. One is about objectives and relates (again, from the OED) "to the generation of income," or "maintained for the sake of profit."[3] The other element concerns the particular social realm encompassed by the economy. Here, the critical word is *materiality*. The OED, for example, refers to "the material resources of a community or state."[4] And Timothy Mitchell (2002, p.82; original emphasis), writing about the historical emergence of current understandings of economy, observes that the economy came to be distinguished from other social realms explicitly "by the fact that it stood for the *material* sphere of life."

Just as there is nothing "wrong" with subject-matter-based definitions of economic geography, neither is there anything necessarily wrong with the OED's definitions of the economy. They are very useful definitions, as far as they go. Rather, the point is that if an economic geography defined by its fixation on, say, "the absolute and relative location of economic activities" is inherently limited *by* such definition, the OED definitions similarly constrain what we can and do think of as economic. For example, according to the OED an activity not mediated by a monetary calculus, not enacted for the sake of income generation, or without an obvious "material" basis could never be considered economic. There *may* be reasonable and defensible grounds for considering such an activity as extra-economic (and instead as, perhaps, "cultural"), and hence as beyond the bounds of economic geographic inquiry. But equally there may not be.

In recent years a growing number of scholars have addressed the question of how, why, and when today's dominant understanding of the economy cohered and ultimately came to assume the particular form it did. Three periods in history, and three extremely important systems of economic representation, were especially important. First, from the late eighteenth century, what became known as "political economy" – the work of Adam Smith, David Ricardo, and others – set about separating the "economic sphere" from other spheres of life, partly by imputing to it, as Shamir (2008, p.5) notes, "its own laws, its own logic of operation, its own conception of the human subject." A century later, neoclassical economics arrived on the scene and sharpened the economy's separateness, rigorously abstracting it from the rest of social life through new statistical techniques and models (Breslau 2003). Yet it was not until the 1930s and 1940s and the birth of national accounting – the calculus which generates estimates for gross domestic product (GDP) – that this separation, at least in the public eye, was formally cemented. By providing estimates for the size of the "national economy," such accounts appeared to furnish conclusive proof that the entity called the economy existed and on its own terms (Mitchell 2002, chapter 3).

These studies of the economy's "emergence" raise various fundamental issues. Who, for instance, has the power to decide where the economy ends and other spheres of life begin? In other words, who decides where the economy's boundaries lie? As individuals, we can all make our own pictures of the economy, of what it does and does not contain, but only some of those pictures prove influential because only they are officially legitimated, institutionalized, published, and reproduced. Today, it is the statisticians and economists working in national and international statistical offices and producing different countries' national accounts that have the power, in the most immediate sense, to calculate how big economies are and, prior to such enumeration, to make the more fundamental assessment of what particular activities should "count" as economic or not (Box 2.1: National Accounts).

Representations, in short, matter, not least when they attribute value – economic or otherwise – to certain people or activities but not to others, and national income accounting is an excellent example. That certain activities, most notably unpaid household work, are conventionally excluded from the accounts' definition of economy serves to reaffirm prevalent beliefs about their lesser worth and status and thus maintain unequal relations of power (Waring 1999). Similarly, albeit conversely, the historical trend toward greater *inclusion* of other activities – most notably financial services – has had equally profound implications, increasing their presumed value (Christophers 2011). Moreover, such effects take on a heightened significance when one considers that the differentiation between what is in and what is outside the economy is not only arbitrary but often severs activities that in practice are inextricably linked. Household labor, inasmuch as it enables the reproduction of paid labor and the market economy, is again an exemplary case.

Recall also the OED definition that stipulates profit generation. If only for-profit activities are regarded as properly "economic," then entire economies – socialist ones – would presumably *not* be "economies." This tells us something very interesting, that the OED definition of the economy is an explicitly capitalist one. The broader point we want to make is that *all* ways of picturing and talking about "the economy" and the "economic" are soaked

Box 2.1 National Accounts

National accounts estimate the size of a country's economy – its total economic output. The most well-known national accounting metric is GDP, which is an estimate for the total value of products and services output generated within the country in question. Another important metric is GNP (gross national product), which values the output generated not *within* that country but by citizens (individual and corporate) *of* that country. Here already we can see that national accounts, in cutting up the world and its economy according to explicitly spatial classifications, are a decisively economic geographic artefact.

The accounts produce not one but three GDP estimates, using three different methods; the single official GDP figure that is published and communicated is derived by triangulating between the three estimates. One is generated using the "product" method. Here, the market value of all "final" goods and services – those purchased directly by consumers – is totaled and the value of "intermediate" goods and services – those used either for resale or for further production – is deducted to avoid double counting. The second estimate uses the "income" method, which involves summing incomes across the primary sectors of the economy: wages for labor, profits for companies, and so on. The final method is the "expenditure" approach, which looks at monetary value spent rather than produced (method 1) or earned (method 2). The formula is quite simple (even if its calculation is anything but):

consumption + investment + government spending + net exports

National accounts may sound boring and unimportant, but even if the former is true, the latter is not. These numbers, GDP and the like, are enormously consequential. All governments, at least in capitalist societies, are judged by their record on economic growth. Many governments posit delivery of growth as central to their credibility and electability. How do we know whether an economy is growing or not? The national accounts tell us. They also tell us which parts of the economy – resource industries or the creative industries, pharmaceuticals or software – are delivering this growth, and in what proportions.

There have always been critics of national accounting and its uses, even from its earliest days as a formal, state-sponsored statistical enterprise in the decades immediately prior to and after World War II. But these criticisms have multiplied and deepened in recent years. They are significant and well taken, and include: the fact that certain crucial activities (e.g., unpaid household work) are ordinarily excluded and thus devalued; the fact that GDP does not capture phenomena that arguably matter as much as or more than economic growth (e.g., wellbeing or sustainability); and the fact that the dominance of GDP gives too much power to numbers and to those who produce them. Recent examples of critiques in this vein include Fioramonti (2013) and Philipsen (2015).

in values and prejudices and as such are ripe for critical interpretation (see Chapter 1). Here again is Massey (2013, pp.6-7): "The whole vocabulary we use to talk about the economy, while presented as a description of the natural and the eternal, is in fact a political construction that needs contesting."

We suggest it is best to think of the economy's scope and composition as an interplay between representation on the one hand and social organization on the other. After all, socialist economies fail the OED test not only because the OED has decided, per this definition, that economic = profit-seeking, but also because in those (non-)economies the production and consumption of goods and services is organized on a not-for-profit basis. The point is that while people and institutions with power make decisions about what is "economic," they also make decisions about the organization of goods' and services' production and consumption that result in some activities qualifying as "economic" and some not.

Consider the situation in a "mixed economy" where some goods and services, generally provided by profit-generating firms, are paid for, and some are provided free of charge (often by the state). If we abide by the OED definition, only the former are "economic." But plenty of goods and services could, and do, swing either way. Think of education, or healthcare, or even banking. Often provided gratis by governments as a "public good" utility, they equally well could be – and increasingly are being – provided by companies operating in markets on a profit-oriented basis. Here the question of who has the power to decide when such a service is "economic" and when it is not is arguably much less important than the parallel highly political question of who has the power to decide whether markets (and profits) should be involved in the delivery of that service in the first place (Christophers 2013). As Figure 2.1 shows, the answer to the key existential question – to be or not to be economic – depends on both decisions.

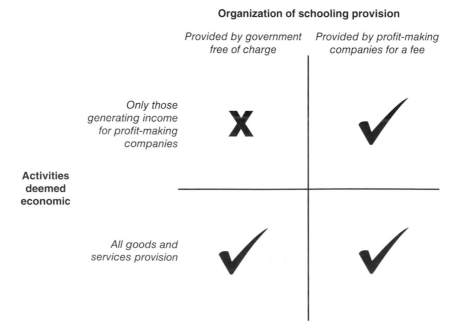

Figure 2.1   Is schooling an "economic" service?

All of this really matters for economic geography and for how we perceive, practice, and produce it. After all, if economic geography constitutes a geographical approach to the economy, and if the definition of the economy is both contingent *and* deeply implicated in social relations of power, it clearly matters a great deal how the "economic" in economic geographic research is delineated. It will affect economic geography's intellectual purchase, its scholarly credibility, and its practical effects. What different understandings of "economy," then, have typically featured in and shaped economic geographic scholarship, and with what implications? The rest of this section addresses these critical questions.

2.3.2   The "economic" in economic geography

Economic geography has traditionally worked with and (re)produced a relatively rigid and frankly conservative conception of "the economy." Indeed, if we wanted succinctly to encapsulate the basic nature of that economy it would be close to the OED definition. The economic in traditional economic geography is conceived as largely formal, commercially oriented, and materially (industrially) based. Moreover, it has typically been characterized by a set of marked functional, geographic, and gender assumptions. The economy is presumed to be about production (where capital meets productive labor), the spaces of the Global North, and men.

Since the early 1980s, however, there has been a discernible "refiguring [of] the economic in economic geography" (Thrift and Olds 1996). The question of the scope and substance of the particular complex of processes and structures that warrants and requires economic geographic examination has been widely and forcefully readdressed. This has led to a refiguring of the economic on multiple fronts. Three of them merit particular attention.

First, there has been a broadly based endeavor to make the object(s) of economic geographic inquiry more gender-inclusive: to render the "economic" less male. There is now a fruitful ongoing interaction between economic geography and feminist theory. Here is not the place to discuss the wide range of critical insights generated by that interaction (see Chapter 4). Our particular interest is how such research has refigured what *counts* as "economic." In contrast to national accounting methods, scholars at this intersection have argued for a wider definition of the economy that includes what was previously excluded, and to "blur" the conceptual boundaries conventionally placed between the (male) economy and its (female) other (Mitchell, Marston, and Katz 2004).

Second, the economic in economic geography has also been rendered more inclusive in spatial terms. As Sue Roberts (2012) has argued, historically there was a clear divide between economic geography and so-called development geography. That division coincided closely with the nominal divide between the Global North on the one hand and the Global South on the other. Economic geography was the geographical study of the economies of the Global North. Development geography was the study principally of the countries of the Global South, which by definition lacked (proper) "economies" in that they were not yet "modernized." Recent economic geographic scholarship, though, has increasingly torn down that divide. It insists that the Global South is no less "worthy" of economic geographic study than the Global North, the processes by which goods and services are produced and consumed there being no less "economic" (e.g., Murphy and Carmody 2015).

Lastly, the range of activities and environments considered legitimate subjects of economic geographic inquiry has also proliferated in view of a broad rethinking of deeply rooted conceptual dualisms featuring "economy." Those dualisms include, "economy and culture," "economy and nature," and "economy and society." Each of these binaries has been subjected to searching scrutiny, with important implications for economic geography. One of them is the subject of Chapter 11, which examines the erosion of the traditional distinction between the economic and the "natural" world. The putative boundary between economy and "culture" has also been questioned. This questioning, much of it taking place *in* economic geography, and often referred to as the "cultural turn," has had profound implications *for* economic geography (Crang 1997; Chapter 3.6). In particular, Thrift and Olds (1996) pointed to the burgeoning critical attention that is now given to processes "which until quite recently would have been counted as outside the normal orbit of economic geography, yet which illustrate the extraordinary difficulty of separating out something called 'the economic' from 'the social' or 'the cultural' or 'the political' or 'the sexual' or what have you" (p.312).

Nevertheless, for all this multidirectional refiguring – captured by Thrift and Olds as economic geography "'stretching' its own definition so as to make itself more inclusive" (p.313) – we think a dose of circumspection remains in order. Unlike Martin and Sunley (2001), for whom the project of "rethinking the 'Economic' in Economic Geography" had, in a sense, (already) gone too far, we are inclined to believe that in many respects it has not yet gone far enough. On close inspection, the "economy" figured in and through economic geography remains in large measure traditional and conservative, and as such problematic. As Lee (2006, p.413) observes, "the arbitrary notion that the economic is somehow separate and autonomous, rather than constituted through multiple social relations and conceptions of value, persists." And there is more.

First, the cultural turn, notwithstanding its closing of the gap between economy and culture, has ultimately tended to leave "the economy" intact. For example, the cultural turn in economic geography has often meant "embedding" the economy within the larger culture. The difficulty with this formulation, however, following Mitchell (2008, pp.1117–1118), is that the economy is still treated as possessing "some essential form." However embedded in "ties of friendship, affection, altruism, morality, control, culture or other apparently non-economic relations," the economy is still deemed separable in order for it to be embedded. For Mitchell, as for Lee, *genuinely* rethinking economy does not mean posing questions about "differing degrees of embeddedness." It means rethinking the projects and practices by which the economy is constituted. Mitchell's work aside, there is precious little economic geography of such an ilk (cf. Barnes 2008).

Second, if we think of the economy as a totality of interconnected "moments," economic geography continues to offer a partial and limited reading of those moments. In the work of scholars such as Louise Crewe (2000), a focus on consumption has been successfully added to the traditional emphasis on production. But as several commentators have recently noted (e.g., Christophers 2014), much less has been written from an economic geographic perspective about markets and about processes of exchange, market-based or otherwise. Lee (2002), moreover, suggests that even adding the exchange "moment" to the mix might not be enough because our overall figuring of the economy would remain stilted. He calls, therefore, for a far more sweeping re-imagining. "Economies," he insists (p.336), "are more than mere moments of consumption, production and exchange or even a sequence of such moments: they are circuits of material reproduction involving the continuous flow of values and energy from consumption via exchange to production, via exchange to consumption and so on"; and they "must entail the possibilities of drawing value from there and then as well as from here and now, and of projecting from past into future spaces and times."

Finally, there are J.K. Gibson-Graham's multiple provocations (Chapter 3.6). In repeatedly arguing that economic geography remains tethered to unbending and confining conceptions of the economic, the gauntlet Gibson-Graham has thrown down is probably the most challenging of all. She maintains that the major problem is not one of scope, or of *what* is included (or not) within "the economy" (household labor practices, the Global South, "informal" economies, and so on). She sees the problem rather as *how* this inclusion is effected – on what terms. Attempts such as those elucidated above to "add to the picture of what contributes to the production of goods and services … do not necessarily help us think differently about the economy" (Cameron and Gibson-Graham 2003, p.151). This is because truly valorizing activities such as household labor requires recognizing them as having value equal to "traditional" economic activities; merely labeling them "economic" is not in itself sufficient. And she says that this lip service is what economic geography's "refiguring" has generally amounted to. Economic geographers have certainly expanded the realm of the economic. But the activities newly admitted to the fold, those previously considered non-economic, are still not meaningfully valorized; these "'added in' sectors, though recognized and counted, remain locked in the subordinate, under/devalued position vis-à-vis the core economy" (p.151).

In her own seminal work, therefore, Gibson-Graham has sought to think *differently* about the (geographical) economy, and from two main, linked perspectives: feminist and non-capitalist. "Feudalisms, slaveries, independent forms of commodity production, household economies and other types of economy," by Gibson-Graham's way of thinking, "may be seen as coexisting in a plural economic space – articulated with and overdetermining various capitalisms rather than necessarily subordinated or subsumed" within them (1995, pp.278–279). This is certainly in part a project of "adding in" – elsewhere, in discussing non-paid activities of sharing, bartering, volunteering, and so forth, Gibson-Graham (2008, p.617) emphasizes that such "'marginal' economic practices and forms of enterprise are actually more prevalent, and account for more hours worked and/or more value produced, than the capitalist sector" – but it is much more than that too.

In this radical refiguring Gibson-Graham has not been alone. A comparable trail has been blazed by researchers writing economic geographies of the Global South that not only add the Global South to economic geography's canvas but, more significantly, advance alternative ways of thinking "economy" in the process. Stimulated by influential and overlapping deconstructions of Eurocentric social science such as Edward Said's famous critique of *Orientalism* (1978) and the Subaltern Studies historical-theoretical tradition (Guha and Spivak 1988), studies such as those by Hart (2002), Chari (2004), and Gidwani (2008) ask what the economy might look like – and what Western terms such as "capitalism" and "globalization" even mean – once the Eurocentric blinkers are removed. These are not easy issues to grapple with, and we have barely scraped their surface here. But, in the work of scholars such as these and Gibson-Graham, the "economic" in economic geography is being given perhaps its most substantial and significant makeover in several decades.

2.4   Geographies of the Economic

2.4.1   Putting flesh on the definitional bones

We have seen, then, that as a category the economic is considerably less straightforward than one might imagine. We shall now consider what it means in turn to adopt and exercise an analytical attentiveness to, and an explanatory emphasis upon, the substantive implication of space, place, scale, landscape, and environment *in* (broadly conceived) economic processes – to practice and produce, by our definition, economic geography.

What do we mean to begin with by analytical attentiveness and explanatory emphasis? Attentiveness can be thought of as vigilance. It means configuring one's investigational antenna in such a way as to ensure that certain signals are *not* missed. How this attentiveness can be effected and how it can be rendered *analytical* are obviously crucial questions – of theory (Chapter 5) and methodology (Chapter 6). But the essential point stands: economic geography requires being on the lookout for the imprint of the geographical on the economic, albeit without closing one's mind to other signals or pre-imposing a "geographical" (mis)reading *on* such signals. This latter aspect clearly speaks also to the concept of explanatory emphasis. What economic geography is definitely not is an unyielding insistence on explaining everything about the economic world in geographical terms or as a function *of* geography. Rather, economic geographic scholarship *emphasizes* geographical factors in its explanation, while not neglecting other factors.

These considerations have three profound implications for "doing" economic geography, and indeed – because the two are not necessarily coterminous – for "being" an economic geographer (which many scholars, including ourselves, would profess to be).

First, while one may set out to do economic geographic research, in the end one's analysis may ultimately lack a distinctively geographical component. However attentive one might be to "the geographical," it might not be *substantively implicated* (on which more below) in the economic processes under examination.

Second, and following, economic geography is unavoidably intertwined with other ways of knowing and explaining the economic world. Doing economic geography, and being an economic geographer, means trying to decipher economic processes that not only might at one level or another defy geographical explanation, but which will almost certainly (also) require other tools and modes of explanation – be they derived from mainstream economics, from political economy, or from somewhere else. Chapter 4 explicitly explores this ineluctable terrain of theoretical and methodological cross-fertilization.

And third, being an "economic geographer" means being convinced on the basis of one's own research and/or that of others that geography helps explain the economic world at large – that, economically, geography matters (Massey and Allen 1984). But an economic geographer is always open to the possibility that specific economic processes, or their components, may not require a specifically geographical explanatory dimension. And she does not claim "the geographic" as geography's private property; she is not a "spatial separatist." Given the significance of geography to the economy, *all* economic sciences are, or at least should be, concerned with geographic relations. An economic geographer is simply someone who privileges this concern, who *emphasizes* that geography matters, knowing others are less wont to do so.

This notion of the "materiality of geography" is what we refer to as the *implication* of space, place, scale, landscape, and environment in economic processes. By "implicated" we mean that the geographies in question *affect* the processes in question. Whether it is inflation, globalization, raw material extraction, industrial waste disposal, or whatever else, geography is integral rather than peripheral or incidental to the form and outcomes of the process. It is an active ingredient. We add the word "substantive" to make a key further point. This is, that the effects of the geographic are not trivial or marginal. They are significant to the degree of being necessary to explanation. Or, to look at things from an alternative perspective, one might say that the form and outcomes of economic processes in which space and other factors are *substantively* implicated would be significantly different were it *not* for those factors. Geography makes a (big) difference. As such, meaningful understanding and explanation of such processes requires explicit attention to and consideration of their geographical dimensions.

This leaves just one final component of our definition to be elucidated: the nature of the "space, place, scale, landscape, and environment" upon the implications of which economic geography focuses. What are these things? This question, unsurprisingly, has no simple answer beyond saying that they constitute collectively the essential "stuff" of human geography. Probably at least as much ink has been spilled on each such concept and its meaning as on "economy."

At this point we provide no attempt at definitions or identification of general usages of these terms in economic geography. Definitions would be impossible. And usage in economic geography is in many respects what this book as a whole is about. The meaning of a term like "place" only really becomes apparent and consequential in the context of its utilization. How economic geography understands and invokes each key geographic term will become clearer in the following chapters. For now, therefore, we limit ourselves to a preliminary sketch of each term's basic contours, which we think is a helpful foundation from which to proceed (Box 2.2: The "Stuff" of Human and Economic Geography).

One other important point remains. Some readers may be wondering about the "region," or about "territory." Are these not also the "stuff" of geography? Indeed they are. It is important to be clear, then, that we are neither neglecting such phenomena/concepts nor gainsaying their centrality to geographic processes and knowledge. However, we did want to keep the array of geographic terms invoked in our definition to manageable proportions. And while in our view space, place, scale, landscape, and environment all potentially denote substantially *different* – although interrelated – ideas and properties, we believe the concepts of region and territory can be accommodated by some combination of the factors already on our list.

2.4.2 Economic geography in practice: two examples

Given that much of the rest of the book is concerned precisely to discuss critically work that demonstrates geography's manifold implication in economic processes, the objectives of the remainder of this section are modest. They are to illustrate the basic idea of "implication" by showing what it looks like in practice, with two examples. We provide a flavor of what economic geography offers, both to make the discussion less abstract, and (we hope!) to whet the reader's appetite.

Box 2.2   The "Stuff" of Human and Economic Geography

All social sciences have their own key terms and concepts, those by which they come, to one degree or another, to be recognized – you're talking about class, you must be a sociologist; you're talking about culture, so I am guessing you are an anthropologist; and you're talking about utility, which makes you an economist. Human geography has many such concepts; it doesn't "own" them, because other social scientists use them too, but it *does* make particular play of them. We have singled out five: space, place, scale, landscape, and environment – it would be odd to find an article in a human geography journal that did not refer to at least one of these. Our interest in this book is in how they are put to work specifically in economic geography.

Space: This is arguably the broadest, most general, and sometimes vaguest of geography's key concepts. It typically refers to the spatial arrangements of a society or economy, the ways in which different locations are connected to one another; a total lack of connection would represent one (extreme) spatial configuration, tight and rapid connection – perhaps facilitated by advanced transportation and communications technologies – another. Economic geographers are interested in the spatial configuration of *economic* processes. For example, is a particular economic process characterized by spatial fluidity and deep integration of the economies of different locations, or is space more punctuated, more slowly traversed? And (how and why) does this matter?

Place: This concept refers to the individual locations stitched together by spatial relations. Place is particular, unique. Economic geographers are interested in all sorts of issues relating to place. What type of economy do we find in place A, and why is it so different from that in place B? How significant is the economy of one place – Chicago, say – to the wider regional (Midwestern) or national (US) economy? And what combination of wider economic processes serves to produce the unique economic outcomes that we find in particular places?

Scale: This denotes spatial extent: local, regional, national, international, or global. Is the market for automobiles a global one – that is, do retailers operate and customers search globally – or does it comprise a series of smaller national or intranational markets? Do we have local, national, or supranational currencies, that is, what is the scale of the monetary system? More fundamentally still, how do different economic processes *constitute* different scales as relevant spheres of economic organization? *Why*, in other words, is the national or "the global" an important scale of economic inquiry? What processes made it so? All such questions are of central concern in economic geography.

Landscape: What is a human or social landscape? The critical consideration is *visibility*: a social landscape is what we see on the land as a result of the history of human intervention, for instance different forms of land use and different types of built environment. Economic landscapes therefore represent the visible manifestations of economic processes, for instance the different landscape forms associated with different economic activities (agriculture, mining, manufacturing, and so on).

Economic geographers study the processes that create these landscapes and, no less importantly, the ways these landscapes shape the economies that develop on and through them.

Environment: The environment is physical "nature," adapted and transformed, of course, by millennia of human intervention. Economic geography is concerned with all manner of different relations between economy and environment – from the physical dependence of a vast array of economic processes on primary environmental resources of various kinds (timber, water, minerals, etc.), all the way to the creation of modern financial markets in environmental "goods" (e.g., wetland preservation credits) and "bads" (e.g., carbon emissions allowances).

Our two examples were chosen partly because they are exemplary of economic geography as we have defined it, and partly because they concern an area of the economy – and of economic geography – not specifically addressed in significant depth elsewhere in the book (although see Chapter 12): labor processes.

The two studies in question – Jamie Peck's *Work-Place* and Don Mitchell's *The Lie of the Land* – were published coincidentally in the same year, 1996. We first consider *Work-Place*, one of the few exceptions to the "rule" noted above that the economy of/in economic geography has neglected markets. Peck's is a major study of how labor markets *work*. It examines the form, operations, and outcomes of those socioeconomic processes that we refer to collectively *as* labor markets, and which comprise at their core the search by firms for employees and workers for wage labor. Peck's account is theoretical in the sense (discussed at length in Chapter 5 of the present book) that it seeks to explain labor markets at a macro, generalized level, and with reference to underlying structural factors. Those underlying factors consist primarily of institutional forces and power relations.

Yet, as Peck admits, there was already a substantial political economic literature – one he refers to as segmentation or structured labor market theory – that explained labor markets precisely in terms of social structuration and institutional mediation. In what sense, then, was that literature lacking, and how did *Work-Place* aim to embellish it? The answer lay in the geographies of labor markets, which the existing literature essentially overlooked. The great achievement of *Work-Place* was to show how geographical factors were *substantively implicated* in labor market processes, thus also demonstrating that the existing literature was insufficient to fully explain such processes.

How does it do so? The book is dense and multilayered, and repays a full reading, but it is possible to articulate its main insights by clearly distinguishing the central argument from more superficial readings of labor market geographies. The latter readings, as Peck notes, tend simply to rehearse the banal truism that "the concrete form of labor markets varies over space" (p.262) – that those markets display geographical variability. On our terms such a finding does not even qualify as economic geography. Economic geography is about the implication of geography in economic processes, not merely spatial variance in economic outcomes, which could very well result from non-geographical factors (e.g., different institutions or policies).

Work-Place, by contrast, analyzes "the spatiality of the processes themselves, not just the unevenness of their outcomes" (p.265). Moreover, it demonstrates that this spatiality is *material* because "labor market processes are themselves transformed by the way that they are … stretched out over space" (p.262). More exactly, and calling on one of the geographical concepts missing from our "primary" list, Peck shows that labor markets are "territorially constituted" (p.263): they assume particular spatial forms that *matter* to their processual dynamics. The particular spatial/territorial form that Peck emphasizes in his account, partly because it tended to be neglected by other commentators, is the local. Insofar as their scale (an adequate substitute, here, for territory?) shapes their operation, then, Peck insists that local labor markets must be seen as "more than data units or study areas. Fundamentally, they represent the scale at which labor markets work on a daily basis" (p.263).

Our second study, *The Lie of the Land*, deals with an altogether different set of labor processes, in an altogether different context. Whereas Peck focuses on contemporary Anglo-American labor markets, Mitchell examines the history of migrant labor and agribusiness in California in the first half of the twentieth century. But his book is no less powerful a demonstration of the significance of the geographic to the dynamics of the economic.

Mitchell's book stands out in economic geography insofar as it is concerned with arguably the least discussed of the geographic phenomena/concepts on our list: landscape. Landscape, Mitchell maintains, is implicated in economic processes just as substantively as, say, space or place. And he substantiates this claim with reference to landscape both as a material, on-the-ground reality and as a discursive artefact.

This second point warrants exemplification. For Mitchell, the "stuff" of geography takes metaphorical or discursive as well as material forms. Economic geography in principle is just as concerned with the significance of the former as that of the latter. This is not to say that economic geography *in practice* has paid as much attention to the discursive as it has to the physical; it demonstrably has not, which is part of the reason why studies that do so, such as by Henderson (1998) and Goswami (2004), are so striking. In our view, however, this dearth is much more a symptom of a failure of economic geography than a sign that geographical representations are generally insignificant. They are not. The ways that space, place, scale, environment, and landscape are represented – in fora ranging from art to the media, from literature to planning policy, from military strategy to financial market regulations – make a difference to the constitution and playing-out of economic processes. Mitchell's book is noteworthy for showing just this.

In many ways Mitchell's core arguments are even trickier to summarize than Peck's. Landscape, Mitchell argues, is two things: a physical reality (the worked land) and a representation (a work of art). In both cases, it was substantively implicated in labor processes in early twentieth-century California. For example, Mitchell shows that the landscape imagery of California was infused with racial ideologies, which "naturalized" particular, iniquitous forms of labor relations.

Mitchell then adds two further levels of analysis. First, he shows that there was a vital reciprocity to the above relation. Labor processes shaped landscapes (physical and representational) even as they were shaped by them. His chapter 3, for instance, details the effects of the "subversive mobility" of migrant labor on both landscape morphologies and ideas. Second, Mitchell demonstrates the inadequacy of treating physical and representational landscapes separately. The reason, of course, is that they are themselves interlinked, and thus their reciprocal relation to labor processes can only be comprehended in the light of such linkages. Here, Mitchell follows in the sizeable footsteps of Denis Cosgrove (1984), who showed that the idea of landscape more generally is inseparable from the history of capitalist commodification, of labor *and land*, to which it belongs. The key contradiction of the classic landscape representation was its attempt to make the landscape appear "natural" or unworked. *The Lie of the Land* does something similar to Cosgrove in linking labor, landscape-as-morphology, and landscape-as-representation. Consequently it is a work of economic geography with considerable sophistication and power.

2.5   Principal Contemporary Narratives of Economic Geography

We have developed here an expansive definition of economic geography that allows, necessarily and intentionally, for a wide variety of scholarly economic geograph*ies*. In the next chapter, we examine the principal forms that economic geography has assumed historically. To complete the current chapter we want to focus explicitly on the present. If ours is a general and generic answer to the chapter's core question – "What is economic geography?" – what more specific answers to the same question are furnished explicitly or implicitly by the most prominent approaches to practicing and producing economic geography presently in circulation? In considering these narratives, our primary aim is to start to give the reader a clearer sense, not so much of what economic geography *can* in principle look like (our generic answer), but of what it *does* today look like in practice.

We also have an important secondary aim, which is to provide a direct and meaningful segue into the chapters which follow. Those chapters critically examine and explain economic geography through a series of different but connected lenses: its histories; its relations (with mainstream economics and with other social sciences); the theories and methods it brings to bear; its practicing "communities"; and its central substantive topics of analysis. Our argument in the current section of the present chapter is that the various principal contemporary narratives that we identify all mobilize distinctive *assemblages* – different conjunctural comings-together – of those topics, methods, theories, and so on that the book's later chapters examine in more depth. Our intention here is to isolate and describe at a headline level dominant current narratives, nodding to – but without elaborating upon – their theoretical underpinnings, communities of practice, topical interests, and so forth. The rest of the book, by elucidating the wider currents of economic geography, will enable the reader to situate and understand these principal approaches *in context*.

We identify four approaches, four distinctive narratives, within economic geography. We provide a brief overview of each, and compare them in Figure 2.2. That diagram raises as many questions as it provides answers, and for this reason we want to make three very brief points about our categorization before proceeding. First and most obviously, it is not comprehensive. Economic geography spills over any attempt categorically to constrain it. Second, economic geography clearly could be "cut" in innumerable ways, and differently from how we do so here: as *different* assemblages, for instance, or strictly on the basis of, say, theoretical disposition. But we think and hope our own cut is illuminating and helpful. And third, the boundaries between the narratives and approaches we identify are porous in several different senses, which our static figuring of them unfortunately does not reveal. For one thing, there is overlap between them: the narrative we term "Geographies of Business" has links and commonalities with both "Geographies of Capitalism" and "Geographical

Narrative ("economic geography is …")	Key themes	Key theoretical underpinnings	Selected key protagonists
Geographies of Capitalism	• Accumulation and value • Circulation of capital • Power • Social relations of production • Uneven development	• Political economy • Feminism • Postcolonialism	• David Harvey • Doreen Massey • Linda McDowell • Jamie Peck • Allen Scott
Geographies of Business	• Firms • Industrial clusters, districts, and regions • Innovation • Knowledge and learning • Networks	• Business studies • Institutional and evolutionary economics	• Ron Boschma • Gordon Clark • Meric Gertler • Ann Markusen • Henry Yeung
Geographical Economics	• Agglomeration economics • Spatial equilibrium • Trade costs • Urban economics	• Mainstream economics • Regional science	• Paul Krugman • Henry Overman • Andrés Rodríguez-Pose • Michael Storper • Anthony Venables
Alternative Economic Geographies	• Context • Discourse • Performativity	• Cultural economy • Poststructuralism	• John Allen • Christian Berndt • J.K. Gibson-Graham • Roger Lee • Andrew Leyshon

Figure 2.2   What is economic geography?

Economics." For another, some of those individuals we identify as key protagonists have made contributions in more than one area, often at different points in their research careers; Michael Storper, for instance, could arguably be "placed" within any of the three aforementioned approaches.

2.5.1   Geographies of Capitalism

Much of the contemporary scholarship falling within our definition of economic geography is concerned to understand the implication of geographies of various types in the central economic processes of capitalist society. Such scholarship focuses, then, on what are widely seen to be capitalism's cardinal economic dynamics, including but not limited to capital–labor relations, the circulation of capital (in and through money, commodity, and productive circuits), processes of capital accumulation, the social relations of credit and debt, and the state and regulation. Studies in this vein range from the macro (e.g., Harvey 2010; Mann 2013) to the much more micro, but they typically share the conviction that capitalism represents a coherent and expanding mode of socioeconomic production and reproduction with characteristic social structures, relations, and outcomes, and that it needs to be understood, theorized, and explained as such.

This narrative of, and approach to, economic geography is inspired primarily by political economy of various stripes. Such political economy, however, typically does not comprise the popular contemporary approaches known as "comparative" and "international" political economy, although these assuredly do feature in some economic geographic work.

Instead, it tends to be more longstanding political economic traditions that economic geography draws on and, in turn, strives to speak critically to: political economy from *before* the late nineteenth-century revolution in economic thought that spawned mainstream economics broadly as we know it today; and the political economy which subsequently emerged *alongside* mainstream economics in the first half of the twentieth century. From the first such era, Marx is especially influential for economic geographic thinking; from the second, the dominant figure is arguably Karl Polanyi. Whoever is the theoretical inspiration, however, and whatever is the particular strain of political economy, economic geography's signal endeavor is to aim to add geographical rigor and sensitivity to existing explanatory frameworks. Geographers argue such frameworks are limited by their underappreciation, or even outright neglect, of geography's substantive implication in capitalist economic processes.

Yet geographical political economy – as we might style the approach associated with this narrative of economic geography (Sheppard 2011) – in its current guise is also indebted to other sources of inspiration. If, as many working in this area would insist, political economy concerns itself above all with questions of socioeconomic power and its social-relational structuring, the insights of more contemporary traditions of thinking – concerned with axes of social differentiation other than class – must play no less a formative role than those derived from a Marx or a Polanyi. In economic geographic studies, as intimated earlier in this chapter, feminist theory has been particularly influential. But even as it reveals the stark gendering of economic processes and spaces, the work of scholars such as Pratt (2004) and McDowell (2008) continues to speak centrally to the theme of "Geographies of Capitalism." Rather than offering a separate narrative, and a separate answer to the "What is economic geography?" question, these scholars are better seen as answering that question in related but more expansive and inclusive ways. So, too, are those writing economic geographies shaped by postcolonial rather than by feminist theory but with capitalism's characteristic geographies similarly their abiding concern (e.g., Gidwani 2008).

2.5.2   Geographies of Business

The suggestion that "Geographies of Business" represents a materially different answer to this chapter's title question than "Geographies of Capitalism" might strike some readers as a curious one. Is capitalism not *about* business – its production of goods and services, its competitive environments, its revenues and costs, its growth and profitability imperatives? Economic geography written from the "Geographies of Business" perspective would answer: yes, it is. Economic geography written from the "Geographies of Capitalism" perspective would answer: yes, it is, but it is about so much more than that, too.

Economic geography in the "Geographies of Business" guise therefore displays a markedly different, and considerably narrower, scope than the first of our four stylized approaches. It provides a distinctively *partial* perspective on capitalism and its geographies insofar as it isolates one of capitalism's principal organizational forms – the firm – and makes it the privileged and abstracted locus of analysis. Its practitioners are more interested in how firms seek to differentiate themselves from one another, secure competitive advantage, and thus commercially succeed (or fail) than they are in understanding the firm in relation to the wider constitution and dynamics of capitalism as a mode of social (re)production – hence their interest in issues such as (corporate) learning and innovation, in the business studies literature, and in the work of that great "prophet of innovation" Joseph Schumpeter (McCraw 2010). Meanwhile, the geographies they investigate concern principally the spatial configurations of productive organization believed to encourage or discourage processes of innovation, learning, and competitive differentiation – industrial districts, learning regions and, above all else, clusters – rather than the spatial configuration of capitalist economies, in all their multidimensionality and heterogeneity. Hence their typical appeal to Alfred Marshall (Amin and Thrift 1992), not Marx.

If "Geographies of Business" narrows the frame within and through which "Geographies of Capitalism" views the economy, then why not simply treat the former as a subset of the latter? Why identify it as a *different* narrative and approach? Partly this is to do with theoretical inspiration. Political economy is almost wholly absent. In its place there is a turn to evolutionary economics (Boschma and Lambooy 1999), especially in its Schumpeterian variants, and to (new) institutional economics (Martin 2000), both of which frame the economy in decidedly different ways from political economy as invoked in economic geography. More importantly, albeit relatedly, "Geographies of Business" asks very different questions *about* the economy and its geographies. There is, for instance, negligible consideration of social (still less environmental) outcomes, which many of those working in the "Geographies of Capitalism" tradition not only place front and center but offer strong normative (generally negative) viewpoints about. Indeed, social relations, and the possibility of treating them as relations of power, scarcely feature: if studies positing economic geography as the geography of business treat the economic world as relational at all, the relations they are concerned with tend to be strictly those within and between firms and among the entrepreneurs who drive them, relations ordinarily figured as "networks" of one form or another.

2.5.3   Geographical Economics

The third dominant narrative of economic geography today is one which posits it as a geographical version of economics – of, specifically, mainstream (typically labeled "neoclassical") economics (Chapter 4). The key themes addressed by those adopting such an approach are not necessarily so different from those which recur under the two previous approaches. In fact, they represent something of a mix of the two. With "Geographies of Business," "Geographical Economics" shares a concern with industry clustering; with "Geographies of Capitalism," it shares a concern with uneven geographical development at the international scale. What *are* substantially different, however, are the disciplinary training and identity of the primary practitioners and the methods employed by them. We will take these in turn, although in reality they represent two sides of the same coin.

Although in Figure 2.2 we have included in our sample list of protagonists within "Geographical Economics" two scholars trained as geographers (Rodríguez-Pose and Storper), the fact is that the leading practitioners are nearly all economists who are trained in mainstream economics and working in economics departments. They tend to see their research therefore as adding geography to the geographically *in*sensitive corpus of knowledge that traditional mainstream economics represents – a "wonderland of no dimensions," in the felicitous phrase of one of the great forerunners of today's "Geographical Economics," the regional scientist Walter Isard (1949, p.477). The label that today's economists have chosen for this work of spatializing the aspatial, and what they see themselves *doing*, is "new economic geography" (e.g., Krugman 1998).

Meanwhile, the term "Geographical Economics" is the term usually applied to the work in question by those working on economic questions from the other side of the disciplinary divide, namely, geography. This alternative label reflects a certain suspicion of the so-called new economic geography and of the appropriation of disciplinary turf to which that term gestures. Partly by dint of their disciplinary affiliation, and partly by dint of their methods (see Chapters 4 and 6), Paul Krugman and other "geographical economists" are ultimately not regarded as doing "proper" economic geography. Hence the semantic distancing. We, however, disagree. As we have been at pains to argue in this chapter, we do not think economic geography is best defined as a bounded discipline, or as a set of prescribed methods, or as a set of limited thematic questions. If one defines economic geography as an attentiveness to and emphasis on the materiality of geography to the economic, one must include the "new economic geography," whatever label one affixes to it.

The methods issue is, nevertheless, pivotal. It is here that the truly significant differences between "Geographical Economics" and our first two (and fourth) approaches to economic geography lie. All the other approaches use primarily qualitative methods, albeit sometimes with quantitative methods mobilized on a limited basis. "Geographical Economics," by contrast, is fundamentally quantitative in nature, relying like mainstream economics more widely on formal modeling techniques.

Yet to identify the differences in terms merely of qualitative versus quantitative is to oversimplify (Chapter 6). First, there is the question of explanation. In "Geographical Economics," models, and the quantitative relations they express, are deemed *to* explain. For other types of economic geography, quantitative findings generally can do no more than describe, with explanation requiring qualitative analysis. Second, "Geographical Economics" and its models rely on assumptions that other types of economic geography refute. To be sure, some of the wider assumptions ordinarily invoked in mainstream economics, such as perfect competition and constant returns to scale, have been relaxed by the "new economic geographers." Yet more deep-seated assumptions – not least regarding the possibility and analytical power of general equilibrium and the derivation of aggregate behavior from individual maximization – remain. These set "Geographical Economics" definitively apart in the wider terrain of economic geography.

2.5.4 Alternative Economic Geographies

The contemporary economic geographic scholarship we want to highlight under this fourth and final heading offers a very different answer to the "what is" question. Indeed, difference is arguably its central narrative property. Although those practicing such economic geography would likely recoil from the very idea of definition, if they were pushed they might begin by saying that economic geography needs to be defined in tension to the three approaches already discussed. "Alternative Economic Geographies" is what those three are not, or, better, what they are not *only*. If the three previous approaches are the "dominant" or "leading" narratives, there needs also to be a type of economic geography to challenge or at least unsettle them and to achieve something different.

But it would be a very limited reading to leave things there. If the work we have in mind here is certainly "alternative" by being different, it is also alternative in another, more identifiable sense: it is alternative in terms of being explicitly non-*mainstream* in approach and subject matter. For example, the focus is generally not the spaces of large-scale commercial industry. Rather, it is those of Lee's (2006) "ordinary economy," which expand the very nature of the economy that geography takes as its object. As Leyshon, Lee, and Williams (2003, p.x) put it in introducing a volume of essays on just such "alternative economic spaces," the concern is with "the possibilities of diversity and economic proliferation," and even more expansively with "the perpetual opening out and transformation of social life." Those essays, and other comparable economic geographies, explore how "people create and implement ways of practicing economic life shaped and directed through sets of social relations differentiated from – and in some cases opposed to – mainstream relations."

If such geographies share a commitment to exploring alternative socioeconomic spaces, or spaces of "alterity" (e.g., Amin, Cameron, and Hudson 2003), they also share a methodological and theoretical pluralism again best described as non-mainstream. Often stimulated by poststructuralism (with its own distinctive take on "difference"), and both contributing to and shaped by the "cultural turn," "Alternative Economic Geographies" throws a sharp light on the *commonalities* rather than differences among our first three narratives, which can all appear rationalist, totalizing, even imperious. And where work in any of those three traditions seeks to shed such qualities – as it very much does in the work of a Pratt (2004) or a Gidwani (2008), for instance – it arguably aligns more with the narrative we are concerned with here.

This narrative, never actually singular, questions rationalism, embracing the uncertainty of metaphor (Barnes and Curry 1992). Relatedly, it is alive to the power of representation and of the social "discourses" that frame what is and is not – can and cannot be – spoken and written with confidence and authority in different historical and geographical conjunctures. It rejects the taken-for-granted, including the taken-for-grantedness of the economy, asking instead who "performs" the economy into being, and why. It also repudiates totalizing narratives, emphasizing instead the irreducibility of context. Pickles (2012) argues that the cultural turn, a catalyst of "Alternative Economic Geographies," was in fact a turn more than anything else *to* context. Perhaps above all, this approach denies the possibility or desirability of closure, explanatory or political. Economic geography needs to show humility. It must be open-ended, just like the socioeconomic spaces it studies.

2.6   Conclusion

In this chapter we laid the groundwork for the rest of the book. We have done so by considering a very elementary question: "What is economic geography?" Economic geography comes in many shapes and sizes, and in all these forms it provides its own answer – implicitly or explicitly, minimally or maximally – to that titular question. Those answers vary greatly. What fundamentally links them, we have argued, is not objects or methods of analysis or disciplinary location so much as an insistence on the materiality of the geographic in understanding the dynamics of economic life.

This, however, is arguably all that *is* constant and consistent about economic geography. Theories, methods, practices, even principles display a veritable multiplicity to which the following chapters will now seek to impart a sense of order, meaning, and critical assessment. Central to this critical perspective is a commitment to questioning the taken-for-granted. Things – including things in the contemporary world of economic geography – are not as they are accidentally. They are that way because they were *made* that way, even allowing for the historical imprint of the hand of contingency. The next chapter examines how economic geography became what it is today by excavating its own geographical histories.

Notes

1 It bears pointing out that this entry was written by one of us (Barnes).
2 http://www.oed.com/view/Entry/59393 (#11) (accessed July 7, 2017).
3 http://www.oed.com/view/Entry/59384 (#4c) (accessed July 7, 2017).
4 http://www.oed.com/view/Entry/59384 (#4a) (accessed July 7, 2017).

References

Amin, A., and Thrift, N. 1992. Neo-Marshallian Nodes in Global Networks. *International Journal of Urban and Regional Research* 16: 571–587.

Amin, A., Cameron, A., and Hudson, R. 2003. The Alterity of the Social Economy. In A. Leyshon, R. Lee, and C.C. Williams (eds), *Alternative Economic Spaces*. London: Sage, pp.27–54.

Barnes, T.J. 2008. Making Space for the Economy: Live Performances, Dead Objects, and Economic Geography. *Geography Compass* 2: 1432–1448.

Barnes, T.J., and Curry, M.R. 1992. Postmodernism in Economic Geography: Metaphor and the Construction of Alterity. *Environment and Planning D* 10: 57–68.

Boschma, R.A., and Lambooy, J.G. 1999. Evolutionary Economics and Economic Geography. *Journal of Evolutionary Economics* 9: 411–429.

Breslau, D. 2003. Economics Invents the Economy: Mathematics, Statistics, and Models in the Work of Irving Fisher and Wesley Mitchell. *Theory and Society* 32: 379–411.

Cameron, J., and Gibson-Graham, J.K. 2003. Feminising the Economy: Metaphors, Strategies, Politics. *Gender, Place and Culture* 10: 145–157.

Castree, N., Kitchin, R., and Rogers, A. 2013. *A Dictionary of Human Geography*. Oxford: Oxford University Press.

Chari, S. 2004. *Fraternal Capital: Peasant-Workers, Self-Made Men, and Globalization in Provincial India*. Stanford: Stanford University Press.

Christophers, B. 2011. Making Finance Productive. *Economy and Society* 40: 112–140.

Christophers, B. 2013. Mad World? On the Social Construction of Economic Value. https://antipodefoundation.org/2013/06/03/mad-world/(accessed June 14, 2017).

Christophers, B. 2014. From Marx to Market and Back Again: Performing the Economy. *Geoforum* 57: 12–20.

Coe, N., Kelly, P., and Yeung, H.W.C. 2007. *Economic Geography: A Contemporary Introduction*. Oxford: Blackwell.

Cosgrove, D. 1984. *Symbolic Formation and Symbolic Landscape*. London: Croom Helm.

Crang, P. 1997. Cultural Turns and the (Re)constitution of Economic Geography. In R. Lee and J. Wills (eds), *Geographies of Economies*. London: Arnold, pp.3–15.

Crewe, L. 2000. Geographies of Retailing and Consumption. *Progress in Human Geography* 24: 275–290.

Fioramonti, L. 2013. *Gross Domestic Problem: The Politics Behind the World's Most Powerful Number*. London: Zed Books.

Gibson-Graham, J.K. 1995. Identity and Economic Plurality: Rethinking Capitalism and 'Capitalist Hegemony.' *Environment and Planning D* 13: 275–282.

Gibson-Graham, J.K. 2008. Diverse Economies: Performative Practices for Other Worlds. *Progress in Human Geography* 32: 613–632.

Gidwani, V. 2008. *Capital Interrupted: Agrarian Development and the Politics of Work in India*. Minneapolis: University of Minnesota Press.

Goswami, M. 2004. *Producing India: From Colonial Economy to National Space*. Chicago: University of Chicago Press.

Gregory, D., Johnston, R., Pratt, G. et al., eds. 2009. *The Dictionary of Human Geography*, 5th edn. Oxford: Wiley-Blackwell.

Guha, R., and Spivak, G.C. 1988. *Selected Subaltern Studies*. Oxford: Oxford University Press.

Hart, G.P. 2002. *Disabling Globalization: Places of Power in Post-Apartheid South Africa*. Berkeley: University of California Press.

Harvey, D. 2010. *The Enigma of Capital and the Crises of Capitalism*. London: Profile Books.

Henderson, G. 1998. *California and the Fictions of Capital*. Oxford: Oxford University Press.

Isard, W. 1949. The General Theory of Location and Space-Economy. *The Quarterly Journal of Economics* 63: 476–506.

Krugman, P. 1998. What's New about the New Economic Geography? *Oxford Review of Economic Policy* 14: 7–17.

Lee, R. 2002. 'Nice Maps, Shame about the Theory'? Thinking Geographically about the Economic. *Progress in Human Geography* 26: 333–355.

Lee, R. 2006. The Ordinary Economy: Tangled Up in Values and Geography. *Transactions of the Institute of British Geographers* 31: 413–432.

Leyshon, A., Lee, R., and Williams, C.C., eds. 2003. *Alternative Economic Spaces*. London: Sage.

Mackinnon, D., and Cumbers, A. 2007. *An Introduction to Economic Geography: Globalization, Uneven Development and Place*. Harlow: Pearson Education.

Mann, G. 2013. *Disassembly Required: A Field Guide to Actually Existing Capitalism*. Edinburgh: AK Press.

Martin, R. 2000. Institutional Approaches in Economic Geography. In E. Sheppard and T.J. Barnes (eds), *A Companion to Economic Geography*. Oxford: Blackwell, pp.77–94.

Martin, R., and Sunley, P. 2001. Rethinking the "Economic" in Economic Geography: Broadening our Vision or Losing our Focus? *Antipode* 33: 148–161.

Massey, D. 1997. Economic/Non-economic. In R. Lee and J. Wills (eds), *Geographies of Economies*. London: Arnold, pp.27–36.

Massey, D. 2013. Vocabularies of the Economy. In S. Hall, D. Massey, and M. Rustin (eds), *After Neoliberalism? The Kilburn Manifesto*. London: Soundings.

Massey, D., and Allen, J., eds. 1984. *Geography Matters!: A Reader*. Cambridge: Cambridge University Press.

McCraw, T. 2010. *Prophet of Innovation: Joseph Schumpeter and Creative Destruction*. Cambridge, MA: Harvard University Press.

McDowell, L. 2008. Thinking through Work: Complex Inequalities, Constructions of Difference and Trans-national Migrants. *Progress in Human Geography* 32: 491–507.

Mitchell, D. 1996. *The Lie of the Land: Migrant Workers and the California Landscape*. Minneapolis: University of Minnesota Press.

Mitchell, K., Marston, S., and Katz, C., eds. 2004. *Life's Work: Geographies of Social Reproduction*. Oxford: Blackwell.

Mitchell, T. 2002. *Rule of Experts: Egypt, Techno-politics, Modernity*. Berkeley: University of California Press.

Mitchell, T. 2008. Rethinking Economy. *Geoforum* 39: 1116–1121.

Murphy, J., and Carmody, P. 2015. *Africa's Information Revolution: Technical Regimes and Production Networks in South Africa and Tanzania*. Oxford: Wiley-Blackwell.

Peck, J. 1996. *Work-Place: The Social Regulation of Labor Markets*. New York: Guilford Press.

Philipsen, D. 2015. *The Little Big Number: How GDP Came to Rule the World and What to Do about It*. Princeton, NJ: Princeton University Press.

Pickles, J. 2012. The Cultural Turn and the Conjunctural Economy: Economic Geography, Anthropology, and Cultural Studies. In T.J. Barnes, J. Peck, and E. Sheppard (eds), *The Wiley-Blackwell Companion to Economic Geography*. Oxford: Wiley-Blackwell, pp.537–551.

Pratt, G. 2004. *Working Feminism*. Philadelphia: Temple University Press.

Roberts, S.M. 2012. Worlds Apart? Economic Geography and Questions of 'Development.' In T.J. Barnes, J. Peck, and E. Sheppard (eds), *The Wiley-Blackwell Companion to Economic Geography*. Oxford: Wiley-Blackwell, pp.552–566.

Said, E. 1978. *Orientalism: Western Conceptions of the Orient*. New York: Random House.

Shamir, R. 2008. The Age of Responsibilization: On Market-Embedded Morality. *Economy and Society* 37: 1–19.

Sheppard, E. 2011. Geographical Political Economy. *Journal of Economic Geography* 11: 319–331.

Thrift, N., and Olds, K. 1996. Refiguring the Economic in Economic Geography. *Progress in Human Geography* 20: 311–337.

Wallerstein, I. 2003. Anthropology, Sociology, and other Dubious Disciplines. *Current Anthropology* 44: 453–465.

Waring, M. 1999. *Counting for Nothing: What Men Value and What Women are Worth*. Toronto: University of Toronto Press.

Chapter 3

Inventing Economic Geography: Histories of a Discipline

> "Reality isn't there yet; it has to be brought forth or produced; and this is the duty and stake of writing."
>
> Tom McCarthy (2014, p.21)

3.1 Introduction

It might seem odd to put "inventing" in front of "economic geography" in this chapter's title. Hasn't economic geography existed forever? It has no need to be invented. It is not like the incandescent light bulb or anything. We suggest in this chapter it is, though. Economic geography is very much an invention. Further, like lots of inventions, once it entered the world, the world was reshaped. Following Tom McCarthy, writings by economic geographers have helped bring forth or produce a new reality.

Material acts that we now think of as economic geographic, those meeting the definition of economic geography that we gave in Chapter 2, have existed for a very long time (Sahlins 1972). Perhaps Eve picking fruit from the tree of knowledge in the Garden of Eden was the inaugural economic geographic act. This act, as well numerous others, could not at the time be described as economic geographic, however. This is because the very idea of economic geography was not yet invented. When did that happen?

That's difficult. Possible contenders include: 1925, when the flagship journal of the subject, *Economic Geography,* was first published at Clark University in Worcester, Massachusetts

---

Economic Geography: A Critical Introduction, First Edition. Trevor J. Barnes and Brett Christophers.
© 2018 John Wiley & Sons Ltd. Published 2018 by John Wiley & Sons Ltd.

(see Chapter 7); or 1893, when economic geography courses were first taught at Cornell University and the University of Pennsylvania (Fellmann 1986); or 1882, when the German geographer Wilhelm Götz distinguished between commercial and economic geography (Sapper 1931); or 1826, when *Der isolierte Staat* (*The Isolated State*), for some the first classic treatise in economic geography, was written and privately published by the German landowner and farmer Johann Heinrich von Thünen (1783–1850) (von Thünen 1966 [1826]).

These dates at least narrow the invention of economic geography to the nineteenth and early twentieth centuries. They suggest there was something about that period – at least in Western Europe and North America – that was ripe for the emergence of the discipline. Again there is a parallel to the process of invention. To invent, say, the incandescent light bulb, as Thomas Edison did in 1879, required a peculiar combination of circumstances to make it possible including knowledge of electricity, the means to generate electricity and transmit it across long distances to users, and a research laboratory containing instruments, materials, and trained technicians and scientists. It was the conjuncture of these historically contextual factors that made the incandescent light bulb conceivable, shaping its later form and capabilities. Moreover, once the incandescent light bulb was invented, it brought forth and produced a new reality: radiant night-time city streets; brightly lit homes after dark; and illuminated factory spaces enabling round-the-clock production.

The same holds for the invention of economic geography. It is not possible in a single chapter to identify all the historical contextual factors that made the invention of economic geography conceivable. There are some obvious candidates, however. One was the prior existence of discursive entities such as "society," "economy," and "nation." They were necessary conceptual building blocks for economic geography. Without the idea of the economy, for example, it would be hard to imagine how economic geography could have been invented. Likely it wasn't until the seventeenth and eighteenth centuries that the idea of a separate, independent sphere, the economy, was conceived (Dumont 1977; Buck-Morss 1995) (see also Chapter 2). Another was the emergence, first in Western Europe then in North America, of social science as a form of institutionalized inquiry concerned with understanding and analyzing social processes broadly conceived. Under its umbrella were such disciplines as economics, anthropology, sociology, politics, and by the end of the nineteenth century, economic geography. That these disciplines were institutionalized during this period was connected to a changed European and North American nation-state increasingly bent on controlling, managing, and organizing its population, as well as the populations of colonies. Recognized by Michel Foucault (1977) as biopower, the state strove to control and manage its population by drawing upon and deploying knowledge acquired from the new social sciences. Yet another factor was the growth of universities in Western Europe and North America. Their expansion was prompted in part by a rising middle class that could afford to send its children to be educated in, among other subjects, economic geography. And yet one more factor was the rampant program of European colonization particularly from the late nineteenth century ("the Scramble for Africa"). European imperialism induced hitherto unprecedented levels of global economic specialization, commerce, transportation, and movements of people. Made possible in part by inventions like Edison's, it was also made possible, as we will argue, by the invention of new academic disciplines like economic geography.

The invention of economic geography was thus hardly innocent. It emerged from a concatenation of social, political, geopolitical, economic, and institutional interests.

Those interests were shaped and motivated by, among other things, various forms of power, economic inequality, racial prejudice, and vaulting imperial ambition. If there is any part of economic geography that requires a critical introduction it is its history. Further, that same critical sensibility is also needed to examine the effects of the discipline as it went out into the world, whether as a university class, or as a specialized text, or as a specific kind of practice or technique.

There is one other point. Inventions like the incandescent light bulb and academic disciplines like economic geography require much hard work to be realized. Their completion is fraught and precarious. Edison famously said, "genius is one per cent inspiration and ninety-nine per cent perspiration." There was a lot of perspiration in inventing both the light bulb and economic geography. Neither task was for the weak or faint hearted. They involved bringing together and enrolling various kinds of resources, trying to persuade them to work together collectively to produce a stable entity – whether a light bulb that could remain lit for more than a few minutes, or an academic discipline that generated distinctive knowledge. In neither case was the invention instant, but instead unfolded hesitantly over time, with false starts, dead ends, and frequent regret (Edison said, "I haven't failed. I have just found 10,000 ways that won't work"). The history of economic geography is a history of hard work, of intellectual sweat, but also of foiled efforts, frustration, and disappointment.

The chapter is divided into five short sections, each turning on a different historical context that profoundly influenced the shape and content of the economic geography that emerged in that period (and explaining also the pluralizing within the chapter title). The first concerns the institutional emergence of economic geography during the late nineteenth century within the context of colonialism and associated burgeoning global commerce. The second is about the interwar period in which regionalism – defined as the systematic description and comparison of regions – is dominant. The third focuses on the period immediately after World War II when economic geography mimics natural science, seeking formal theory, rigorous empiricism, iron-clad explanation, and instrumental intervention. The fourth is about the emergence of a radical political approach, political economy, which unfolds from the early 1970s during a period of increasing (global) industrial turmoil. The last is the discipline's poststructural phase, which began sometime in the 1990s and is associated with a shift and reconception of both the economic and the role of economic geographers.

There are at least two qualifiers. In setting out our five periods from oldest to most recent we are not implying any big P (absolute) progress. To establish Progress there must be a set of independent and objective criteria for measuring it. No such criteria exist. While within each period there are local criteria for marking progress, those criteria are radically different across the different periods. Those differences mean there is no basis for asserting big P progress. The other qualifier is that ours is *a* history – or a series of overlapping histories – of economic geography and not *the* history of economic geography. There can never be *the* history of economic geography precisely because of the point just raised: that every account, including ours, is necessarily stuck inside the bubble of its own historical setting. We are also constrained by our restricted language abilities and the limits of our knowledge of traditions in economic geography other than Anglo-American ones. This chapter is concerned primarily with the invention of English-speaking, Anglo-North American economic geography.

3.2 In the Beginning: Commercial Geography

The very first economic geography textbook written in English was George Chisholm's *Handbook of Commercial Geography* published in 1889 (Figure 3.1). It was a book with legs, going into 20 editions, with its last republication as recent as October 2011. By then its author had been dead for more than 80 years.

Chisholm was born in Scotland. He attended Edinburgh University, and later worked for the Edinburgh publisher W.G. Blaikie & Son on such projects as *The Imperial Dictionary*. He then moved to London, becoming a member of the Royal Geographical Society (RGS) in 1884. He made his living primarily from writing and editing geographical textbooks, gazetteers, and atlases. From 1896 he supplemented his income by lecturing on commercial geography at the University of London's Birkbeck College (Wise 1975; MacLean 1988).

Photo by Elliot and Fry.

Figure 3.1 George G. Chisholm (1850–1930). Source: *The Scottish Geographical Journal.*

At age 58 Chisholm came home, having been appointed Lecturer at the newly formed Department of Geography, University of Edinburgh.

The *Handbook* was written while Chisholm was still in London, and was later used by him as a text for his Birkbeck College extension course for mature students. They took the course primarily to better their prospects. They believed that detailed knowledge of British imperial commerce, the subject of Chisholm's course and *Handbook*, might enable them to share in its fruits. In 1889, the year the *Handbook* appeared, Victorian British imperialism was in its heyday, with the United Kingdom the workshop of the world. That project was about not only acquiring territories – coloring the world map pink – but also economically transforming them to benefit Britain. That transformation often turned on the production and trade of particular commodities that were either indigenous to the colony, or were introduced from outside by the colonists. The classic example was the cultivation of tea in India. Although not indigenous to the subcontinent, the value of tea production went from nothing in 1850 to becoming the country's most valuable cash crop by the close of the nineteenth century. Commodity-based imperial commerce framed every page of Chisholm's book. It was found in his meticulously compiled statistical appendix, in his many tables, in his myriad detailed descriptions of individual commodities, and in his maps, especially insert maps separately glued to the book's interior binding. For example, the map insert of India showed that once that subcontinent was laid out on a piece of paper (Figure 3.2), its spaces could then be literally written on by the British colonial project: tea in the Northeast, opium in the Northwest, coffee in the Southwest, and wheat, millet, cotton, and oil seeds in the center.

Also extending through the pages of Chisholm's book was another imperial project, a disciplinary one. It was to colonize an academic space for the new subject of economic geography. The *Handbook* became the Anglo-American starting point for a brand new subject. For anyone who doubted the new discipline one could literally show its substance by pointing to Chisholm's book (it was over 500 pages long and weighed over a kilogram). Economic geography was no longer a weightless idea, or intellectual distinction, but in the form of Chisholm's volume it was something tangible that could be held in the hand. Once published and in circulation, it was physical proof that economic geography as a discipline possessed substance. It took a lot of work, though, much of it mundane and routine rather than inspirational (following the Edisonian tradition). Chisholm spent enormous amounts of time finding and assembling data in tabular form, searching down every last fact, and constructing detailed maps (albeit often drawn not by him but by another RGS member, F.S. Weller). He was less keen on theory, however, and in one hyperbolic moment was even tempted "to wish … th[e] love of pure theory to the devil" (quoted in Wise 1975, p.2). Chisholm's practices might appear ordinary, tedious even,[1] but they contributed to the reproduction of imperialism, while at the same time helping to invent a new subject.

For Chisholm the intellectual rationale for economic geography was given by environmentalism: the idea that the physical environment influenced the kind of economic activity carried out at a given place. Chisholm (1889) wrote on page 1 of *The Handbook*:

Economic geography should be concerned with examining places in terms of the natural advantages that they possess for the supply of a particular market. By natural advantages are meant such as these – a favourable soil and climate, the existence of facilities for communication external and internal so far as these lie in the nature of the surface and physical features, the existence of valuable minerals in favourable situations and so on.

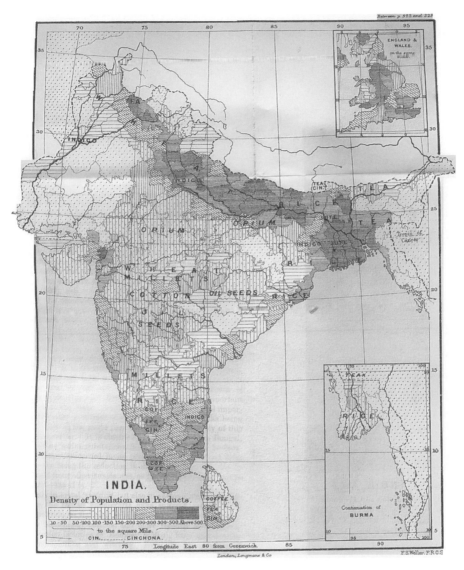

Figure 3.2   "India: Density of population and products." Chisholm 1889, between pp.322–323.
Source: *The Handbook of Commercial Geography*, 1st edition.

For Chisholm there was something about the natural environment that made each place uniquely suitable to undertaking particular types of economic activities. That position became more extreme, with explicit racist overtones, in the hands of a number of environmental determinists working in economic geography at the turn of the century. Environmental determinism was racist because it suggested that only racial characteristics, set especially by climate, directed the ability of a people to carry out particular activities, including making a living. Perhaps the most notorious of all environmental determinists was the Yale University-based geographer Ellsworth Huntington (1876–1947), the first American to use the term economic geography (Box 3.1: Ellsworth Huntington and

Box 3.1   Ellsworth Huntington and Environmental Determinism

Ellsworth Huntington. Source: Images of Yale individuals ca.1750–2001 (inclusive). Manuscripts & Archives, Yale University.

Ellsworth Huntington argued that the level of economic development of a given place was determined by the climate regime found there. His 1915 book *Climate and Civilization* included a chapter on "Work and weather" (Huntington 1915, chapter 6). He calculated that mental work efficiency was maximized if mean seasonal temperatures remained above 38°F (3.3°C), and physical work efficiency was maximized if mean seasonal temperatures did not exceed 65°F (18.3°C) (Huntington 1915, p.129). His conclusion was that mental and physical work occurred most efficiently in the world's temperate regions, which just happened to be occupied by white populations (that is, if you forget the aboriginal inhabitants who in any case had been mainly decimated by earlier European colonists). People in North or Central Africa or South or East Asia, or Central or South America could never make it economically because the climate regime under which they labored made them labor so badly. The early American economic geographer J. Russell Smith fully approved, writing to Huntington to say that the latter's "chart showing the relation of human output to temperature" was "real geography" (quoted in Livingstone 1994, p.143, fn. 35).

In 1924 Huntington extended his analysis. He mapped the variable global levels of mental and physical efficiency, now combined as a single measure, "climatic energy," and correlated with a second index measuring the spatial distribution of levels of civilization drawn up by 120 "experts" who included J. Russell Smith (Figure 3.3). In Huntington's mind, the results affirmed the case for the "The White Man's Burden," the title of an 1899 poem by Rudyard Kipling that expressed the idea that non-European populations could never generate autonomous development and civilization. Both needed to be brought to them by "white men" who had already achieved development and civilization.

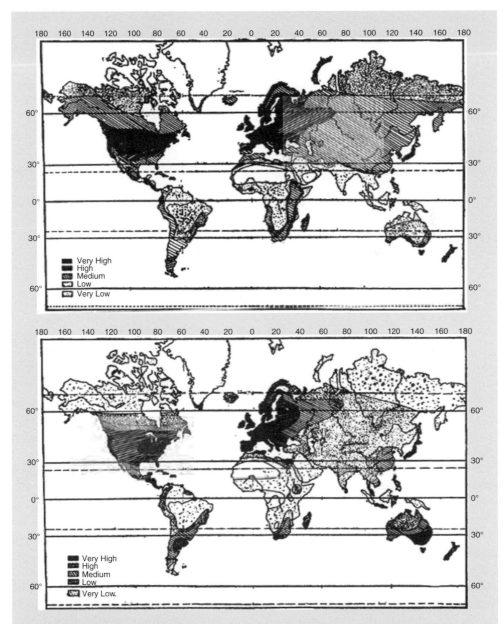

Figure 3.3 According to Huntington, the heights of civilization, found in Western and Northern Europe, the North East, Mid-West, and West Coast of America, and the East Coast of Australia and New Zealand, correspond to the heights of "climatic energy."
Source: Huntington 1915, p.200.

Environmental Determinism). Huntington, in fact, invited Chisholm to participate in his environmental determinist project to map and rank the world's civilizations on the basis of the climatic regime in those places. But while Chisholm was an imperialist, he was no racist. He refused Huntington's request, saying that he had "a peculiar incapacity for forming [such] judgments" (quoted in Livingstone 1994, p.143). That said, all those numbers Chisholm had compiled, maps he had drawn, and facts he had reported helped bring forth and produce the reality of British imperialism.

Economic geography did not simply describe the late nineteenth-century imperialist world in which it arose, but was complicit with it, contributing to its maintenance and expansion. Furthermore, insofar as the discipline relied on environmental determinism as a causal mechanism for explaining geographically divergent levels of the economy (uneven development), it engaged in a barely disguised racism bent on justifying European imperialism and white settler colonialism. If ever there was a need to bring a critical sensibility to economic geography it is to understand these politically and morally shaky institutional beginnings.

3.3   Economic Regionalism

Even before World War I began in Europe in 1914, there were signs of change in economic geography. There was a movement away from the focus on global commodity production that preoccupied Chisholm to a focus on discrete regions. In 1914 the American economic geographer Ray Whitbeck argued in a critical review of Chisholm's book that there was "a distinction between a textbook of commerce and industry and one on commercial and industrial geography" (Whitbeck 1914, p.540). For Whitbeck (1915–1916, p.197) the emphasis should be geographical: "the unit should be the country and not the commodity."

A focus on the region had long antecedents in the discipline, going as far back as the Ancient Greeks, who along with the earth (*geos*) and place (*topos*) also recognized the centrality of the region (*choros*) to geographical study (Lukermann 1961). That the region resurfaced as the unit of inquiry in economic geography during the early part of the twentieth century was also a result of other contemporary factors, often going to shifting relations of power both outside and inside the academic world. The slowing of colonialism (for example, the Scramble for Africa had ended by 1914), and the effects of the Great Depression of the 1930s on world trade and commerce, rendered Chisholm's scheme focused on global commodity production less pressing. There were also internal sociological reasons, of which the most cited was the embarrassment from the discipline's earlier affair with environmental determinism. A turn to regional study represented if "not greener pastures" then at least "smaller safer pastures" (Mikesell 1974, p.1). The discipline of geography itself was also becoming more professionalized and institutionalized within a university setting. As it did so, the stakes became higher, with individual academic reputations hinging on the successful promulgation of original ideas and discourses such as the region.

Practicing a regional economic geography meant deploying a fixed typological scheme to classify the salient facts of a given region. For example, for the region British Columbia, put Vancouver and Victoria under the category "Cities," put Coastal Mountains and Fraser River under the category "Topography," and put wood products industry and motion picture industry under the category "Economy." The same classification scheme would then be used to characterize neighboring regions such as Alberta or Washington State. Once all regions were described, that is, when all boxes were filled in creating a single large typological grid, a region's distinctiveness became immediately obvious. By running your eye along any row – "cities," "topography," "economy" – the uniqueness of a region was apparent straightaway by comparing the entry of one cell with another.

For example, Ray Whitbeck and his University of Wisconsin colleague Vernor Finch practiced this form of economic regionalism in their joint textbook *Economic Geography*, published in 1924. It provided "a large and varied body of detailed facts" (p.4) organized by political region beginning with the United States and Canada. Those facts were neatly ordered under an identical fourfold typology for each of the regions investigated: "agriculture," "minerals," "manufacture," and "commercial trade, transportation, and communications."

The regionalist perspective represented a marked shift in the economic geographic discourse, changing the object of enquiry. The focus of the discipline was no longer the single commodity and its travels but a fixed geographical entity, the region, defined by a list of geographical facts organized by predetermined categorical boxes. In 1939, the American geographer Richard Hartshorne (1899–1992), in his book *The Nature of Geography*, provided a philosophical and methodological rationale for this regional approach.

Hartshorne conceived the region as consisting of a unique combination of objective geographical elements. That conception meant, however, that the traditional natural scientific explanation based on physical laws could not apply (see Chapter 5). Scientific laws pertain only to classes of phenomena that are homogeneous. Because one atom of, say, hydrogen gas is like any other, one can generalize from experiments on a sample of hydrogen atoms about the effect of an outside force, for example, heat or pressure, to all hydrogen atoms. Because of homogeneity, it becomes possible to universalize, to state a law – Boyle's Law, for example. Regions are not like hydrogen atoms, though. The mixture of constitutive elements forming them makes each region unalike, leading them to react differently to the same cause. Consequently, one can't make law-like generalizations about regions. As Hartshorne (1939, p.446) summarized, "We arrive, therefore, at a conclusion similar to that which Kroeber has stated for history: 'the uniqueness of all historical phenomena…. No laws or near laws are discovered.' The same conclusion applies to the particular combination of phenomena at a particular place." Geographers, therefore, cannot do any of the things that natural science accomplishes by its recourse to physical laws; that is, to explain, to predict, and knowingly to intervene. In the Hartshornian conception of the region, geographers only describe. "Regional geography, we conclude, is literally what its title expresses: … [I]t is essentially a descriptive science concerned with the description and interpretation of unique cases …" (Hartshorne 1939, p.449).

In retrospect, the timing of Hartshorne's argument for an economic geography based on the descriptive study of unique regions could not have been worse. In 1939, when Hartshorne published his book, social sciences and even some humanities in the United States were on the cusp of fundamental transformation. They were about to be changed into quasi-natural sciences, seeking everything that Hartshorne's conception of regions was unable to offer: general explanation, prediction, and knowing intervention. After World War II, this new conception of "naturalized" social science took hold. Geography was initially a hold-out. For at least a decade or so the old regionalist paradigm was maintained, held in place in part by an "elite club" of senior male geographers (including Hartshorne) that policed the discipline (Butzer 1989, p.5). From the mid-1950s onward that club began to lose its grip, a result in part of criticisms from a younger generation who were never members, and in part of a changing academic and political context. Slowly at first, but then much more quickly, there was a shift away from the old regional geography to a new kind, called later "spatial science." Regions were still discussed, but conceived utterly differently, not as unique places, as lists of discrete facts, but as abstract spaces to be rearranged, manipulated, formally represented, and used as a tool to achieve instrumental planning objectives. Economic geography became one of the contributors to bringing into being a new post-World War II world.

3.4   Spatial Science

3.4.1   World War II and the Cold War

The larger context for this change comprised macro-scale geopolitical conflicts, World War II followed by the Cold War. The Allies won World War II, it was claimed, because of their better science. Social sciences, including geography (Barnes and Farish 2006), played a part, but the physical sciences were very much to the fore. As a result of radar, calculating machines like the Colossus used at Bletchley Park to break the German Enigma code, ENIAC (Electronic Numerical Integrator and Computer) used at the ballistics firing range in Aberdeen, Maryland, and most spectacularly the development of the "Bomb," the Allies secured military victory. Because of the success of the physical sciences, social sciences were increasingly modeled on them. Social science began to mimic especially "Big Science," defining itself by interdisciplinary research teams, solving practical problems, and deploying rigorous analytical methods and objective data. During the Cold War these features became even more entrenched and pervasive, and even some humanities, such as philosophy, linguistics, and literary criticism, followed suit (Schorske 1997).

The other factor in the emergence of spatial science was a much more interventionist state. From World War II the state began aggressively using its considerable power and resources to regulate, plan, and control in order to realize not only military but also domestic socioeconomic objectives. One of the seeds of that change was the publication in 1936 of the economist John Maynard Keynes's book *The General Theory of Employment, Interest and Money* (Chapter 1). In effect that volume was a how-to manual for government intervention. It identified buttons needing to be pressed and levers pulled for governments to produce particular economic outcomes. The book, in combination with the state's demonstration of its effectiveness as a planner during World War II, helped create and maintain what was called, at least in the United States during the Cold War, the "military-industrial complex." It was an assemblage of tight interlinkages among the state, private business (mainly large corporations), and often universities. Within this world of Keynesian military-industrial planning, economists devised a burgeoning array of models, measuring techniques, theoretical precepts, and predictive tools to accomplish state intervention and control. In doing so they were inventing a new kind of economics. As we will see, the younger generation of geographers who were never part of the "elite club" of regionalists drew upon similar tools and similar aims. They aimed to invent a new kind of economic geography that could intervene practically in the spatial economy, to be another element, albeit minor, in the Keynesian military-industrial complex.

This disciplinary move both toward a natural scientific sensibility and cozying up to an enormously powerful military-industrial state called out for critical political interrogation. An entirely new politics of knowledge based on scientific objectivity was claimed, rooted in a masculinist rationality (masculinist in that those putting forward the approach were overwhelmingly men, but also in that the approach asserted purity, universalism, and dominance). Along with that went also a new politics of state accommodation, with the discipline carrying out its bidding. Very little critical interrogation occurred, however, and when it finally did the movement was almost at an end anyway.

3.4.2 The "quantitative revolution"

Younger geographers who were never part of the club, at least in the United States, initially gathered at two principal sites: the University of Iowa, Iowa City, and the University of Washington, Seattle. It was from Iowa that the first explicit attack on Hartshorne and economic geography's regionalism was launched. In 1953, Fred K. Schaefer (1904–1953), a socialist political refugee from Nazi Germany, and inaugural member of the Department of Geography at Iowa, published in the discipline's flagship journal, *Annals of the Association of American Geographers*, "Exceptionalism in Geography" (Schaefer 1953). It was a sweeping denunciation of Hartshorne's position, calling instead for a scientific approach to geography based upon the search for geographical laws. Schaefer died before his article even appeared in print, and was unable to defend himself from Hartshorne's (1955) subsequent blistering attack. But the article became a rallying point for a younger generation of economic geographers intent on practicing the discipline differently, as "spatial science."

To do so required the techniques, logic, and vocabulary of science. Here a colleague of Schaefer's at Iowa, the Head of Department, Harold McCarty (1901–1987), played a critical role. While suspicious of Schaefer's grand philosophizing, and rankled by his combativeness, McCarty concurred with Schaefer that economic geography should move away from regionalism and become more scientific. That meant two changes.

First, the use of theory, but not any old theory (see also Chapter 5). It needed to be theory of the kind found in the physical sciences. It should both explain (to identify the causes of an event) and predict (to anticipate an event before it occurs). To achieve both ends, theory should be expressed as a set of logically connected abstract terms, as found, for example, in Isaac Newton's paradigmatic scientific theory of gravity:

$$G = M_1.M_2/d^2$$

Here a set of abstract terms, G (gravity), M (mass), and d (distance) are logically related using a mathematical formulation. Mathematics serves as a purified form of connecting the terms rationally.

Second, the use of empirical data and statistical techniques of analysis (see also Chapter 6). Theory as it is defined above is expressed as a set of only logical relations among terms. For it to have explanatory or predictive purchase theory must be empirically connected to the world; that is, the abstract terms of theory must be operationalized as real-world statistical variables. This second step involves constructing variables, gathering data about those variables, then measuring, calculating, and undertaking statistical analysis.

Both these steps, designing theory and carrying out statistical analysis, defined spatial science. At Iowa, McCarty thought that economic geographers were suited to carrying out only the second of these tasks, data gathering and statistical analysis, not the first. He thought theory development should be handed over to economists. Consequently, McCarty and his students at Iowa carried out in the name of spatial science only statistical analysis, principally drawing on correlation and regression techniques to measure spatial association in industrial geography (McCarty, Hook, and Knox 1956; Barnes 1998).

At the other site, the University of Washington, there was less reluctance to theorize. There were two catalyzing faculty members, Edward Ullman (1912–1976) and William Garrison (1924–2015). Then there were the graduate students, later nicknamed the "space cadets," who serendipitously arrived in Seattle all around the same date, *annus mirabilis* 1955. Garrison offered the first ever advanced course in statistics in a US geography department in 1954. It wasn't only numbers that were important but also machines. There were the large, cumbersome electrical Friden calculators, but more significant was the even larger, more cumbersome, computer. In an early advertisement for the department, Donald Hudson (1955) boasted about the departmental use of an IBM 604 digital computer, also a national first. The programming technique of so-called plug wiring, involving plugging wires into a circuit board, was crude and inefficient, but it helped define and consolidate the new vision of the discipline based on science and the latest technology. The theory came from a rediscovered branch of German economics, location theory (see Box 4.2: The German Location School). Here newly translated books on urban economies by August Lösch (1954 [1940]) and Walter Christaller (1966) proved invaluable, enabling the "cadets" to link their empirical analyses, especially of metropolitan Seattle, to formal theory and the construction of mathematical models.[2]

We already begin to see the shape of a different kind of economic geography emerging. It was clearly not Hartshorne's field-based, typological description of the region. The new economic geography was primarily undertaken at a desk, involving calculators, computers, graphs, numbers, and spreadsheets; it employed an increasingly abstract set of terms taken from economics and physics (Newton's gravity equation was a special favorite); it concerned itself with finding causes and explanations, and not simply with classification; and it was not content with written descriptions of the unique, but focused on the logical and numerical analysis of the general. Economic geography was becoming a fully-fledged social science stressing the social over the natural and scientific analysis over "mere" description; today's "Geographical Economics" (Chapter 2) contains powerful echoes of this type of postwar economic geography. More generally, as economic geographic discourse and practices were reinvented, an entirely different economic geographic world emerged. It was defined by Euclidean and non-Euclidean space, geometrical axioms, Greek symbols, and regression lines.

But it was hard work, and often contested. This can be seen in Britain where a similar process unfolded, albeit slightly later. There "the revolution" was associated especially with Peter Haggett, who was at Cambridge and later Bristol University, and Richard Chorley, a physical geographer at Cambridge. They were given the moniker the "terrible twins" because of their aim to shake up the discipline by moving it to a scientific approach. In Haggett's (1965) case that approach was largely worked out within the context of economic geography. He argued that through a process of abstraction economic geographers should construct an ideal simplification of reality – a model – and use statistical methods to test it against the real world.

Further reinforcing economic geography's trajectory toward spatial science during the mid-1950s was a related and allied movement, regional science (see also Box 4.1). It was the child of the energetic and ambitious American economist Walter Isard (1919–2010). His purpose was to add spatial relationships to the hitherto "wonderland of no dimensions" that constituted the economist's world. The strength of Isard's theoretical treatise *Location and Space Economy* (1956) (immediately used by Garrison at Washington for his economic geography seminar), was such that by 1958 Isard had established the first Department of Regional Science at the University of Pennsylvania. During these early years there was theoretical, personnel, and institutional cross-pollenization between regional science and economic geography producing mutual benefit. It was a relationship always pregnant with potential conflict, however, because the two projects claimed the same academic turf. Isard's attitude that economic geographers were simply hewers and bearers of data for regional scientist theoreticians didn't help, despite the fact that it was close to McCarty's original view.

By 1960 these different elements – the new economic geography in the United States and the United Kingdom, and regional science – came together to form a coherent network. It was characterized by a specific set of geographical nodes such as Iowa City, Seattle, Philadelphia, and Cambridge (UK). They were connected by: the circulation of papers – early on graduate students at Washington initiated their own Discussion Paper series and sent it to kindred souls around the world; flows of money – especially important in the United States were grants from the Office of Naval Research that favored the large-scale, collaborative, and practical projects pursued by new economic geographers (Pruitt 1979); and the movement of people – visitors were brought in, sometimes from far afield, and as graduate students obtained their PhDs they carried the revolutionary spirit with them to departments where they were hired.

Its practical-mindedness as an approach, along with cutting-edge techniques and technologies, aligned the discipline in the United States at least with state and even military ends (Barnes and Farish 2006), certainly with state and military research funding. There was almost no internal political criticism of that alignment, however, or of the new conception of objective knowledge that the discipline proffered – abstract, statistical, mathematical, and model-based – that relied less on being in the field than being in the sterile spaces of a computer center or in a room full of mechanical calculators. What were the politics of being enrolled in the military-industrial-academic complex, though? And what were the politics of an objective knowledge that seemingly derogated subjectivity, morality, and the rich messiness of everyday spaces and places that earlier geographers at least tried to represent? Some of those critical questions began to be posed by what came next in economic geography, a concerted critical disciplinary sensibility, radical economic geography.

3.5 Radical Economic Geography

From the 1970s spatial science came under increasing attack. By then it was clear that science and a scientific methodology could not do everything. The world appeared increasingly broken, alienated, and traumatized. There was an environmental crisis (the first Earth Day was April 2, 1970). Civil rights campaigns, especially in the United States, mounted mass rallies, protest marches, and various acts of civil disobedience to overthrow the historical yoke of institutionalized racial segregation and discrimination. From the early 1960s second-wave feminists criticized and took action against an entrenched male patriarchy and privilege. Urban riots broke out all over the world, the one in Paris in May 1968 almost toppling the French government. And in places like Vietnam, Cambodia, and Laos, science and technology were used not to improve the human lot but violently to worsen it. Science was no magic solution providing a happy-ever-after ending. With that realization at least some within social science cast around for other approaches that offered more persuasive bases for understanding the then current messy, fractured, and riven world.

By the early 1970s there were also signs that it wasn't business as usual for the economy. Apart from the occasional downturn, the postwar period in North America and Western Europe had been a Fordist "Golden Age" with full employment and rising productivity and wages. From the early 1970s things started to unravel, however. The Golden Age became progressively tarnished by high rates of inflation, stagnating productivity, falling profits, sluggish employment growth, and slowing investment. By the end of the decade, the economy was in full-blown crisis, and this took a distinctive geographical form. Longstanding sites of manufacturing, such as the US North East and Mid-West, or the Midlands and the North of England, experienced rapid, and in some cases catastrophic, declines in investment. Millions of industrial workers were laid off, and manufacturing towns and cities became shadows of their former selves. The manufacturing belt became the rust belt. The flip side of "deindustrialization" was the rise of new manufacturing spaces. Some were high-tech clusters in places like Silicon Valley, California, or Cambridge in the United Kingdom. The vast majority, though, were newly established export processing zones located in the Global South. The manufacturing carried out there was often by one large multinational corporation or another, drawing on cheap, frequently female labor, and engaging in mass production of component parts or in assembly.

In coming to grips with this turbulent and torn world, economic geographers increasingly moved away from the models of spatial science cast, for example, as equations of the Newtonian law of gravity, or as the geometrically ordered landscapes of German location theory. Instead, they drew on social theory. Their argument was that to understand the social world and its churning character the starting point must be with an analysis of society itself. That was where spatial science went wrong. It was based on physical science, which while appropriate for understanding the natural world was not up to the task of understanding the social world.

Of the various social theories available for understanding the social world, economic geographers gravitated especially to Marxism. It averred that the economy was inseparable from society and politics, indissolubly joined. From the early 1970s, a number of economic geographers took up the analysis of a Marxist perspective in a movement dubbed radical economic geography.

3.5.1   David Harvey's Theory of Geographical Accumulation

David Harvey was and remains the best-known interpreter of Marx's writings for economic geography. His work is voluminous, as well as wide-ranging both theoretically and substantively. While we discuss his work here, it is also threaded throughout the book, especially prominent in Chapters 5 and 10.

Our focus is one of Harvey's (1975) earliest and most important contributions: his theory of accumulation, that is, his theory of capitalist investment. It is a foundation for much of his subsequent theorizing. It is so fundamental because accumulation is so fundamental, going to the heart of the capitalist economy, to what makes it succeed and fail. Harvey's theoretical contribution is to show that economic geography is integral to the accumulation process. Sometimes economic geography furthers the accumulation process, warding off crisis; at other times, it brings accumulation to a crashing halt, producing destruction and mayhem.

Marx conceived accumulation as the engine of capitalism. "Accumulate, accumulate! That is Moses and his prophets," writes Marx (1992 [1867]) in the first volume of *Capital* (chapter 24). Without accumulation, the economic system would grind to a stop. Factories would close, workers would be out of work, and no goods would be produced. Accumulation is not autonomous, though, existing in its own economic bubble, separated from society and polity. Rather, the reverse. Marx emphasizes that accumulation occurs only because of the exploitation and oppression of one social class, the workers, by another social class, the capitalists. Without that exploitation and oppression there would be no surplus value (profit), and without surplus value there would be no accumulation. In this sense, Marx believes (as does Harvey) that capitalist accumulation rests on social and political injustice; it is built into capitalism's very operation and maintenance.

To Marx's theory of accumulation Harvey (1982) adds an economic geographic sensibility. That sensibility enters into the very frame of Marx's theory, transforming it. Harvey contends that a fundamental limit to the accumulation process is space itself. Accumulation does not happen on the head of a pin, but in real places and across wide-ranging spaces. Capitalists must continually find ways to overcome the barrier of space that otherwise constrains their ability to create surplus value (profits), and hence to accumulate. They do so by making use of new technologies of transportation and communication that tear down spatial barriers, "the annihilation of space by time," and in the process creating brand new configurations of economic geography.[3] There is thus an underlying imperative within capitalism to change the spaces of accumulation. But here is the paradox. To change those spaces, to annihilate space by time, requires the destruction of already established places, their devaluation and write-down. Hence there is a continual tension within the geography of accumulation – Harvey (1986, p.150) calls it "knife-edge" – between the forces of geographical inertia that keep intact the value of past existing investments in a place, and the forces of geographical change bent on destroying such places by annihilating space by time. For a period, at least, there will be stability, a "spatial fix" to use Harvey's term (see also Chapter 5). But sooner rather than later the landscape is torn asunder. The forces of the annihilation of space by time overcome the tendency to geographical inertia. Old spaces are destroyed and new spaces created. The only constant is "the restless formation and reformation of geographical landscapes" (Harvey 1986, p.150).

When Harvey articulated this thesis during the late 1970s and early 1980s it seemed a perfect description of, and explanation for, the central economic geographic event of the time – the destruction of old manufacturing spaces (deindustrialization) and the creation of new ones sometimes half a world away (new manufacturing spaces). It appeared a brilliant response to the context of changed economic geographic times.

3.5.2   Doreen Massey's *Spatial Divisions of Labour*

Another brilliant response from within radical economic geography was Doreen Massey's, found in her landmark book *Spatial Divisions of Labour* (1984). Massey's book represented a sea change in the discipline. There was "Before Massey," and "After Massey." One aspect of the change revolved around gender. Hitherto economic geography had been overwhelmingly masculinist in subject matter, style of representation, and membership. Massey broke all those molds. Another was that she opened up economic geography both to new content (home life was as important as work life, for example) and to new forms of theorization (theory could be written in colloquial prose and not necessarily as Greek letters).

Massey, like Harvey, was concerned with three elements within capitalism: accumulation (called by her "social reproduction"); the relation between capital and labor (class conflict); and space. How she put those elements together was quite different from Harvey's approach, however. Furthermore, despite her socialism and the deep influence of Marx, she generally shunned using Marx's texts in her own (Harvey's approach). Also unlike Harvey, she made her argument through empirical cases, particularly examples drawn from the massive industrial restructuring taking place in England during the 1970s and 1980s, which gave her work added purchase and potency.

Massey conceived accumulation as a set of distinct rounds of investment unfolding over time. As for Harvey, accumulation for her was not only temporal but geographical. Each round of investment produced a distinct geography, a "spatial division of labour." The idea of the division of labor was an old one going back to one of the first great economists, Adam Smith, and introduced in his 1776 book *An Inquiry into the Causes of the Wealth of Nations* (see also Chapter 12). Smith was the first to recognize that industrial production was predicated on a high degree of worker specialization, a division of labor, with each laborer undertaking a different specific work task. Massey argued that because accumulation necessarily occurs within space, there was also a distinctive spatial division of labor. Some places, say, accumulate investment in iron and steel furnaces; their spatial division of labor is heavy industry. Other places, say, accrue investment in car assembly plants; their spatial division of labor is manufacturing automobiles.

Massey further argued that the type of investment carried out in a place shaped the social relationships established there. Her example was South Wales in the United Kingdom. For at least a century up until the 1980s South Wales experienced a particular form of accumulation associated with heavy industry, coal mining, and iron and steel production. Massey contended that this type of industry produced a particular form and configuration of social relationships: a masculinist patriarchal culture in which men worked and women were expected to stay at home; left-wing politically active communities allied with unions and to the (left-wing) British Labour party; and tight-knit social and cultural solidarity reinforced by Methodist chapel going (Figure 3.4).

Figure 3.4   Massey's spatial division of labor thesis. Round 1 of accumulation: investment (the spatial division of labor) shapes geography.

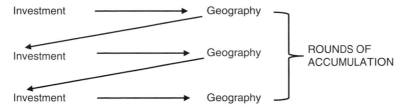

Figure 3.5   Rounds of accumulation: investment (the spatial division of labor) shapes geography, and geography shapes investment (the spatial division of labor).

So far the economy does all the work in Massey's account. Geography is the consequence of the determining force of economic investment. Ultimately Massey is a geographer, though. "Geography matters!" in her famous phrase. At a certain juncture, she argued, geography, which was initially determined by the form of investment, comes to prominence, and begins to influence the type of accumulation pursued in the next round of investment. It is no longer simply a one-way flow of determination, but two-way, a reciprocal relationship. Geography reacts back on the accumulation process, now shaping its form (Figure 3.5).

Using the previous example, by the late 1970s foreign-owned electronics firms started to invest in South Wales (representing the next round of investment in Massey's scheme, Figure 3.5). Following Massey's argument, those firms invested in South Wales because of the changed economic geographic character created in the previous round of accumulation – in this case, by heavy industry. Heavy industrial investment had (inadvertently) produced a pool of women who were now a perfect potential labor force for electronics firms. Because of patriarchy, many women in the region had never worked in the formal economy before – they were "greenfield labor." Consequently, they had never been unionized, and tended to be relatively pliable, especially under male supervision. Furthermore, because of patriarchy those female electronics workers were also expected to continue to provide domestic labor at home, and so could not commute far. They were a captive labor market unable to seek out jobs elsewhere, which made them even more attractive as employees. Moreover, during this same period of the late 1970s and early 1980s, the state – the principal employer of male workers in coal mining and iron and steel – began withdrawing investment from South Wales's heavy industry following the logic of neoliberalism introduced by Margaret Thatcher's Tory government that came to power in May 1979 (see Chapter 1). The resulting protracted miner's strike (1984–1985) also pushed women to work in the electronics industry to increase family income. The geographical characteristics of South Wales therefore changed radically, setting up the region for the next round of investment (Figure 3.6).

By providing the mechanism by which geography influences future forms of investment (the spatial division of labor), Massey demonstrates that economic geography is a critical disciplinary inclusion. Economic geography enters the very bones of the economy. If you fail to understand how economic geography matters, you fail to understand some of the central events of our time.

In sum, both Harvey and Massey imported a critical sensibility into theories of economic geography. They did so in part because, during the late 1970s and early 1980s when they made their contributions, they were on the front line of economic change. Seeing the acute and traumatic consequences of economic restructuring and deindustrialization as it

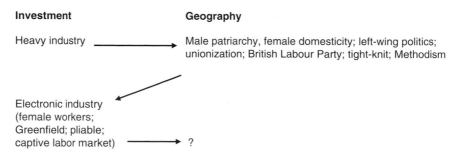

Figure 3.6   The reciprocal relationship between investment and geography: the South Wales case.

occurred, Massey and Harvey both moved away from economic geography's formerly ordered, static, abstract, and remote mathematical model-based accounts (spatial science). Instead, they looked for explanations that captured the dynamism, crises, inequalities, and particularities they now saw unfolding unrelentingly over space and place. For both, there was no better critical explanation of what they saw than Marxism. If today spatial science lives on in one form or another in "Geographical Economics," then "Geographies of Capitalism" (Chapter 2) remains the heartland of the radical economic geography that Harvey and Massey initially helped to create.

3.6   Poststructuralism and the End of Capitalism (As We Knew It)

The last major form of economic geography we discuss emerged sometime in the mid-1990s (see also the related discussion of "Alternative Economic Geographies" in Chapter 2). Like other movements we reviewed, it was provoked partly by substantive economic geographic changes on the ground, and partly by new ideas about how to represent them.

The changed world was a more complicated capitalism, and perhaps even more ruthless. The complexity derived in part from the processes of crisis and restructuring that Harvey and Massey theorized. At least in the Global North the old form of industrial capitalism was increasingly hollowed out (deindustrialization, Chapter 1), replaced by various forms of the service economy, the products of which were frequently intangible, requiring brain work rather than brawn to produce. In this new world, the relevance of Marx's nineteenth-century political economic analytical categories, especially around class and production, seemed distant. Parts, at least, of the Global South were also transformed, becoming new manufacturing centers. Here Marx's nineteenth-century writings had renewed purchase. Select regions in the Global South, like the Pearl River Delta in Southern China, or along portions of the border in Mexico, or in Southeast Asia in the industrial triangle embracing southern Malaysia, Singapore, and Indonesia, contained cavernous factories, the size of Ford's Rouge plant or even larger, filled with an enormous lumpen proletariat, and subject to long work hours, poor working conditions, and low pay. They were the same working class that Marx had described some 150 years earlier. The Global South, in turn, was connected to the Global North by complex, elongated, and overlapping production chains, held in place by behemoth multinational corporations whose revenues were often larger than the economic output of many Global Southern countries. This larger system that had evolved seemed at once not only inscrutable (how did it all fit together?) but massive and overwhelming, seemingly impossible to change.

As capitalism became more globalized, seeping into even the world's remotest nooks and crannies, it also appeared increasingly brutal and pitiless. Under neoliberalism that first took hold in the Global North in the 1980s, and in the Global South by the 1990s, strict market discipline was enforced (Chapter 1). Gone was coddling, the Nanny State. Instead, it was dog-eat-dog competition, and market efficiency of the starkest and most cold-blooded form. There was no alternative, at least according to Margaret Thatcher. Given the fall of the communist Soviet Union and its satellite states in 1989, and the adoption of market reform by the Chinese Communist Party during the 1980s, it seemed as if she might be right.

Joined with these changed economic geographic conditions was a changed intellectual framework, poststructuralism. Its origins were with a set of Parisian intellectuals of the 1960s, of whom perhaps the most important was Michel Foucault. Originating in studies of philosophy, cultural history, literary criticism, and psychiatry, by the 1970s some of poststructuralism's key ideas had spread to the Anglo-American humanities. And by the 1980s they reached the social sciences, entering economic geography sometime during the 1990s.

Poststructuralism was in part a reaction to classical Marxism (and represented by an even earlier generation of Parisian intellectuals such as Jean-Paul Sartre). The poststructuralists were critical especially of Marxism's economism, its tendency to make the economy, and within the economy the material antagonistic relation between social classes, workers and capitalists, the cause of everything. Within Marxism, all explanation seemed to come down to the effects of social class. In contrast, poststructuralists argued that matters were more complex and messy. There was more at stake than class, such as other social dimensions like gender, race, and sexuality. None could be reduced to social class. Rather, they had an integrity of their own, they were part of culture, which had some degree of autonomy from the economy. Under classical Marxism, though, culture was not independent. It lay within what Marxists called the superstructure, determined by the economy (the base or infrastructure). This determinate relationship was denied by poststructuralists, who even on occasion suggested the relation could be reversed, with the economy shaped by culture. No wonder there were some bitter disputes between Marxists and poststructuralists.

Their different approach did not imply poststructuralists were satisfied with the status quo, however. Their project was also intensely political, as concerned with changing the world as Marx's. But they thought how the world was to be changed, and what it should be changed to, was very different from Marx's prescriptions.

3.6.1   J.K. Gibson-Graham and poststructural economic geography

In economic geography, these changed substantive circumstances on the ground, as well as changed ideas from poststructuralism, were forcefully given expression in the work of Kathy Gibson and Julie Graham from the early 1990s. For them poststructuralism went all the way down to their self-identification as authors. In poststructural fashion they signaled the complexity of their own authorial identity by writing under a single name, J.K. Gibson-Graham.[4]

As with the broader movement with which she identified, Gibson-Graham wrote partly in opposition to classical Marxism, and classical Marxism's best known exponent in economic geography, David Harvey. This provoked debate between the two factions, although it was never an outright war. And while there were some testy moments, increasingly economic geographers tried to combine, or integrate, or at least acknowledge the arguments of the other (whatever the other was).

Gibson-Graham herself started as a classical Marxist. She believed that because of contradictions in capitalism of the kind that David Harvey acutely revealed, workers would rebel, overturning the larger system. Expropriators would be expropriated, and a new system, communism, would rise phoenix-like from capitalism's ashes.

But in the mid-1990s she had an epiphany. Marxism would never change the world because theoretically it presented capitalism as implacable, an inviolate entity driven by indomitable forces. She called this conception a capitalocentric view, the belief that capitalism was constituted by the insurmountable power of capital. Further, it was a view reinforced by populist portrayals of globalization (see Chapter 8). Those too represented the capitalist world economy as all-encompassing, rooted, with nothing that could budge it. For this reason, Gibson-Graham argued, while Marxists might say that they were revolutionaries, their capitalocentric depiction of capitalism in fact militated against the possibility of revolution. You threw in the towel before you even began. Gibson-Graham did not want to throw in the towel, though. She wanted to recognize the economy's complexity and force, but at the same time she did not want to fall into the paralysis of capitalocentrism. She also did not want to give up on Marx's own injunction that the point is to change the world and not merely to interpret it (eleventh thesis in Karl Marx's (1976 [1888]) *Theses on Feuerbach*). How could she do that?

It was through poststructuralism, and especially the version forwarded by feminists (discussed further in Chapter 6). Poststructuralism provided a means to rethink the nature of capitalism so that it was no longer as intimidating and unchallengeable. That was the significance of Gibson-Graham's 1996 book title *The End of Capitalism (As We Knew It)*. To change capitalism, to end it, we must learn to know it in another way. Poststructuralism enabled that new kind of knowing. It undid capitalocentrism. Rather than representing capitalism as a single essential force, determining everything else, it was portrayed as varied and diverse, constructed by joining different elements, many of which were not traditionally economic at all. Rather, many were cultural, and the turn to poststructuralism in economic geography thus overlapped with a larger "cultural turn" in the discipline to take more seriously the cultural specificity and "embeddedness" of all economies. In any event, Gibson-Graham contended that poststructuralism allowed one to move away from the iron cage of capitalocentrism, and instead to conceive the economy as piecemeal, varied, fragile, and vulnerable. Once capitalism was thought about in those terms, it was less powerful and scary. It could be chipped away, resisted bit by bit.

This goes to another difference between poststructuralism and Marxism. For Gibson-Graham capitalism was to be changed incrementally, not, as traditional Marxists thought, in one single apocalyptic revolutionary moment. Gibson-Graham's model of revolution was of the kind pursued by second-wave feminism (see Chapter 6). During the early 1960s patriarchal society likely seemed as impossible to change as capitalism, deeply entrenched and seemingly here for ever. But gradually, through small local revolutionary acts, patriarchal society changed. Of course, there was and is still huge room for improvement. The point, though, was that the improvements that occurred were not brought about by a single big revolution, but rather by many small everyday revolutionary acts. Gibson-Graham thought capitalism could be similarly changed. It would occur gradually. Over time small changes would add up to a big change.

What is it that these changes are to realize? It is not the grand utopian future society envisioned by Marx. Instead, for Gibson-Graham these changes are directed toward realizing diverse forms of the economy that already exist. Here Gibson-Graham draws on what she calls the iceberg model (Figure 3.7). Its tip is pure capitalism: free markets, competition, large corporations, myriad financial flows, wage labor markets, and so on. But below pure capitalism there is 90% of the economy that is generally invisible, and which is at best semi-capitalist, if capitalist at all. This 90% is what Gibson-Graham calls diverse economies; economies that operate on different principles than found under pure capitalism. Gibson-Graham's political hope is to move the diverse economies that lie (invisibly) below the water line to above the waterline. They become mainstream. They become the new and improved economy. This will not happen all at once, with the death knell of capitalism sounded overnight. But over time, she believes, small acts produce major transformation.

Figure 3.7   Gibson-Graham's iceberg model of the economy. From Community Economies Collective, drawn by Ken Byrne. Source: Gibson-Graham, 2006, p.70.

Here Gibson-Graham draws on another poststructural idea: performativity. It is the notion that our knowledge is less a mirror of the world than a means to intervene and to change it, to make it a different world. Capitalism's power derives not from any inherent properties that define it but because certain ideas about its power are believed by people who then perform them, make them real through their acts. If people can be persuaded to believe and enact different ideas, however, the world can be altered. For example, once people are encouraged to believe that there are other ways to perform the economy, for example, through co-ops, then a co-op economy will come into existence. For Gibson-Graham, those alternatives already exist, except they are below the waterline, not visible. By enabling people to see those alternatives and to perform them, the unseen diverse economy below the waterline can become the seen diverse economy above the waterline. In her work, Gibson-Graham uses her own academic research, which she calls action research, to persuade people to engage in alternative economic performances (Chapter 6). This strategy is the opposite to what has gone on in the name of economic geography for most of its existence. The research process within economic geography for over 100 years was hands off. Research described the object of inquiry. It did not attempt directly to change it. In contrast, Gibson-Graham uses the performance of the research process to transform the very object she is studying. She wants the research to bring into being new forms of diverse economies, and, like feminism over the last 60 years, to spur wide-scale social transformation.

Poststructural economic geography from its very outset is a deeply critical project. Gibson-Graham uses the research process to change what is researched; that is, to perform into existence a new and, for her, improved reality. This is different from Harvey's and Massey's critical projects. They use their research in economic geography in order to understand better the capitalist reality, to ferret out its secrets. They too want to change capitalism, but neither believes that their research in and of itself will change the world. It might reveal capitalism's contradictions, which might then lead to change. But their research itself does not initiate that change. Gibson-Graham suggests her work does, however. She believes in the critical potential of economic geography more than anyone else has ever done before.

3.7   Conclusion

The history of economic geography has been one of continual invention and reinvention. It is one of the most restless social sciences. Other social sciences have revered past figures, sacred methods, hallowed canons, and sacrosanct texts. Economic geography appears to do none of these things (see also Chapter 5). Its history has been one of pell-mell change, intellectual openness, eclecticism, and pluralism, which sometimes verged on the chaotic and anarchic. Inconstancy is its only constant.

But it hasn't been entirely a free spirit. One of the main themes of the chapter is to show that the intellectual content of the discipline is in part anchored by its historical period. This is perhaps clearest in Chisholm's work during the late nineteenth and early twentieth centuries. His portrayal of the geography of the commodity is a mirror reflection of the commercial interests found within British imperialism during the same era. That tethering may be more difficult to discern for the most recent form of economic geography we described, Gibson-Graham's poststructural approach, but it is there. Her work emerged partly from her frustration that there seemed nothing one could do about the latest turn of capitalism, bound up as it is with neoliberalism and globalization. But she wanted to do something, to intervene, and to show how that might be possible. Although Gibson-Graham's relationship to her context is different than Chisholm's is to his, as with that of Chisholm, we can't understand her work, its inspiration and purpose, as well as its form, without understanding the larger historical context in which it was written.

It is not only larger events that have been important to economic geography but larger ideas. This has been key in all the historical vignettes above. Not that many of the ideas were original to economic geography. The discipline has been the great appropriator of other disciplines' concepts, begging, borrowing, and stealing. Again this goes to the openness of the subject. As is clear from our history, almost anything can be its intellectual fodder. This has led to charges by some outside of the discipline of a hopeless eclecticism, or a slapdash approach, or a fuzzy logic, or a lack of rigor. Likely all of these charges are true to a certain extent. But there are benefits to such a loosely conceived discipline. The lack of a canon, or a strict methodology, or a clear definition of theory has historically given economic geography an ability to roam where there are interesting problems, and not be straightjacketed, or waved off an issue because disciplinary police have ruled it is not economic geography. With that freedom has also gone much energy, vigor, and dynamism, suffusing the discipline, providing it with momentum and purpose.

That doesn't mean it is perfect, going to perhaps the most important theme in this chapter and book, the need for tenacious critical scrutiny. It is so easy to whitewash the past, to make history have a happy ending. That is what textbooks often do, to omit the nasty bits, to tell only of the triumphs. In contrast, what we tried to do in our textbook's chapter is to tell the imperfect history of economic geography, nasty bits and all. Nietzsche famously said, "what doesn't kill me, makes me stronger." Our hope is that our history makes you stronger.

Notes

1 Chisholm recognized the tedium involved in at least his form of economic geography, although he discounted it: "If ... there is some drudgery in the learning of geography, I see no harm in it" (Chisholm quoted in MacLean 1988, p.25).

2 Although the published version of Carl Baskin's English translation of Walter Christaller's *Central Places in Southern Germany* did not appear until 1966, a mimeographed version of the same text – it was Baskin's PhD thesis at the University of Virginia – was in circulation from 1957.

3 The phrase "annihilation of space by time" was originally coined by Marx in his *Grundrisse*, the collection of notes he kept for writing *Capital*. It referred to a fundamental impulse in capitalism to reduce the so-called turnover time, the period between starting and finishing production, by using new technologies of communication and transportation. Those technologies enable faster exchange between differently located points of production; that is, they shrink the world ("time-space compression" as Harvey 1989, chapter 16, also calls it). The distances between places, space, is being annihilated by increased speed, time. The idea was picked up by David Harvey, especially in his book *The Condition of Postmodernity* (1989), and joined with his larger theory of accumulation.

4 The untimely tragic death of Julie Graham in 2010 did not halt the continued writing of J.K. Gibson-Graham.

References

Barnes, T.J. 1998. A History of Regression: Actors, Networks, Machines and Numbers. *Environment and Planning A* 30: 203–223.

Barnes, T.J., and Farish, M. 2006. Between Regions: Science, Militarism, and American Geography from World War to Cold War. *Annals of the Association of American Geographers* 96: 807–826.

Buck-Morss, S. 1995. Envisioning Capital. *Critical Inquiry* 21: 435–467.

Butzer, K.W. 1989. Hartshorne, Hettner and The Nature of Geography. In J.N. Entrikin and S.D. Brunn (eds), *Reflections on Richard Hartshorne's The Nature of Geography*. Washington, DC: Occasional Publications of the Association of American Geographers, pp.35–52.

Chisholm, G.G. 1889. *Handbook of Commercial Geography*. London and New York: Longmans, Green, and Co.

Christaller, W. 1966. *Central Places in Southern Germany*, trans. C.W. Baskin. Originally published in German in 1933. Englewood Cliffs, NJ: Prentice-Hall.

Dumont, L. 1977. *From Mandeville to Marx: The Genesis and Triumph of Economic Ideology*. Chicago: University of Chicago Press.

Fellmann, J.D. 1986. Myth and Reality in the Origin of American Economic Geography. *Annals, Association of American Geographers* 76: 313–330.

Foucault, M. 1977. *Discipline and Punish*. London: Allen Lane.

Gibson-Graham, J.K. 1996. *The End of Capitalism (As We Knew It): A Feminist Critique of Political Economy*. Oxford: Blackwell.

Gibson-Graham, J.K. 2006. *Postcapitalist Economies*. Minneapolis, MN: University of Minnesota Press.

Haggett, P. 1965. *Locational Analysis in Human Geography*. London: Edward Arnold.

Hartshorne, R. 1939. *The Nature of Geography. A Critical Survey of Current Thought in Light of the Past*. Lancaster, PA: AAG.

Hartshorne, R. 1955. "Exceptionalism in geography" re-examined. *Annals, Association of American Geographers* 45: 205–244.

Harvey, D. 1975. The Geography of Capitalist Accumulation: The Reconstruction of Marxian Theory. *Antipode* 7: 9–21.

Harvey, D. 1982. *Limits to Capital*. Chicago: University of Chicago Press.

Harvey, D. 1986. The Geopolitics of Capitalism. In D. Gregory and J. Urry (eds) *Social Relations and Spatial Structure*. London: Macmillan, pp.128–163.

Harvey, D. 1989. *The Condition of Postmodernity: An Enquiry into the Origins of Cultural Change*. Oxford: Blackwell.

Hudson, D. 1955. University of Washington. *The Professional Geographer* 7: 28–29.

Huntington, E. 1915. *Civilization and Climate*. New Haven, CT: Yale University Press.

Huntington, E. 1924: *The Character of Races as Influenced by Physical Environment, Natural Selection and Historical Development*. New York: Charles Scribner's Sons.

Isard, W. 1956. *Location and Space Economy*. London: Wiley.

Keynes, J.M. 1936. *The General Theory of Employment, Interest and Money*. London: Macmillan.

Livingstone, D.N. 1994. Climate's Moral Economy. Science, Race and Place in Post-Darwinian British and American Geography. In A. Godlewska and N. Smith (eds), *Geography and Empire*. Oxford: Blackwell, pp.132–154

Lösch, A. 1954. *The Economics of Location*, 2nd edn, trans. W.H. Woglom with the assistance of W.F. Stolper. Originally published in German in 1940. New Haven, CT: Yale University Press.

Lukermann, F.E. 1961. The Concept of Location in Classical Geography. *Annals of the Association of American Geographers* 51: 194–210.

MacLean, K. 1988. George Goudie Chisholm 1850–1930. In T.W. Freeman (ed.), *Geographers Bibliographical Studies*, volume 12. London: Infopress, pp.21–33.

Marx, K. 1976. Karl Marx Theses on Feuerbach. In F. Engels (ed.), *Ludwig Feuerbach and the End of Classical German Philosophy*. Originally published in German in 1888. Peking: Foreign Languages Press, pp.61–65. http://www.marx2mao.com/M&E/TF45.html (accessed July 4, 2017).

Marx, K. 1992. *Capital: A Critique of Political Economy: Volume 1*, trans. by Ben Fowkes. Originally published in German in 1867. Harmondsworth: Penguin.

Massey, D. 1984. *Spatial Divisions of Labour: Spatial Structures and the Geography of Production*. London: Macmillan.

McCarthy, T. 2014. Writing Machines. *London Review of Books*, 36: 21–22.

McCarty, H.H., Hook, J.C., and Knox, D.S. 1956. *The Measurement of Association in Industrial Geography*. Iowa City: Department of Geography, University of Iowa.

Mikesell, M.W. 1974. Geography as the Study of the Environment: An Assessment of Some New and Old Commitments. In I.R. Manners and M.W. Mikesell (eds), *Perspectives on Environment*. Washington, DC: Association of American Geographers, pp.1–23.

Pruitt, E.L. 1979. The Office of Naval Research and Geography. *Annals of the Association of American Geographers* 69: 103–108.

Sapper, K. 1931. Economic Geography. *Encyclopedia of the Social Sciences*, volume 5. New York: MacMillan, pp.626–629.

Sahlins, M.D. 1972. *Stone Age Economics*. Chicago: Aldine-Atherton.

Schaefer, F.K. 1953. Exceptionalism in Geography: A Methodological Introduction. *Annals of the Association of American Geographers* 43: 226–249.

Schorske, C.E. 1997. The New Rigorism in the Human Sciences, 1940–60. *Daedalus* 126: 289–309.

Von Thünen, J.H. 1966. *Von Thünen's "Isolated State,"* trans. C.M. Wartenberg, edited and with an introduction by P. Hall. Originally published in German in 1826 as *Der isolierte Staat*. Oxford: Pergamon Press.

Whitbeck, R.H. 1914. Review of J. Russell Smith's *Industrial and Commercial Geography*. *Bulletin of the American Geographical Society* 46: 540–541.

Whitbeck, R.H. 1915–1916. Economic Geography: Its Growth and Possibilities. *Journal of Geography* 14: 284–296.

Whitbeck, R.H., and Finch, V.C. 1924. *Economic Geography*. New York and London: McGraw-Hill.

Wise, M.J. 1975. A University Teacher of Geography. *Transactions, Institute of British Geographers* 66: 1–16.

Chapter 4

Economic Geography and its Border Country

4.1 Introduction

One of Britain's most famous twentieth-century critical public intellectuals, Raymond Williams, grew up literally in border country, in his case, along the boundary between Wales and England.[1] He was betwixt and between other borders too. He was raised in a working-class family (his father was a railway signalman), but he became Professor of Modern Drama at one of England's elite universities, Cambridge, "teaching the enemy" as he once put it. He was a Welsh speaker, an avid promotor of the Welsh literary tradition, but he taught English language classics – novels, poetry, and plays – to undergraduates. And while he believed that Wales was colonized by the English from the thirteenth century when the English King, Edward the Conqueror, first invaded, Williams's adult life was spent not on the oppressed Celtic Welsh margins but at the heart of the imperial Anglo-Saxon center. Raymond Williams's stretched and displaced existence between the various borders within which he lived was shot through with tensions, frustrations, anxieties, contradictions, and sometimes even anger. But in the end he always managed to negotiate successfully the different borders that defined his life. Rather than hobbling him, living in the border country, crossing and re-crossing starkly different worlds, became for Williams a source of vitality, purpose, creativity, and critical inspiration.

This chapter is about the border country of economic geography. It has two main aims. The first is cartographic: to map the boundaries economic geography shares with cognate social sciences. Economics is perhaps the most important of those adjoining disciplines, but we also discuss five others: anthropology, cultural and gender studies, environmental

Economic Geography: A Critical Introduction, First Edition. Trevor J. Barnes and Brett Christophers.
© 2018 John Wiley & Sons Ltd. Published 2018 by John Wiley & Sons Ltd.

studies, political science, and sociology. None of these other disciplines do the same thing in the same way as economic geography, but there are at least family resemblances that make their work more or less intelligible and potentially useful for a critical project.

This goes to the second aim of the chapter: to explore the utility of moving across the borders between economic geography and neighboring social science disciplines to engage in critical inquiry, just as Raymond Williams did. Borders represent liminal space; that is, a space that lies on the threshold between one thing and another, and so is neither. It is precisely this in-betweenness, liminality, which we will suggest provides prospects for an enhanced critical economic geographic sensibility. On the one hand, it necessarily disrupts and subverts, calling into question fixed, accepted, and naturalized truths – the cultural critic Homi Bhabha (1994, p.5) talks about a "disruptive inbetweenness." On the other hand, liminal space represents a potentially creative space. In liminal space normal disciplinary constraints are relaxed – the space does not belong to one discipline or another. Old binaries and hierarchies consequently no longer apply. As a result, you can do things normally not permissible, think thoughts normally not thought, and produce new meanings, social relations, and identities not seen before. The border country's openness becomes a space conducive to creativity, innovation, and novel social experimentation.

That's not to deny the difficulties of working within such a space. Disciplines, as their very name implies, impose discipline. Their internal spaces and boundaries are often tightly policed, subject to surveillance and restriction. A failure to conform to those internal rules can produce a range of reactions: bafflement ("What do you mean?"); rebuke ("You are wrong"); derision ("You idiot"); and banishment ("Get out of here"). Consequently, at least for those working within some disciplines, going into the border country can be hazardous, like crossing the Wall between East and West Berlin during the Cold War, or the 38th parallel separating North and South Korea.

Our argument, though, and in line with Raymond Williams, is that while there may be some risks, there are also potential rewards from working in the border country. Trafficking across boundaries can trigger, as we will show for economic geography, original experimentation and invention; new enabling vocabularies; creative theoretical frameworks; fruitful forms of academic debate and discourse; and fresh methodological strategies.

The intent of this chapter is to provide examples of various successful interactions that have occurred in the border country between economic geography and adjacent social sciences. That cross-border traffic, we suggest, was possible in part because of the existence of "trading zones." Like free trading zones that exist in some countries, such sites permit a free exchange of ideas from one discipline to another along with their productive comingling.

We divide the chapter into three main sections. The first defines disciplines (cf. Chapter 2), and reflects on the general meaning of disciplinary boundaries, the border country, and the sites, or trading zones, where ideas cross boundaries and mix and join. The second and longest section provides a schematic map of economic geography's border country, identifying the different disciplinary spaces against which economic geography abuts. We focus especially on economic geography's trading relation with one of those disciplines, historically the most important, economics. We suggest, though, that over the past three to four decades, much of economic geography has shifted attention away from trade with mainland orthodox economics, neoclassicism, to an archipelago of loosely grouped alternative economic approaches, heterodox economics. This is followed by brief reviews of economic geography's exchange relations with other social sciences: anthropology, cultural and gender studies, environmental studies, political science, and sociology. The final section discusses three specific examples of critical work stemming from occupation of the border country.

4.2   Institutions, Disciplines, Borders, and Trading Zones

Social science disciplines were not created in heaven, given by the hand of God, but were invented at particular earthly historical moments in response to social, political, and economic circumstances (see Chapter 3). To say disciplines were invented does not mean, as we argued in Chapter 3, that they were insubstantial and ephemeral. The social sciences that first emerged during the late nineteenth and early twentieth centuries, which included economic geography, have remained robust and durable. In spite of enormous and significant changes occurring in the world since their appearance roughly 150 years or so ago – two world wars, innumerable political revolutions, unprecedented technological change, profound physical transformations of the planet – the social sciences continue to thrive, even their original labels and subject divisions remaining. They appear as bulwarks of the contemporary academy, resilient and dogged.

4.2.1   Institutions, disciplines, and boundaries

Academic disciplines never exist as only pure, detached bodies of knowledge (see also Chapters 2 and 7). They are always integrated within a social institutional structure, most often centered on the university. This means that the sustainability of any discipline is not simply the consequence of the inherent rightness or truth of its ideas and research. It also depends on a set of social relationships embedded within both the institutional structure in which the discipline sits and those structures found in outside institutions that buttress it. Sufficient backers (allies) with enough power and resources to uphold the discipline's value have been present for the roughly 150 years that economic geography has existed as a university subject. Had the discipline not met the institutional conditions of adequacy then, in Darwinian fashion, it would have gone the way of the dodo, and you would not be reading this book. Survival is never guaranteed, however, and history is littered with disciplines (some of which were even relatives of economic geography) that fell by the wayside, victims of social institutional failure (Box 4.1: The Rise and Fall of Regional Science).

The initial institutionalization for economic geography was crucial. As we saw from Chapter 3, it emerged on the scene in the late nineteenth and early twentieth centuries because universities were willing to take a chance on it. They created new departments (such as at the University of Edinburgh, which inaugurated its Department of Geography in 1908); hired staff (like George Chisholm); purchased books (for example, Chisholm's own *Handbook of Commercial Geography*); bought specialized furnishings and apparatus (wall maps, map cases, magic lanterns [early projectors]); provided space in buildings (at the University of Edinburgh, Geography was given room at the "Old High School"); and otherwise spent money to foment a new line of academic inquiry. Universities may have been prodded to take this action by other institutions such as government departments, Royal Commissions, royal learned societies, business associations, or arms of the military. The university was the immediate moving force, however.

Yet another important element solidifying a discipline, also social in origin, is the process of self-identification by its practitioners. The philosopher Ian Hacking (1995) uses the phrase "looping effects" to describe how made-up social categories gain purchase, become real, as people change their identity to inhabit them. In this case, the made-up category of a social science discipline becomes one basis by which individuals identify themselves: "I am an economist," or "I am a sociologist," or "I am an economic geographer." One's very sense of self comes to coincide with the discipline. Moreover, as this process of self-identification unfolds, the discipline's social structure becomes internalized by its practitioners. That structure determines how one operates within a discipline, making it distinctive, setting it apart from other disciplines. As the economic geographer Erica Schoenberger (2001, p.368) reflects, a discipline's interior social structure

> defines, for example, who has authority to speak or how one learns to do the work of the discipline. It underwrites a community of practice with its own understandings of professional development and career ladders. Every field knows which are the departments to come from or to go to, which are the hot topics, which theories can be mentioned in polite company, and in which journals one has to publish. Every field has its own "circuits of fame" with citations and references passed along like Kula shells in a complex process of recognition and social reinforcement.

The larger point is that a social science discipline is a complex entity. While it is a container for a particular type of knowledge, it is always more than a container, more than a type of knowledge. A discipline is lived (see Chapter 7). It represents a set of dynamic practices that intersect among other things with elements of the physical world, with levels of social hierarchy, demarcation, and power, and with personal identity and a sense of self. Such complexity partly explains why disciplines have such substance and import; why dividing lines are so hard fought, protected, and policed; and why sanctions are meted out for transgression. But sometimes there are good reasons for transgression, that is, to work in the border country.

At least, this is suggested in Thomas Kuhn's (1970) famous book, *The Structure of Scientific Revolutions* (supposedly the most cited academic book of the twentieth century). Kuhn contended that while working within a well-defined discipline is sufficient for carrying out ordinary science (his phrase is "normal science"), it is sometimes necessary to move outside of a discipline to achieve novelty and creativity. Kuhn thought normal science was primarily about "mopping up," that is, diligently filling out the various implications of a new idea once it was coined. In Kuhn's vocabulary, normal science was about remaining within a prevailing "paradigm," the stable matrix of values, assumptions, methods, and exemplars that constituted a scientific community's guiding framework (for more on Kuhn, normal science, and paradigm, see Box 7.1: Paradigm). For Kuhn the principal home of a paradigm was a discipline. It was where normal science occurred. But there were times, said Kuhn, when existing paradigms no longer worked, and needed to be changed. Specifically, whenever there were persistent and striking anomalies between what our paradigms told us should be in the world, and what we actually saw in the world, we needed to think outside the box; we needed to find a new paradigm. Instead of practicing normal science we needed to practice revolutionary science (see also Chapter 5 on "revolutionary theory"). To do this, Kuhn suggested, one should move outside the internal constraints of a discipline, and the concomitant blinkers of normal science. One should go into the liminal space of the border country.

Box 4.1 The Rise and Fall of Regional Science

Walter Isard. Source: Walter Isard papers, #3959. Division of Rare and Manuscript Collections, Cornell University Library.

A kindred discipline to economic geography, regional science, provides a bracing example of the potential fragility of a social science discipline when institutionalization breaks down. It shows how disciplines are institutional creatures of their time and place. If they fail to adapt to changing conditions, they can easily share the fate of the dinosaur.

In 1954, seemingly born fully formed Athena-like from the head of one man, the American economist Walter Isard, regional science developed an impressive set of formal mathematical theories and empirical models of, especially, the postwar American spatial economy. It combined ideas from economics, geography, and urban and regional planning. For its first 30 years it was on a disciplinary tear, expanding geographically across the world, as successful as any American multinational during the same period. That success was partly a consequence of its institutionalization that anchored the discipline. In 1958, Isard persuaded an Ivy League University, the University of Pennsylvania (Penn), to provide a home for both the brand new Department of Regional Science and its affiliated Regional Science Association. That process of institutionalization assembled different kinds of resources at the McNeil and Observatory Buildings on Penn's campus in Philadelphia. Those resources included various human bodies, such as Isard's – he was Chair of the Department and Professor – and those of young assistant professors, some of whom were economic geographers, such as Allen Scott and Michael Dacey, as well as graduate students who came from around the world, including Doreen Massey, Michael Dear, and Neil Smith. There were also large inflows of money that greased institutional wheels, derived from such sources as student fees, association and journal dues, US government research grants, philanthropic foundations such as Carnegie and Rockefeller, and remuneration for contract research from non-profits such as RAND. Finally, there were machines, essential for the big data-driven models in which regional science specialized. Penn's Moore School for Electrical Engineering in 1946 had built ENIAC (Electronic Numerical Integrator and Computer), the world's first electronic general purpose computing machine (completing more mathematical operations in its lifetime than had been completed previously in the entirety of human history). Regional science was thus perfect for postwar America, providing incisive explanation, documentation, analysis, and advice during an unprecedented period of urban, regional, and national expansion.

Sometime from the late 1970s, however, regional science began to experience a reversal of fortune, losing its various institutional backers, including: former individual proponents (even former employees, such as Allen Scott, for example); past supportive disciplines such as economic geography; funders of its research such as the US government; and high-level university administrators who previously offered it a comfortable home. The discipline no longer resonated and engaged, and so, like the Roman Empire, it declined and fell. In 1994 the University of Pennsylvania closed the Regional Science Department; its funding was cut; it lost its building; it did not train students; and its faculty were either let go or moved laterally to other departments. Regional science may have still existed as a body of ideas, but without an appropriate accompanying institutionalization it became increasingly lifeless.

For Kuhn, as well as for Raymond Williams, the border country represented potentially fertile intellectual soil. It was where the usual disciplinary constraints were relaxed, where it was possible to talk to others who worked with different paradigms. Consequently, there were opportunities to create by joining unfamiliar theories and methods from other disciplines with the familiar ones of our own.

4.2.2   Trading zones

The historian of science Peter Galison (1998) suggested that such opportunities might be realized by creating "trading zones." These are sites enabling the exchange of ideas, concepts, methods, and techniques among disciplines that otherwise do not normally interact. Admittedly, it is hard to put to one side one's normal disciplinary prejudices, often so ingrained. That's what a trading zone requires, however. It involves loosening one's usually tightly held disciplinary convictions, and finding a balance between keeping one's identity as an economic geographer while also acceding to voices from other disciplines.

Trading zones are a useful device in thinking about the border country. The border country is a space where different disciplines are contiguous, but where there is not necessarily interaction. That is where a trading zone is useful. It is a connection, a passageway, a trail across the border, allowing some people from one side of the divide to traverse and interact with people on the other side. For Galison (1998, p.47), key to that interaction is establishing a mutual language allowing members of different communities to communicate. Galison suggests as a model pidgin or Creole, which are made up of a mixture of different languages, improvised, partial, subject to change, but nevertheless enabling the work of exchange to occur. Such a language is a form of bodging. It is not perfect, but it gets the job done.

4.3   Mapping Economic Geography's Border Country

Over its history, economic geography has compared itself in different ways to the different social sciences on its borders: sometimes with envy, at other times with cocky superiority, and at yet other times with shrugged indifference. Likewise, the border traffic between geography and its neighbors has varied from light to heavy, from one-way to two-way, and the permeability of the different borders that this traffic crosses ranges from soft to hard. Some borders seem like the European Union's Schengen zone, where you hardly realize that you have crossed into another country. The natives on the other side are almost like you, maybe with only a slight accent, and on occasion making unusual word choices. Other borders, however, are like the former Iron Curtain, requiring the correct entrance papers in triplicate, with still the possibility of interrogation in a small windowless room, followed by a visa denial.

For most of their history economic geographers have been curious to explore their border country, to cross disciplinary boundaries. Maybe it is in the genes of a geographer. For example, in an early paper George Chisholm (1910) drew on the work of the economic sociologist Alfred Weber to explain industrial location in Scotland. Over the past 35 years, the inclination "to read around," as Gibson-Graham (1996) put it, has significantly increased. Economic geographers have become ever-more promiscuous in the books that they bring home (Gibson-Graham label themselves "theory sluts"; 2006, p.xi). Before 1980, though, economic geographers were inclined to monogamy. They remained more or less loyal to a single social science partner, economics.

4.3.1   Economics

You would think that economics should be economic geography's most important trading partner in the border country. It's the economy, stupid. But it has been often a troubled and hesitant relationship. Initially there was little contact. The intellectual projects were too dissimilar. That changed during the early postwar years, with some economic geographers becoming smitten by mainstream economics, attempting to make their discipline like it. More recently, there has been a falling out, with standoffishness, and bad mouthing (nearly always from the geographers). Nevertheless, some economic geographers continue to harbor warm feelings about economics, and at a few select sites a geographical version of orthodox economics is practiced.[2]

Once upon a time
Making the relation between economic geography and economics complicated is that there is more than one kind of economics. There is a dominant (likely hegemonic) form, neoclassicism. Mainstream or orthodox economics, as neoclassicism is also called, originated during the late nineteenth century in Western Europe. Initially there were different schools, but they all postulated that if rational producers and consumers met in a free market, through the well-oiled movement of prices the end result would be both optimality (maximization) and equilibrium. No consumer or producer could be better off (they were at their optimum), and no consumer or producer had any incentive to change their position (they were in equilibrium). Proof of equilibrium and optimality was through mathematical deduction.

Starting with a set of precisely defined abstract assumptions, such as rational choice (*homo economicus*), optimality and equilibrium were logically derived by rigorously working through the equations that connected initial postulates.

Economic geography arose during exactly this same late Victorian period. Initially there was almost no contact between the two disciplines, however. Economic geography and economics were so different. Early economic geography was defined by encyclopedism (Chisholm's agenda of dull rote memorization) and a single-minded emphasis on the environment, whereas neoclassical economics was defined by abstract mathematical proof and a single-minded emphasis on the market. They were ships passing in the night. Border traffic was absent not because of internal restrictions but because neither side had any reason to visit the other.

Postwar romance

That changed in 1940 when the American economic geographer Harold H. McCarty (1940, p.xiii) argued for full-blooded interdisciplinary trade: "economic geography [should] derive its concepts from the field of economics, and its method largely from the field of geography," he contended. In McCarty's scheme, economic geographers would do the manual labor, hewing and drawing geographical data, gathering and storing facts, while economists would do the brain work, the mental heavy lifting of developing theory. In the trading zone, economic geographers would offer data, and economists, in return, theory.

The impulse for economic geography to draw on economic theory became stronger from the mid-1950s with the wider disciplinary move to spatial science (Chapter 3). It was a period in which economic geographers made repeated border crossings into mainstream economics, bringing back whatever they thought would best make economic geography into its twin. They brought back assumptions, for example, *homo economicus*; theories, for example, the theory of the firm; and perhaps most of all mathematical-statistical techniques, for example, linear programming, game theory, Monte Carlo simulation, and input–output analysis. In contrast, economists showed almost no interest in economic geography. On the rare occasions economists left their "wonderland of no dimensions" (Isard's 1956, p. 25, unflattering term for the aspatial world in which economists operated) they developed their own framework rather than drawing on existing economic geographic scholarship. For example, when Paul Samuelson (1954), possibly the most famous American neoclassical economist, developed his theory of international trade he fancifully conceived transportation costs as like a melting iceberg ("the iceberg model"), ignoring other less fanciful conceptions previously worked out by economic geographers.

The problem was that economic geographers provided no value added when they reworked what they took from the economists. They gave back only simplified economic theory with spatial subscripts. Economists were underwhelmed, explaining the one-way traffic. Another reason was that there was no parallel in economic geography to the economist's mathematical training and expertise. Economic geography's tradition was resolutely qualitative, literary not mathematical. The first university classes in statistics in US geography departments were not inaugurated until the mid-1950s, and the first statistical geographical textbooks were not published until the 1960s. Economic geographers were still in their mathematical infancy, barely able to speak the language, certainly unable to contribute to formal economic theorizing.

There were two areas to which postwar economic geographers contributed, but in both cases they were only sort of in economics; certainly they were not central to the mainstream. The first was location theory, associated with a loosely connected, eclectic set of German scholars writing over the course of more than a century (Box 4.2: The German Location School). They included the Prussian autodidact, landowner, and farmer Johann von Thünen (1783–1850), inventor of the concentric model of agricultural land use; the economic sociologist Alfred Weber (1868–1958), developer of the locational triangle that calculated the optimal location for any given manufacturing plant; and the duo Walter Christaller (1893–1969), a geographer, and August Lösch (1906–1945), an economist, who each independently formulated central place theory, a general theory of the urban location of economic activities set on a grid of hierarchically nested hexagons. Unlike borrowed work from mainstream economists, location theory from the start was inherently geographical, not requiring the addition of spatial subscripts. It could be taken off the shelf and immediately applied. It was through empirical application and contextualization that economic geographers made their contributions, whether as studies of land use and rural settlement (von Thünen as interpreted by Chisholm 1962), or the location of iron and steel mills in the United Kingdom (Warren 1970), or the distribution of urban services in Iowa and Nebraska (Christaller and Lösch as interpreted by Berry 1967).

Box 4.2   The German Location School

Location theory in economic geography is a varied body of theory and techniques concerned with explaining and sometimes predicting the location of both individual and aggregate economic activities. As a body of work it is most associated with a set of German scholars who first began writing in the early nineteenth century, dubbed the German Location School (Blaug 1979; Barnes 2003).

The first member of the school, and likely most well-known, was the Prussian landed aristocrat and independent scholar Johann von Thünen. In 1826 he privately published *Die isolierte Staat* (*The Isolated State*) which presented a ring-like "concentric model" of agricultural land use. It rigorously set out and empirically demonstrated a set of general spatial relationships between land use, commodity prices, transportation costs, and land rents. Both the empirical testing (resting on meticulous, laboratory-like farm records kept by von Thünen) and the theorization (Alfred Marshall, the first professor of economics in England, said, "I loved von Thünen above all of my masters"; Marshall and Pigou 1925, p.360) were remarkably sophisticated. In effect, von Thünen invented the technique of deductive modeling. Calling what he did *Form der Anschauung* (the intuitive form), it involved starting with simplified assumptions, gradually relaxing them, and then using a formal deductive method (which included applying calculus) to derive the consequences. The second theorist, albeit still not a card-carrying economist, was Alfred Weber. His work was born out of Germany's increasingly vibrant urban-industrial economy of the second half of the nineteenth century. In 1909 he published *Über den Standort der Industrien* (*About the Location of Industries*), which presented the conceptual technique of the "locational triangle" (later taking material form as the Varignon frame, a wooden triangle fitted with ropes, pulleys, and weights).

The aim was to isolate the relative pull of different locational factors in determining the optimal location (that is, cost-minimizing and profit-maximizing) for a given industrial manufacturer. Weber's thesis was not as empirically exact or as dense as von Thünen's, but it compensated by its formal presentation. Weber enlisted the help of the brilliant German mathematician Georg Pick, who later assisted Einstein in formulating his theory of relativity, to write the equations of his locational model. Finally, there was the geographer Walter Christaller and the economist August Lösch, who each independently developed central place theory (Christaller's term). It was a general theory of the spatial distribution of economic activities set across an entire urban hierarchy. For both writers, those activities were located on a set of overlapping different-sized hexagonal grids, creating an idealized honeycomb landscape of economic spaces. Both Christaller and Lösch were also concerned to use their theory to change the world, to make it better. Lösch (1954 [1940], p.4) famously said that "the real duty of the economist was not to describe our sorry reality but to improve it." Christaller, though, deployed his theory to assist the Nazis, making at least his definition of what it meant to "improve" reality counter-intuitive if not grotesque.

The other area, and related to location theory, was an account of industrial districts. It was an idea originating with England's first professor of economics, Alfred Marshall (1842–1924). An industrial district is a tightly bounded industrial area specializing in a single type of good like clothes or furniture. Firms within the district are highly specialized, undertaking only one or a limited set of tasks within the overall production process. To produce the final good, firms within an industrial district must therefore interact. Marshall (1919) theorized that propelling the formation of industrial districts were agglomeration or external or increasing economies (see also Chapters 10 and 12). These were cost savings passed on to each firm from the very act of agglomerating, from joining and congregating within an industrial district. The implication was that the more firms agglomerated, the greater would be cost savings, and thus the more attractive the district would be as a place to locate for yet even more firms. It was a virtuous cycle: the more there is, the more there will be.

The contribution of economic geographers to work on industrial districts and agglomeration was complicated, and goes to the next phase of the story in the relationship between economic geography and economics. While Marshall's (1919) idea of an industrial district was taken up at least implicitly by economic geographers as early as the 1940s (e.g., in Michael Wise's 1949 work on the gun and jewelry quarters in Birmingham), it was not integrated into a theoretical account. That did not come until economic geography began its break with mainstream economics from sometime around 1980. Then industrial districts were folded into the larger theoretical account provided by post-Fordism (see Chapters 1 and 12). In turn, that became a key category within a non-neoclassical approach that increasingly came to structure the discipline.

The break-up
The break between economic geography and economics occurred partly because of a disconnection between the world and mainstream economics' version of it, and partly for reasons of method. The mainstream view was that through the smooth movements of the

market long-run optimality would be achieved, realizing Dr Pangloss's best of all possible worlds. That was not the world of the 1980s, however (see also Chapter 1). By then there had been almost a decade of "stagflation" in the Global North, that is, a stubborn economic downturn associated with high unemployment, low investment, declining economic productivity, sluggish growth in real incomes, and spiraling prices. Even worse, deindustrialization, the systematic hollowing out and abandonment of manufacturing in the Global North, also had begun to bite. Within five years it would ravage long-standing industrial regions, especially in Northern England (Manchester lost a half million jobs) and the Northern US states (the manufacturing belt became the rust belt). This economy was patently neither in equilibrium (more like in free fall), nor operating optimally (more like ruinously). There needed to be a different approach. That came initially from Marxism, a long-time bitter enemy of neoclassical economics that viewed that approach as nothing more than an ideological apology for capitalism, and then later from other non-mainstream economic approaches (collectively labeled heterodox economics).

The other reason for economic geography to decouple from mainstream economics was methodological. Following a more general debate in human geography beginning in the mid-1970s, a determined set of objections were set out against the use of both undiluted mathematical formalism and a statistical-quantitative approach. Both features were the methodological bread and butter of mainstream economics that during the period of infatuation shaped economic geography. The objections made to a mathematical-quantitative approach by some economic geographers from the mid-1970s varied from philosophical, to ideological, to technical, to practical (see Chapter 6). The economic geographic world was too untidy, too ambiguous, too provisional, and too unpredictable to be caught in the steel trap of a singular mathematical logic. Mathematics might be nature's language, but it wasn't society's, or even the economy's, at least the economy with which geographers dealt. Economic geography was dappled and variegated, subject to chance and contingency, mixed up and muddied with the non-economic. It was a plural world requiring not one but many methods. It was yet another reason to sever the relationship with mainstream economics that dogmatically insisted on only one method.

Post-break-up
Thus began the long break-up between economics and economic geography, which widened so much that, as Jamie Peck (2012, p.114) put it, "it was as if a great ocean had come to separate them." From Figure 4.1, it was as if the land bridge that formerly connected economic geography with mainstream economics had washed away, leaving economic geography a separate island. But while it might have broken from orthodox economics, economic geography had not rejected all of economics. Other, alternative, heterodox versions of the discipline existed, some that began life before even neoclassicism (Table 4.1). To extend Peck's metaphor, economic geography became part of that larger archipelago of islands that were held together by both their opposition to neoclassicism and their promotion of an alternative, heterodox economic view (Figure 4.1).

Since the split, economic geographers increasingly have explored the archipelago. The first and the biggest of the islands was Marxist economics. There had been a lone crossing there in the 1920s by the English socialist geographer J.F. Horrabin (Hepple 1999), but from the mid-1970s stopovers became numerous and undertaken by many. The resulting trading zone produced an altogether different type of economic geography. Inhabitants of that

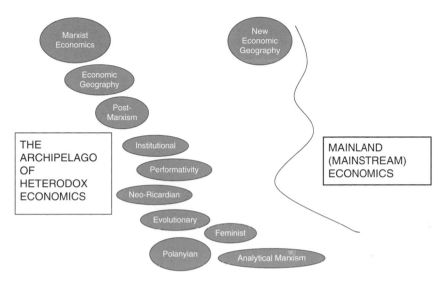

Figure 4.1   The archipelago of heterodox economics.

island portrayed capitalism as dynamic but often destructive. Capitalism's geography was defined by spatially variegated crisis, uneven regional development, precarious "(un) natural" environments, and black spots of violent industrial retrenchment. Notions of optimization and equilibrium were thrown out during the crossing, sinking without trace. The new treasures carried back, such as ideas of "value," "overaccumulation," and "crisis," became central to the changed vocabulary and practices of economic geography (brilliantly represented in Harvey's 1982 work; see Chapters 3 and 5).

There were also other islands to explore. In many ways the recent history of economic geography is the history of those island visitations. They included the islands of French regulationism, as well as institutional, evolutionary, feminist, post-Marxist, analytical Marxist, neo-Ricardian, Polanyian, and performative economics (Table 4.1). In each case, trading zones were opened and ideas exchanged. Compared to neoclassicism, the pluralistic, hybridized, and fuzzily bounded heterodox economic geography that resulted from island trading has been, with some notable exceptions, less abstract, less mathematical, pitched at an intermediate as opposed to a rarefied level of theorization (Chapter 5), and concerned with context and contingency, not with universality and iron-clad mathematical necessity.

There was one rogue island, distant from the archipelago, which maintained strong links with the mainland. It was the island of "new economic geography" or, as we also labeled it in Chapter 2, "Geographical Economics." It was brought into being by the American econo-mist and 2008 Nobel Prize winner Paul Krugman (1995a, p.33), who claimed it first came to him as "a vision on the road to Damascus." That vision was underpinned by the neoclas-sical canon and its two fundamental principles: "maximization (of something) and equilib-rium (in some sense)" (Krugman 1995b, p.75). Also in accordance with neoclassicism, Krugman retained absolute faith in mathematics. "To be taken seriously an idea has to be *something you can model*," Krugman (1995b, p.5; original emphasis) said. But not any old model will do: it must be formalized as "Greek-letter writing" (Krugman 1990, p.ix),

Table 4.1   Forms of Heterodox Economics

Analytical Marxism
An approach that originated during the 1980s, concerned with sorting Marxism into a distinct set of claims that could be analytically scrutinized for their meaning, coherence, plausibility, and truth. Marx was treated as an innovative thinker, but whose ideas required careful logical dissection and development in the light of intervening history, and with the analytical tools available, both conceptual and mathematical, found in contemporary social sciences (see Elster 1985; and in geography, Sheppard and Barnes 1990).

Evolutionary Economics
Evolutionary economics applies theories, models, and concepts from evolutionary biology to the economy. Concepts from evolution such as individual variation, inheritance, selection, competition for resources, adaption, and survival are deployed to explain the economic dynamics of firms, institutions, technological change, and regions. Allied concepts that have also been developed, and which appear to have an especial geographical resonance, are lock-in (once some choices are made, they are very difficult to change), path dependency (decisions in the past determine present trajectory), and resilience (the ability to recover from external environmental shock) (see Nelson and Winter 1982; and in geography, Boschma and Martin 2007).

Feminist Economics
A critical study of all aspects of economics – substantive focus, methodology, history, and philosophy – from a feminist perspective. Part of feminist economics' critical view includes an attack on the androcentric (man-centred) and patriarchal assumptions of mainstream (neoclassical) economics. There is a concerted attempt to include topics specifically germane to the interests of women, such as forms of gender discrimination found in the labor market (from invisible ceilings to differential wages for the same job). The normative purpose of the approach is to improve the lot of women within the economy, including through the use of action research (see Chapter 6) (see Waring 1989; and in geography, Gibson-Graham 1996).

French Regulationist School
Developed from the late 1970s, and most associated with the French economists Alain Lipietz and Robert Boyer, the school argued that there were two critical components to take into account in understanding the dynamics of an economy: the regime of accumulation (the relation between macro investment and consumption), and the mode of regulation (all the institutions, laws, and rules of conduct that shape the economy). The regulationists deployed both concepts to understand, on the one hand, the decline of industrial Fordism in the Global North over the 1970s, and, on the other, the subsequent rise of post-Fordism from the 1980s (also called flexible production or flexible accumulation) (see Lipietz 1987; and in geography, Amin 1998).

Institutional Economics
Institutional economics originally arose with the work of the American maverick economist Thorstein Veblen (1857–1929). Reacting against both an imported neoclassical economics and a domestic industrial system that spawned massive inequalities in levels of wealth and consumption, Veblen constructed a made-in-America theory of his own time and place, institutional economics. Drawing particularly on Darwin, Veblen (1919, p.239) emphasized the close relation between individual economic behavior and evolving institutions (defined as "settled habits of thought") (see Hodgson 1993; and in geography, Martin 2001).

Marxian Economics
The school of economics that traces its origins to Karl Marx's (1818–1883) nineteenth-century analysis of capitalism. Marxian economics is now an enormous and variegated corpus of work. All of it, however, emphasizes the inherent crisis-prone (contradictory) nature of the capitalist mode of production, and its foundation on the exploitation and oppression of labor (the working class) by owners of capital (the capitalist class). Marx believed those contradictions, along with exploitation and oppression, will sooner rather than later produce social revolution, with communism rising from the ashes of capitalism (see Baran 1957; and in geography, Harvey 1982).

Table 4.1 (Continued)

Neo-Ricardian Economics

A school of economics initiated by the Cambridge Italian economist Piero Sraffa (1898–1983) and his slim 1960 monograph, *The Production of Commodities by Means of Commodities*. Sraffa presented a formal model of the economy as a series of input–output equations. He demonstrated that a determinant mathematical solution required reference to a set of historically contingent, non-economic relationships that lay outside the equations. Local context mattered even in economics (see Wolff 1982; and in economic geography, Barnes 1996, chapter 5).

Performativity

Originating in science and technology studies (STS), and especially with the writings of Michel Callon, the performativity approach suggests that economic markets emerge and become real only through their performance. They do not exist prior to it. Instead, the performance, carried out by humans often in concert with various calculative devices, brings into being the new object, the market. Rather than mirroring pre-existing markets, Callon argues that neoclassical theory is a script that people follow to perform the market, to make it come into being. The performative approach thus reverses the normal relation between theory and reality. Theory comes first, producing reality, rather than following it (see Callon 1998; and in geography, Berndt and Boeckler 2009).

Post-Marxism

A loosely bound approach that joins variants of poststructural theory to examine in the broadest definition economic processes. The rigid and exclusive class-centered classical Marxist analysis is rejected. There is no single cause such as the relations of production that in Marx's account propels history forward through a series of crises, and ends with the communist revolution. Instead, there is only overdetermination, causes stacked on causes, which destabilizes and renders incomplete terms such as social class. Ernesto Laclau and Chantal Mouffe (1985) were the first to align themselves to an explicitly post-Marxist agenda. In economic geography post-Marxism is most associated with Gibson-Graham's (2006) (Chapter 3) work on diverse economies that examines such topics as the multinational corporation, mining in the Latrobe Valley, Australia, and the Mondragon Corporation, a federation of worker cooperatives that started in the Basque region in Spain (see Callari and Ruccio 1996, and in geography Gibson-Graham 2006).

Polanyian

An approach to the economy inspired by the mid-twentieth century economic historian and anthropologist Karl Polanyi (1886–1964). Polanyi emphasized the critical importance of cultural-institutional embeddedness to the economy. For the economy to continue, Polanyi argued, it must be buttressed by a set of supporting underlying institutions. If they are not there, as Polanyi showed in *The Great Transformation* (1944), his study of nineteenth-century English industrial capitalism, the larger system can veer perilously close to collapse. It is saved only by reinstalling the missing institutional undergirding (see Dalton 1968; and in geography, Peck 2013).

Krugman's code for mathematics. The problem with the old economic geography, Krugman thought, was that there were not enough Greek letters, and when there were, they were often the wrong Greek letters. Past "modeling efforts of the geographers [were] … murky," he chided (Krugman 1995b, p.87). His solution was to take "the insights of the geographers [but to] integrate [them] … into economics through clever models … [that] meet the formal standards of economists" (Krugman 1995b, p.88). Krugman's epiphany was to realize that he already had the right clever model to fix economic geography. That model was Dixit–Stiglitz, which Krugman earlier tried out on international trade theory (winning him the Nobel prize). Krugman showed that when Dixit–Stiglitz was applied to a set of regional economies it gave a version of the virtuous cycle found in Marshall's industrial districts.

Providing labor was sufficiently mobile, market forces would promote regional agglomeration, generating increasing returns. Consequently, Krugman's regional economic landscape was not smooth but clumpy. Intense concentrations of robust regional economic activity marked by increasing returns would punctuate the landscape, with expanses of anaemic economic space in between.

Although there has been some traffic between the new economic geography and other ("old") parts of economic geography, generally it has been light. And where economic geographers *other* than "Geographical Economists" trade with economics, it is generally not with neoclassical economics of the mainland, but heterodox economics of the archipelago. The problem with neoclassical economics is it has to be its way or no way. That lay behind Krugman's intervention too. Because economic geographers did not do it right, that is, they could not fulfill the formal standards of the economists, Krugman had to do it for them. It goes to neoclassical economics' underlying imperial impulse. The economists' burden was to bring formalism to the dark spaces of the social sciences, such as economic geography, where it was not yet found, or not done correctly.

The relationship of economic geography with varieties of heterodox economics is much different, however. There is no imperial impulse, no single template for doing economics. Instead, as Wendy Larner (2012, p.159) puts it, contemporary economic geography is "based on bricolage and borrowing, inflected with homegrown innovation and ingenuity." Bricolage means do-it-yourself, creating something from the varied and contingent resources at hand. Those include what was found and brought back from other islands of heterodox economics. For this reason, and as we will show in the following chapter, theory in economic geography does not mean a formal model written in Greek letters, but more like a collage, creatively patched together from bits and pieces, some found locally, others from farther afield.

4.3.2   Anthropology

While historically the border country with economics, whether of the mainstream or heterodox kind, has been heavily trafficked, at different times for different durations trading zones have also been established between economic geography and other social sciences.

Admittedly, the interaction with economic anthropology has been limited. In many ways, though, it is a kindred disciplinary spirit, having experienced similar approaches and debates to those experienced by economic geography. Further, the two disciplines share common inclinations to grounded case studies, varied theoretical sources, and explanations pitched at an intermediate level of abstraction. Economic anthropology, however, is much more rigorous than economic geography in its method, ethnography (see Chapter 6), and involves protracted periods of immersion in the field, participating in the practices observed. Some economic geographers claim to practice ethnography. But Elizabeth Dunn (2007) is doubtful, suggesting that this usually means carrying out only an interview, which is different from ethnography because it is much shorter in duration, and involves neither lingering nor joining in.

More successful has been the use by economic geographers of work by anthropologists on material culture. From the beginning, economic anthropologists such as Bronislaw Malinowski (1922) were keen to interpret the cultural significance of material artefacts.

A stick was never just a stick. It could be an agricultural implement, or a prop in a religious ceremony, or an instrument affording its holder the right to speak in public, or a sartorial accessory connoting status. Its meaning depended on cultural context. Similarly, some economic geographers have tried to go beyond mere physical properties of goods when analyzing the components of a commodity chain (Chapter 1). Ian Cook's (2004) work has been formative, ever alert to cultural context. In one of his papers on the geographical travels of the papaya he argued that the fruit was eaten in the United Kingdom not because of its physical characteristics – its sweet taste, its juiciness, its fleshy fibre, its yellow color – but because of its cultural resonances, often unconscious, which gave it meaning, and made it irresistible: Western tropical exoticism, international travel advertisements, Jamie Oliver cooking shows, Delia Smith cookbooks, and the evolving aesthetic tastes of the English middle class.

Perhaps the sprightliest trading zone with anthropology, continuing to yield gains to trade, turns on the work of Karl Polanyi. Polanyi was himself a liminal scholar, brilliantly exploring the border country of several disciplines. The last university department with which he was associated was Anthropology at Columbia University, where he wrote about ancient and primitive economies (Dalton 1968). Polanyi argued that the continued repro-duction of any exchange economy from the very first one – in his reckoning the Ancient Greek household economy – depended on what he termed its embeddedness within a set of cultural institutions. Those institutions would vary from one culture to another. Their importance was in forming a glue that held economy and society together. Without institu-tional embeddedness – disembeddedness – there would be chaos and crisis, an inability to reproduce the exchange economy, with economy and society hopelessly careening apart.

Economic geographers first drew on Polanyi's notion of embeddedness during the 1990s (Grabher 1993). It was a means to move away from the neoclassical theory of the firm that first entered economic geography during the spatial science phase (Chapter 3). That theory conceived the firm as solitary and independent, motivated only by its own ruthless internal drive to maximize the bottom line of profit. By drawing on Polanyi's concept of embedded-ness, economic geographers reconceived the firm as only one element within a larger social and cultural institutional matrix, containing entities both economic and non-economic. This new approach stressed a firm's close linkages both with other firms and with a range of non-industrial institutions such as the state, educational institutes, banks, labor organizations, and various non-profit organizations. That is, the firm was embedded within thick Polanyian, place-specific institutions.

More recently, there has been an even more direct use of embeddedness; or, rather, the idea of disembeddedness. Polanyi (1944) wrote about disembedding in his book *The Great Transformation*. He argued that whenever an economy became unhooked – disembedded – from social and cultural institutions, economic and social crisis followed. In his view, the pure exchange economy was too wild and wayward to sustain itself. It needed always to be brought into check by constraining institutions. If they were no longer there, he thought, sooner rather than later there would be trouble. Economic geographers have made special use of this insight to understand the cascading economic crises gripping capitalism since the emergence of neoliberalism during the early 1980s. They have argued that the principal cause of this economic turmoil was the disembedding of the economy from its hitherto stabilizing institutions. For neoliberalism's agenda from the beginning was to realize a mar-ket pure and simple, deregulated, shorn of any extra-market encumbrances, institutional, cultural, political, or social. Its agenda was disembeddedness. Polanyi would say that was asking for trouble, and trouble has come. The culmination (so far) was the October 2008 financial collapse, when capitalism teetered on the brink of extinction. Even the usually phlegmatic Governor of the Bank of Canada (now the Governor of the Bank of England), Mark Carney, was rattled. "The world was 36 hours from financial Armageddon," he said.

Although Polanyi wrote in the mid-twentieth century, long before neoliberalism, as an anthropologist he appreciated the close relationship between culture, politics, and economy. Economic geographers working in the border country with economic anthropology have recognized and imported his insight into their own work, productively using it to understand the contemporary economic geography of an unreasonable "neoliberal reason" (Peck 2010).

4.3.3   Cultural and gender studies

Cultural and gender studies have separate but intertwined histories both of which date from the 1960s. There are clear shared interests between the two. Both are concerned with the everyday; both tend toward the contemporary (albeit not exclusively); both are methodologically open, pluralist in their inclinations; and both draw on an assortment of critical theoretical perspectives – Marxism, post-Marxism, feminism, poststructuralism, postcolonialism, critical race theory, and queer theory. Like a Venn diagram, there are segments of overlap between the two, and other segments that are separate, subject-specific.

Trading zones between economic geography and gender studies opened from the 1970s, and with cultural studies in the 1990s. They have been two of the most important, certainly the liveliest, border country engagements for economic geography over the past two decades. Several parts of economic geography were radically transformed by these exchanges, including its methods, forms of political activism, and theorization, as well as the content of specific subject areas such as the geographies of labor, work, consumption, and globalization.

Gender studies entered economic geography during the 1970s as second-wave feminism,[3] and during the spatial science phase of the discipline (Chapter 3). As a result, feminism was couched as tables of numbers and statistical equations, showing stark discrepancies in labor market conditions and commuting patterns between men and women, favoring the former. During the 1980s a more vocal and political form of feminism emerged in economic geography drawing on work within socialist feminism. By then political economy was established within economic geography (Chapter 3). From within radical economic geography, socialist feminist geographers raised questions about the social position of women within capitalism, as well as the relationship of capitalism to patriarchy, the system that accorded power and authority to men and justified such a distribution. Doreen Massey's (1984) work on industrial restructuring in South Wales discussed in Chapter 3 partly arose from this context. She demonstrated that patriarchy found in South Wales significantly contributed to the form of industrial restructuring that subsequently unfolded in that region. Male patriarchy turned women in South Wales into a perfect exploitable labor force for American and Japanese multinational electronics firms that began investing there from the 1970s. As a result of patriarchy the South Wales female labor force was: greenfield (they had never worked within industrial capitalism before); pliant (especially in the face of male managers); and spatially trapped (male partners expected their female partners to do a double day, double duty, that is, to work both in a factory and at home undertaking domestic labor thus severely spatially constraining their ability to commute). From the 1990s, interest increasingly turned to issues of gender identity and the body (a consequence in part of the rise of third-wave feminism – see note 3). Linda McDowell's (2013) writings on female workers in the United Kingdom from World War II onward, including various immigrant populations, were exemplary. Following the feminist theorist Judith Butler, McDowell argued that the subject position, that is, the identity taken by the women she interviewed, was fluid, performed in the moment. There were, of course, limits to that performance, at times authoritatively imposed, but as McDowell (and Butler) suggested, transgression was possible, and on occasion it could produce incremental progressive political change (McDowell's work is further discussed below).

Cultural studies began as a "soft" version of Marxism, less concerned with the hard elements of the economy, the "base" or "infrastructure" as they were called. Instead, cultural studies' interest was the spongier surround of everyday life, "the superstructure" as it was termed. Initially, cultural studies interpreted everyday life in relatively narrow political economic terms, as functional, however distant, for the maintenance and continuation of industrial capitalism. For example, cultural studies never interpreted going to school as some youthful idyllic moment of innocence and carefreeness but instead as the first step for entry into the disciplined world of capitalist work life. Even male yobbish acting out at school, cultural studies asserted, was part of the functional plan of capitalism to secure the right kind of labor for the right kind of jobs, thus to ensure the continuation of capitalism (Willis 1977). This soft political economy version of cultural studies did not last, though. Because of the profound changes in the Global Northern economy from the late 1970s turning on ferocious industrial restructuring and deindustrialization, as well as the concomitant burgeoning of service-sector employment both at the bottom (McJobs) and at the top end (the creative class) of the labor market (Chapter 1), the original Marxist inflected nature of cultural studies concerned with a masculine industrial working class lost theoretical purchase. Cultural studies needed to broaden, which it did, analyzing as it does now cultural identity across gender, race, sexuality, and age. Even when class is emphasized in cultural studies it is conceived differently, with a cultural inflexion, and reflected also in an important theoretical shift. Classical Marxist theorists who previously were central to work in cultural studies were moved at least sideways, if not to the back, to allow a new set of theorists in, often with roots in poststructuralism, postcolonialism, and queer studies.

Both gender and cultural studies were critical to the emergence of the "cultural turn" that economic geography experienced from the 1990s (see also Chapters 3 and 5). An animated trading zone between economic geography and these other two disciplines opened up. While the geographers drew on feminist and cultural studies theorists like Judith Butler, Donna Haraway, and Eve Sedgwick, they gave back in exchange reconstituted concepts infused by a spatial sensibility, filled out and shaded by compelling empirical geographical cases. The upshot was a radically different version of economic geography in which, for example, the economic was no longer conceived as separate and autonomous, but utterly mixed up and muddied with other parts of cultural, social, and political life (Chapter 2). There was no essentialized detached economy, nor did it come with its own bounded internal rules and logics. Instead, it was conceived conjuncturally (Pickles 2012). That is a difficult word to define. In the case of economic geography, it means seeing the geography of an economy as one part of a complex balance of relations temporarily stabilized. Defining the precise boundaries of the conjuncture, its exact mechanisms of transformation, the specific relationships among its constituent elements, and the potential sites of intervention for enacting progressive change, all need to be examined *in situ*, that is, within the context, using whatever theoretical means are best suited (Chapter 5).

Cultural studies and feminism, of course, offer a very different approach to economic geography than neoclassical economics. The latter begins with a timeless and pure object, the economy, and with a set of exactly defined, universal explanatory principles that are brought to bear on that object. In contrast, there is nothing timeless and pure about either the conjunctural moment that cultural studies and feminism set out to understand, or the methods and theories used. This is why it is so important for economic geographers to continue to explore the border country.

4.3.4   Environmental studies

Environmental studies, which formally began during the 1950s, was from the start interdisciplinary. It was a combination of physical and social sciences concerned with explaining the relationship between humans and their natural environment.

The significance of the environment as a topic for economic geography has varied historically (see Chapter 11). The first economic geographers, such as George Chisholm and J. Russell Smith, viewed the environment as a pristine stock of raw commodities ripe for extraction. The geographical problem was to locate those resources on maps and gazetteers. Colonialism did the rest. During the interwar years, an interest in the environment meant primarily an interest in agriculture. That attentiveness, though, was mostly typological: was a particular agricultural site best described by the category the "Corn Belt"? Or the "Hog Belt"? Or the "Dairy Belt"? Devastating environmental effects on US agriculture occurring during the same period, such as the Dustbowl when millions of tons of topsoil literally blew away in the wind, were ignored. In the early postwar period, the relationship between economic geography and study of the environment became even more attenuated once spatial science gained disciplinary prominence. Simplifying assumptions necessary to make spatial science's mathematical models tractable – the most well-known was the postulate of an "isotropic plain," the supposition of an infinitely large flat surface undisturbed by resource differentiation – ruled out from the off any serious examination of the environment.

Ironically, the environment as a serious intellectual topic first entered economic geography though its engagement with Marxism. On the face of it, Marx should not have much to say about nature. His focus (and moral contempt) were the "dark Satanic Mills" of nineteenth-century urban-industrial capitalism, places where, as Charles Dickens in *Hard Times* put it, "nature was bricked out." But David Harvey (1974), closely following Marx (and Dickens), argued that nature was yet another of the things that capitalism produced. He did not mean that capitalism literally produced nature from scratch. The Laws of Thermodynamics were never contravened. Rather, Harvey's point was that as soon as capitalism came on to the scene it began to transform existing nature (sometimes called "first nature"; Smith 1984). How nature was exploited, used, thought about, and represented all fundamentally changed. It turned into "second nature" (see Chapter 11 for an extended discussion).

Since then economic geographic studies of the environment have become one of the most vibrant and vital research areas in the discipline, now often carried out under the capacious albeit porous label of "political ecology." That work is set exactly in the border country, joining heterodox economic geography with environmental studies. Initially there was only a Marxist approach. Later, other heterodox traditions were deployed (see Table 4.1). Some of the resulting writing was purely theoretical, but much of it adhered to the tradition of economic geography by combining detailed empirical case studies with theory couched at a mid-level abstraction. The trade with environmental studies was both conceptual (for example, drawing on ecological ideas such as carrying capacity, resilience, and emergence), and empirical (for example, the examination of particular environmental cases such as Oregon forests, Pacific halibut, or Spanish aquifers).

Furthermore, a large portion of this border country work between environmental studies and economic geography was driven by changes in environmental regulation brought about by neoliberalism. Neoliberalism recast nature and in some cases produced new forms. For example, water was treated as a commodity rather than as a public good, and enacted in forms as varied as bottled water menus in upscale Southern Californian restaurants to privatized water companies in England and Wales. Or again, new hitherto unthought forms of nature were freshly minted, such as wetland banks, biodiversity assets, and carbon credit trees. In this work, the demarcation between environmental studies and economic geography was blurred, and in some cases erased. Nature was not separated from economic geography, black boxed, to be dealt with later by someone else, but fully integrated, entering into the disciplinary core (Chapter 11).

4.3.5   Political science

Political science was another late nineteenth-century social science. From early on, American political scientists were concerned to identify and mobilize *scientific principles* to analyze and resolve problems of governance. Even more than in geography, there was a concern for objective empirical data, scientific techniques of analysis, and formal mathematical models and theories. That trajectory has become even more entrenched, with economic geography and political science consequently traveling in opposite directions. Increasingly anything goes in economic geography, while in political science only one thing goes, a scientific, mathematical approach. Perhaps for this reason, it is with political science that economic geographers have most often struggled to establish a trading zone. John Agnew in reviewing the (non-)relation between the two disciplines writes:

> Students of politics potentially have much to offer to a more explicitly politically informed economic geography .... Yet, I can see little or very limited interaction across this intellectual boundary. The centennial issue of the *American Political Science Review* in 2006 ... reviewing the main themes of the journal mention little or nothing of direct relevance to economic geography. (Agnew 2012, p.570)

This is not because the two disciplines have nothing to say to one another. From at least the late 1970s, economic geography, partly influenced by its alliance with Marxism, recognized the importance of the state. Initially the theory of the state used by economic geographers was Marxist, but since then discussion has widened to include multiple theoretical approaches, multiple geographical scales, multiple government functions, and multiple forms of governance (a distinction is usually drawn between "government" carried out by the state, and "governance" carried out by many institutions that might include the state but need not). Some of the most important disciplinary contributions made by economic geographers since the late 1970s have turned precisely on investigating political issues of governance, for example, as they affect industrial districts, or as they define neoliberalism. Here, though, the chief conceptual resources have come not from political science but typically from sociology (discussed in the next section).

The engagement between economic geography and political science, then, has been sporadic, and ad hoc. An incomplete trading zone formed around the study of early 1980s industrial change, particularly the emergence of post-Fordism and its realization as regional agglomerations (industrial districts). It was taken up by the political scientists Michael Piore and Charles Sabel (1984) in their influential book *The Second Industrial Divide*. Their contribution coincided with the investigation by economic geographers of similar new industrial spaces. Piore and Sabel added a novel political spin, claiming that the regional agglomerations presaged a modern form of "yeoman democracy" that was sharply different from the type of politics found among scientifically managed, deskilled workers under the previous Fordist system. Economic geographers contributed detailed, on-the-ground studies of regional industrial agglomerations, along with creative theorizations about their formation. While there was acknowledgment of each other, even some mutual flattery, the union was never consummated.

More satisfying, yet also not fully realized, was potential trade between economic geographers and political scientists around the idea of varieties of capitalism (VoC). This notion was most associated with the political scientists Peter Hall and David Soskice, and their edited book *Varieties of Capitalism* (2001). The gist of the idea was that capitalism came in two main forms: (i) a liberal market version exemplified by the United States and the United Kingdom; and (ii) what was called a coordinated market version and found in countries like Germany and Japan. The difference between the two was the degree to which there was intervention in the market, often by the state, but not necessarily. The liberal market variety of capitalism was characterized by minimal intervention, while the coordinated market version had intervention at every turn. This thesis that there were marked geographical variations in the form of capitalism was then explored by economic geographers. Peck and Theodore (2007) especially saw great potential for trade with political scientists, and in exchange they offered both case studies and potential ancillary conceptions shaped by a full-bodied economic geographic approach. They pushed political scientists to recognize not just two varieties of capitalism but a plethora of them. In Peck and Theodore's language, it is not varieties of capitalism but variegated capitalism. Peck later illustrated the point with Jun Zhang using the case of China (Peck and Zhang 2013). As they demonstrated, the Chinese economy falls betwixt and between a liberal and a coordinated market. As a round peg it cannot be hammered into either of the two square holes given by Hall and Soskice. It requires a different framework, not necessarily one that throws out all of the old, but one which amends it through a geographical sensibility, and which is available by working diligently in the border country.

4.3.6 Sociology

Sociology first emerged as an academic discipline around the same time as political science at the turn of the twentieth century. One of its theoretical pillars, though, was Marx's work, giving it from the start a critical, radical sensibility that has meshed well with economic geography at least over the past 40 years, resulting in several close collaborations on the border country. At times the trading was so close it gave truth to the old saw that geography was sociology with maps, or sociology was geography without maps. Trading has gone on around: Immanuel Wallerstein's world system's theory and the commodity chain analysis derived from it; Pierre Bourdieu's theory of class and his notion of cultural capital; and more recently, Michel Callon's and Donald MacKenzie's theories of market performativity (on which more below).

An early undertaking where economic geographers and sociologists rubbed shoulders was in the so-called localities project of the mid-1980s (Cooke 1989). That project analyzed the spatially variable effects of British deindustrialization in combination with the contemporaneous Thatcherite neoliberal social and political revolution. It asked whether those effects were the same everywhere, or whether they were more marked in some places than in others, and if they were, whether the forms that those effects took varied spatially. The basic geographical unit of analysis was the "locality." It was never definitively defined, but in practice it meant a large urban region. For each locality, teams of researchers were assembled composed of both geographers and sociologists. Their task was to investigate for their assigned locality the peculiar form that larger national structural changes around deindustrialization and Thatcherism induced. It meant border country work. Geographers insisted that place mattered, which was obvious when different localities were compared, while sociologists insisted that broad social processes around class, gender, and race mattered too, which was also obvious from the details brought back from within each locality.

Another fertile encounter also from the 1980s was around theorizing and documenting the transition between the declining older industrial system of Fordist mass production and the new industrial system that came to replace it variously labeled post- or neo-Fordism, or flexible production or accumulation. This became a true interdisciplinary effort involving multiple border excursions, which, apart from economic geographers, included heterodox economists (especially those from the French regulationist school such as Alain Lipietz and Robert Boyer), political scientists (Michael Piore and Charles Sabel), and economic sociologists (Bob Jessop). The broad argument was that the new system operated radically differently from the old one. It used novel computerized methods of production (CAD-CAM) and trained laborers who could undertake a range of work tasks on the shop floor (flexible labor) to produce short-run batches of differentiated products (batch production as opposed to mass production; see Chapter 12).

Within this work of reconceptualization, the state theorist and economic sociologist Bob Jessop was especially prominent, emphasizing particularly the role of government. He used the vocabulary, already introduced in Chapter 1, of the Keynesian Welfare State (KWS), arguing it performed vital roles in maintaining the larger Fordist industrial system by regulating the macro economy, and through the welfare system by providing ideological legitimacy. He also began to sketch out the nature of the state after KWS that he initially labeled the Schumpeterian-Workfare state. Jessop (1993) argued that this was a pared-back state, not bloated, keen to facilitate the creation of markets in all spheres, which it was claimed maximized efficiency and choice. This form of the state was later called neoliberalism. Economic geographers drew on Jessop's formulations and analysis, initially adding to his contributions a discussion of the local state (Peck and Tickell 1995). He reciprocated in the trade adding an explicitly spatial dimension to his work (Jessop and Sum 2006).

One last example concerns a term already discussed, embeddedness, and its further development as the idea of a network. It was associated with one of the most cited papers in twentieth-century economic sociology (well over 30,000 citations), Mark Granovetter's (1985) "Economic Action and Social Structure: The Problem of Embeddedness." By embeddedness Granovetter meant the concrete personal relations and structures (social networks) that individuals use and forge within the economy to facilitate the realization of economic ends. In Granovetter's view we are all socially connected. We can't help it. That's what living in contemporary society means. Consequently, we must view all human acts, including economic acts, as fundamentally set within networks of personal relations. The economic geographer Gernot Grabher (2006, p.165) early on brought Granovetter's idea into economic geography, "afford[ing] a major platform for crossdisciplinary exchange between economic geography and economic sociology." In economic geography it led to burgeoning literatures on networked firms, regions, and industrial districts. In each case, they were analyzed in relation to the larger interconnected network of firms, regions, and industrial districts of which they were component members. No firm, region, or industrial district exists on its own, isolated and hermetically sealed, but is always part of a larger tightly linked collective relational entity, the network.

The history of economic geography's relation to economic sociology demonstrates the benefits of border trade. That does not mean such work should automatically get a free pass. It is important also to be critical of imported ideas, and their application. For example, Jamie Peck (2005) provided incisive criticisms of Granovetter's notion of embeddedness. Working in the border country does not guarantee a better approach, but by provoking new thoughts it at least increases the range of options providing a base for creative and critical interventions.

4.4 Border Life: Three Case Studies

To finish, we flesh out our larger argument about the benefits of border country work by providing three brief case studies. In the best of border work there is mutual trade. Economic geographers may start with ideas taken from an adjoining discipline, but by working them through a specific geographical case, supplementing them with existing or novel geographical concepts, there can be alchemic transformation. A brand new entity is created that is not just derivative. The new object has its own integrity and worth, with the potential to travel, to make a difference.

4.4.1 Linda McDowell on embodied labor

Linda McDowell has spent much of her academic career on the borderlands of feminism, sociology, and economic geography investigating issues around the body and changing patterns of labor and work. McDowell (2007, p.65) says in a semi-autobiographical essay that when she first became an academic economic geographer she was "anxious to assert women's rationality," and as such put issues of the body on one side. Raising issues about corporeality, McDowell (2007, p.65) thought, "only reinforce[d] arguments about women's bodily confinement ... Surely the body could be ignored?" But as McDowell carried out her research on the economic geography of employment and work within the United Kingdom, it was clear it couldn't. The body "as it turned out, was part of the explanation" (McDowell 2007, p.65).

Over more than two decades McDowell has presented a series of meticulous and compelling empirical case studies set in the United Kingdom that show how the body matters at work. Many of the bodies she writes about are women's, but not exclusively. She wrote eloquently about the massive loss of masculine factory jobs ("Life without Father and Ford"; McDowell 1991) in the wake of deindustrialization in the Global North, and in Britain in particular. And that has been followed by a series of studies exploring the new gender order at work, one of which centers on young men who, because they were outperformed at school by young women, often have no work. "Young men are the newly disadvantaged group," McDowell (2007, p.66) writes.

Fordist industrialism was replaced by a postindustrial service sector, turning upside down the old nostrums of work and its associated gender order. For McDowell the important point about service work is that it often requires embodied interactions with customers and with other workers. Labor power alone, muscle, was insufficient to get this work done. The service worker needed to put out emotionally, to put on a cultural performance in which the body of the worker became central to the product sold. How the body was presented could make all the difference: buffed or flabby, groomed or unkempt, confident or timid, well-spoken or tongue-tied.

In carrying out her work, McDowell ranged across a variety of disciplinary borders. For theory she drew on the American feminist scholar Judith Butler and her ideas of bodily performance, as well as the French sociologist Pierre Bourdieu and his notion of cultural capital and distinction (McDowell 2013). Using Butler allowed McDowell to discuss both the detailed performance of the gendered body at work – the learned lines, the props, the costume, and the stage moves – and the forms of discipline (which might occasionally be subverted). Using Bourdieu enabled her to add social class to her analysis. Bourdieu's definition of class is complex, but a central component is cultural taste (forming part of what Bourdieu called cultural capital). Taste specifically, and cultural capital more generally, could be vital for the success of a worker in the service sector in carrying out embodied labor. McDowell's conceptual framework was joined with literatures taken primarily from sociology and political economy describing broader structural economic changes in the Global North that emphasized the growing importance of the service sector. They provided the necessary background for her specific economic geographic studies of particular bodies in particular work places: Latvian women in UK care homes for the elderly after World War II; Afro-Caribbean men working for London Transport during the 1970s; and Oxbridge-educated white British women working in large investment banks in the City of London after the "Big Bang" in 1986 (McDowell 2013). McDowell's border country work produced studies of unusual reach and range, demonstrating close connections across the spectrum of spatial scales from the micro-geography of the individual body to the macro-geography of the British Commonwealth.

4.4.2   Erica Schoenberger on corporate culture and identity

The first half of Erica Schoenberger's career as an economic geographer at Johns Hopkins University in Baltimore involved her studying large American businesses as they strove to survive in the ruthless dog-eat-dog world of corporate capitalism. In an essay reflecting on her research, she says that what most surprised her was just how often such corporations committed "staggering errors of judgment" (Schoenberger 2007, p.28). From her close reading of the financial press that provided inside stories of America's largest firms, she observed corporate misjudgements and mistakes occurring over and over again. Both neoclassical and Marxist economics that she was taught as a student, though, affirmed only corporate rationality. While these two approaches were diametrically opposed on nearly all points, strangely they agreed that capitalists were infallible. But the more she read, the more she saw the opposite, irrationality, egregious foolishness, and colossal blunders. Those at the top of the corporations sometimes made mind-numbingly stupid decisions, worse than irrational, simply barking mad, with large numbers of people and places then suffering sometimes dreadful consequences.

How to explain it? Both her regular go-to guides, neoclassicism and Marxist economics, were obviously inadequate, so she began to search on the borderlands. From specifically exploring anthropology and history, Schoenberger came to realize that the irrational corporate decisions she witnessed, and which gave rise to her "indignant exasperation: What idiots!" (Schoenberger 2007, p.30), were in fact less acts of wanton folly than a mismatch between the form of reasoning in which the decision-makers were steeped and the form of reasoning required for the situation in which the corporation found itself. That insight formed the basis of her book *The Cultural Crisis of the Firm* (1997).

Schoenberger's (1997) borderland explorations are found throughout her volume. A striking example is her use of a historical and anthropological account of the Mamluks, a longstanding fierce Muslim warrior caste, which she uses as an allegory to understand contemporary American corporations. The Mamluks' specialty was charging the enemy on armored horses using lances, bows, and arrows. That was what going into battle meant. That was how a warrior defined himself. It was the culture of male fighting Mamluks. They won many battles that way; that is, until canon and blunderbusses began to be used against them. But for more than 100 years after that they continued their old ways, and were slaughtered as a result. They just couldn't help themselves. Had the Mamluks lost their collective mind? Or was there another explanation that avoided the stark binary of either rationality or irrationality? Anthropologists and historians thought there was, and looked toward culture as an explanation. That became Schoenberger's approach too in understanding the apparent folly of some American multinationals. Using particularly the examples of Lockheed Martin and the Xerox Corporation, Schoenberger brilliantly demonstrated that the wrongheaded decisions they made, and the subsequent crises that unfolded, were not acts of insanity, as she originally thought, but a reflection of the deeper culture and way of life in which key decision-makers were steeped. The problem was that how corporate decision-makers thought of themselves, like the Mamluks, was now out of step with the new reality that they faced. But it was so difficult to think otherwise, going against the grain of everything they formerly believed about themselves, their identity, and the world in which they lived. To see the dilemma of corporate America in these terms, though, required Schoenberger herself to think otherwise. It required going into the border country.

4.4.3   Christian Berndt and Marc Boeckler on the performance of markets

The last trading zone involves work carried out on the performativity of markets under-taken by the German economic geographers Christian Berndt, at the University of Zurich, and Marc Boeckler, at Goethe University in Frankfurt. Their focus has been on the construction of markets, sites that facilitate the buying and selling of goods or services, and where prices are set. Previous economic geographers incorporated the market within their work, especially during the spatial science phase of the discipline. There, a market was conceived as the site, usually marked on a schematic map by a dot or a circle, where goods and services were sold. But while earlier economic geographers referred to the market in their work, they never scrutinized the social and material processes that produced it. Instead, the market was taken simply as self-evident, natural, the end of the story.

For Berndt and Boeckler (2009), though, it is only the beginning of the story. They believe that there is nothing natural about a market. Drawing on border work particularly with the sociology of science (science studies), they argue instead that markets are always the product of concerted and deliberate effort, hard grind, with no guarantees of success. Markets do not appear spontaneously. Often, they are produced by reordering the world so that it acts like (performs) the economist's model of the market. This needs elaboration. Berndt and Boeckler contend that the neoclassical model of the market is in effect a script (along with a set of stage directions) that is then performed by various actors, humans and non-humans, to bring the market into being. This is why they call their approach performa-tive. The model of the market comes first. It is that which is performed, with the world arranged to agree with the model. The model does not passively mirror a pre-existing world, but instead it is active and interventionist, creating a new world in its image.

The sociology of science that inspired Berndt and Boeckler has a long historical pedi-gree, but its most prominent recent form is as science studies. Science studies investigates the multifarious processes by which scientific knowledge is produced. It is both empirical, involving detailed, often historical case studies, and theoretical, developing sophisticated explanatory frameworks, and providing mid-range, hands-on explanatory concepts. Around the new millennium, science studies took up the problem of the economy, and the market in particular. Michel Callon (1998) and Donald MacKenzie (2006) suggested that markets are made possible by the coordinated performance of a varied assortment of human and non-human agents. An agent here means any element necessary to make a market, which can vary from a human body to a pocket calculator to a printout of an Excel spreadsheet. When each of these agents is enrolled within the neoclassical model of the market, that is, when they are persuaded to become "allies," to work together, and to follow the neoclassical script, the market is performed and becomes reality. Callon and MacKenzie also suggest, though, that market performance can be precarious. Allies may desert, interests may change, agents may turn in a duff performance, or forget their lines, or go off script. Then it is curtains.

Berndt and Boeckler (2011) take the basic conception of market performativity from Callon and MacKenzie, and then work it through geographically by using the example of the tomato market, first between Mexico and the United States, and second between Morocco and the EU. The model of the market in this case is the neoliberal one based on neoclassicism. It is the idea that the market should allow free trade between sellers and buyers. As Berndt and Boeckler demonstrate, though, realizing that model takes an enormous amount of work, and not only of agricultural labor and farm equipment, but of bureaucratic labor and its associated machinery. Civil servants along with law-makers energetically intervene providing regulations to allow seemingly for free trade in tomatoes, and thus meeting the neoclassical model of a market. As Berndt and Boeckler argue, though, those regulations involve much fudging and sleight of hand ("overflows" in their vocabulary), one form of which is the deliberate shifting of spatial boundaries. At different times and for different kinds of tomatoes spatial boundaries are made permeable, semi-permeable, and impermeable. The very production of space itself, its delimitation, its division, its control, its policing, becomes part of the fundamental work required to realize a free market. Economic geography is vital to the work of performing a market, but to come to this insight requires traveling in the border country.

4.5   Conclusion

Raymond Williams didn't just live his life on the border country, in his case, between England and Wales, he also pursued it as a broader intellectual venture. During the 1950s he was one of the founders of cultural studies. That project was about constructing a new subject located precisely on the academic borderlands of the humanities and social sciences. In effect, Williams envisaged cultural studies as living within interstitial space, that is, the space between traditional academic subjects, which included English, history, philosophy, anthropology, sociology, and just possibly geography. Cultural studies was radically inter-disciplinary, if not sometimes a bit anti-disciplinary. It inhabited the in-between spaces of traditional academic disciplines, and used that location to disrupt and to criticize ("disruptive inbetweenness"). That was possible because compared to the spaces of tradi-tional disciplines, and the boundaries among them, the border country was far less freighted with obligation and control, less concerned to peddle the standard line (whatever that was), more open to trying out novelty, unafraid to challenge sacred cows, and more willing to experiment with method, with theory, and with representational style. Rather than purify-ing, cultural studies was about messing things up, mashing up, recognizing that the world was untidier than the traditional academic subject categories used to represent it.

   In that light, this chapter explored the work of economic geographers in their border country as they engage in various forms of cross-border trade; that is, establishing trading zones. Their project is not as radical as Williams's cultural studies. Economic geographers are not trying to create a brand new discipline which exists only in in-between spaces. But they have become increasingly adventurous, and like Williams, they believe that a prin-cipal justification for border country work is its critical edge, its ability to unsettle and to dislocate, to denaturalize (cf. Chapter 5).

   Working in the border country, carrying out trade, can be fraught. On occasion it can even be fatal. There are powerful institutional and internal sociological forces that keep disciplines apart, keeping borders closed or at least restricted. Sometimes one must work hard and have a thick skin to cross them. But especially since around 1980, economic geographers have made concerted efforts to do just that, to establish new trading zones. An important moment was the shift in the trading pattern with economics. Until then, the bulk of the interaction – for the most part only one-way from economics to economic geography – was with mainland neoclassical economics. After 1980, more and more of the traffic switched, moving away from the mainland to the various islands that constituted the archipelago of heterodox economics (Figure 4.1). Trade became brisker (and two-way), and was more focused on critique (with orthodox, neoclassical economics often the very object of that criticism). It was not only with a different kind of economics that economic geographers traded. They also stepped out to engage the full range of other social sciences, including Williams's own cultural studies. That engagement was not always successful (although our extended case studies demonstrate the potential high levels of achievement). Perhaps the most interesting feature was the enthusiasm of economic geographers to engage other disciplines, to try to talk across academic boundaries. Samuel Beckett, the playwright and novelist, who lived his own life in an Anglo-Irish-Franco border country, famously said: "Try. Fail. Try harder. Fail better." In exploring their border country, economic geographers have tried harder and failed better.

Notes

1 Raymond Williams (1960) titled his first novel *Border Country*.
2 The best example is the Department of Geography and Environment at the London School of Economics, associated with the new economic geography, or "Geographical Economics" (discussed below, and in Chapter 2). The Department was recently ranked second in the world in the QS (Quacquarelli Symonds) World University Rankings by Subject in 2015 (Geography and Area Studies).
3 Three "waves" of the feminist movement have been generally recognized since the middle of the nineteenth century. First-wave feminism was about attaining for women basic legal and equality rights, especially suffrage (the right to vote). Second-wave feminism, which began around 1960 and lasted through the 1980s, was about extending those rights to include the spheres of reproduction, sexuality, family, and workplace. Third-wave feminism, which started in the 1990s and still continues, is about exploring the full range of differences among women, including around sexuality, race, culture, and religion.

References

Agnew, J. 2012. Putting Politics into Economic Geography. In T.J. Barnes, J. Peck, and E. Sheppard (eds), *The Wiley-Blackwell Companion to Economic Geography*. Oxford: Wiley-Blackwell, pp.567–580.

Amin, A., ed. 1998. *Post-Fordism: A Reader*. Oxford: Blackwell.

Baran, P. 1957. *The Political Economy of Growth*. New York: Monthly Review Press.

Barnes, T.J. 1996. *Logics of Dislocation: Models, Metaphors, and Meanings of Economic Space*. New York: Guilford Press.

Barnes, T.J. 2003. The Place of Locational Analysis: A Selective and Interpretive History. *Progress in Human Geography* 27: 69–95.

Berndt, C., and Boeckler, M. 2009. Geographies of Circulation and Exchange: Constructions of Markets. *Progress in Human Geography* 33: 535–551.

Berndt, C., and Boeckler, M. 2011. Performative Regional (Dis-)integration: Transnational Markets, Mobile Commodities and Bordered North–South Differences. *Environment and Planning A* 43: 1057–1078.

Berry, B.J.L. 1967. *Geography of Market Centers and Retail Distribution*. Englewood, NJ: Prentice-Hall.

Bhabha, H. 1994. *The Location of Culture*. London: Routledge.

Blaug, M. 1979. The German Hegemony of Location Theory: A Puzzle in the History of Economic Thought. *History of Political Economy* 11: 21–29.

Boschma, R.A., and Martin, R., eds. 2007. *Handbook of Evolutionary Economic Geography*. Cheltenham: Edward Elgar.

Callari, A., and Ruccio, D., eds. 1996. *Post-Modern Materialism and the Future of Marxist Theory*. Hanover, NH: Wesleyan University Press.

Callon, M., ed. 1998. *Laws of Markets*. Chichester: Wiley.

Chisholm, G.G. 1910. The Geographical Relation of the Market to the Seats of Industry. *Scottish Geographical Magazine* 26: 169–182.

Chisholm, M. 1962. *Rural Settlement and Land Use: An Essay in Location*. London: Hutchinson.

Cook, I. 2004. Follow the Thing: Papaya. *Antipode* 36: 642–664.

Cooke, P., ed. 1989. *Localities*. London: Unwin Hyman.

Dalton, G., ed. 1968. *Primitive, Archaic, and Modern Economics: Essays of Karl Polanyi*. New York: Doubleday.

Dunn, E. 2007. Of Pufferfish and Ethnography: Plumbing New Depths in Economic Geography. In A. Tickell, E. Sheppard, J. Peck, and T.J. Barnes (eds), *Practice and Politics in Economic Geography*. London: Sage, pp.82–93.

Elster, J. 1985. *Making Sense of Marx*. Cambridge: Cambridge University Press.

Galison, P. 1998. *Image and Logic*. Cambridge, MA: Harvard University Press.

Gibson-Graham, J.K. 1996. *The End of Capitalism (As We Knew It): A Feminist Critique of Political Economy*. Oxford: Blackwell.

Gibson-Graham, J.K. 2006. *A Postcapitalist Politics*. Minneapolis: University of Minnesota Press.

Grabher, G., ed. 1993. *The Embedded Firm. On the Socioeconomics of Industrial Networks*. London and New York: Routledge.

Grabher, G. 2006. Trading Routes, Bypasses, and Risky Intersections: Mapping the Travels of "Networks" between Economic Sociology and Economic Geography. *Progress in Human Geography* 30: 163–189.

Granovetter, M. 1985. Economic Action and Social Structure: The Problem of Embeddedness. *American Journal of Sociology* 91: 481–510.

Hacking, I. 1995. The Looping Effects of Human Kinds. In D. Sperber, D. Premack, and A. James Premack (eds), *Causal Cognition: An Interdisciplinary Debate*. Oxford: Oxford University Press, pp.351–383.

Hall, P.A., and Soskice, D., eds. 2001. *Varieties of Capitalism. The Institutional Foundations of Comparative Advantage*. Oxford: Oxford University Press.

Harvey, D. 1974. Population, Resources and the Ideology of Science. *Economic Geography* 50: 256–277.

Harvey, D. 1982. *Limits to Capital*. Chicago: University of Chicago Press.

Hepple, L.W. 1999. Socialist Geography in England: J.F. Horrabin and a Workers' Economic and Political Geography. *Antipode* 31: 80–109.

Hodgson, G.M. 1993. *Economics and Evolution: Bringing Life Back into Economics*. Cambridge and Ann Arbor, MI: Polity Press and University of Michigan Press.

Isard, W. 1956. *Location and Space Economy*. Cambridge, MA: MIT Press.

Jessop, B. 1993. Towards a Schumpeterian Workfare State. Preliminary Remarks on a Post-Fordist Political Economy. *Studies in Political Economy* 40: 7–39.

Jessop, B., and Sum, N-L. 2006. *Beyond the Regulation Approach: Putting Capitalist Economies in their Place*. Cheltenham: Edward Elgar.

Krugman, P.R. 1990. *The Age of Diminished Expectations: US Economic Policy in the 1980s*. Cambridge, MA: MIT Press.

Krugman, P.R. 1995a. Incidents from my Career. In A. Heertje (ed.), *The Makers of Modern Economics*, volume 2. Aldershot: Edgar Elgar, pp.29–46.

Krugman, P.R. 1995b. *Development, Geography, and Economic Theory*. Cambridge, MA: MIT Press.

Kuhn, T.S. 1970. *The Structure of Scientific Revolutions*, 2nd edn. Chicago: University of Chicago Press.

Laclau, E. and Mouffe, C. 1985. *Hegemony and Socialist Strategy: Towards a Radical Democratic Politics*. London: Verso.

Larner, W. 2012. Reflections from an Islander. *Dialogues in Human Geography* 2: 158–161.

Lipietz, A. 1987. *Mirages and Miracles: The Crisis of Global Fordism*. London: Verso.

Lösch, A. 1954. *The Economics of Location*, 2nd edn, trans. W.H. Woglom with the assistance of W.F. Stolper. Originally published in German in 1940. New Haven, CT: Yale University Press.

MacKenzie, D. 2006. *An Engine, Not a Camera: How Financial Models Shape Markets*. Cambridge, MA: MIT Press.

Malinowski, B. 1922. *Argonauts of the Western Pacific: An Account of Native Enterprise and Adventure in the Archipelagoes of Melanesian New Guinea*. London: Routledge and Kegan Paul.

Marshall, A. 1919. *Industry and Trade*. London: Macmillan.

Marshall, A., and Pigou, A. 1925. *Memorials of Alfred Marshall*. London: Macmillan.

Martin, R.J. 2001. Institutional Approaches to Economic Geography. In E. Sheppard and T.J. Barnes (eds), *The Wiley-Blackwell Companion to Economic Geography*. Oxford: Wiley, pp.77–94.

Massey, D. 1984. *Spatial Divisions of Labour: Social Structures and the Geography of Production*. London: Macmillan.

McCarty, H.H. 1940. *The Geographic Basis of American Economic Life*. New York: Harpers & Brothers.

McDowell, L. 1991. Life without Father and Ford: The New Gender Order of Post-Fordism. *Transactions of the Institute of British Geographers* 16: 400–419.

McDowell, L. 2007. Sexing the Economy: Theorising Bodies. In A. Tickell, E. Sheppard, J. Peck, and T.J. Barnes (eds), *Politics and Practice in Economic Geography*. London: Sage, pp.60–70.

McDowell, L. 2013. *Working Lives: Gender, Migration and Employment in Britain, 1945–2007*. Oxford: Wiley-Blackwell.

Nelson, R.R., and Winter, S.G. 1982. *An Evolutionary Theory of Economic Change*. Cambridge, MA: Harvard University Press.

Peck, J. 2005. Economic Sociologies in Space. *Economic Geography* 81: 129–176.

Peck, J. 2010. *Constructions of Neoliberal Reason*. Oxford: Oxford University Press.

Peck, J. 2012. Economic Geography: Island Life. *Dialogues in Human Geography* 2: 113–133.

Peck, J. 2013. Disembedding Polanyi: Exploring Polanyian Economic Geographies. *Environment and Planning A* 45: 1536–1544.

Peck, J., and Theodore, N. 2007. Variegated Capitalism. *Progress in Human Geography* 31: 731–772.

Peck, J., and Tickell, A. 1995. The Social Regulation of Uneven Development: "Regulatory Deficit", England's South East, and the Collapse of Thatcherism. *Environment and Planning A* 27: 15–40.

Peck, J., and Zhang, J. 2013. A Variety of Capitalism … with Chinese Characteristics? *Journal of Economic Geography* 13: 357–396.

Pickles, J. 2012. The Cultural Turn and the Conjunctural Economy: Economic Geography, Anthropology, and Cultural Studies. In T.J. Barnes, J. Peck, and E. Sheppard (eds), *The Wiley-Blackwell Companion to Economic Geography*. Oxford: Wiley-Blackwell, pp.537–551.

Piore, M., and Sabel, C. 1984. *The Second Industrial Divide: Possibilities for Prosperity*. New York: Basic Books.

Polanyi, K. 1944. *The Great Transformation: The Political and Economic Origins of Our Time*. New York: Farrar and Rinehart.

Samuelson, P.A. 1954. The Transfer Problem and Transport Costs II: Analysis of Effects of Trade Impediments. *The Economic Journal* 64: 264–289.

Schoenberger, E. 1997. *The Cultural Crisis of the Firm*. Oxford: Blackwell.

Schoenberger, E. 2001. Interdisciplinary and Social Power. *Progress in Human Geography* 25: 365–382.

Schoenberger, E. 2007. Politics and Practice: Becoming a Geographer. In A. Tickell, E. Sheppard, J. Peck, and T.J. Barnes (eds), *Politics and Practice in Economic Geography*. London: Sage, pp.27–37.

Sheppard, E., and Barnes, T.J. 1990. *The Capitalist Space Economy: Geographical Analysis after Ricardo, Marx and Sraffa*. London: Unwin Hyman.

Smith, N. 1984. *Uneven Development: Nature, Capital and the Production of Space*. Oxford: Blackwell.

Sraffa, P. 1960. *Production of Commodities by Means of Commodities: Prelude to a Critique of Economic Theory*. Cambridge: Cambridge University Press.

Veblen, T. 1919. *The Place of Science in Modern Civilisation and Other Essays*. New York: B.W. Huebsch.

Waring, M., ed. 1989. *If Women Counted: A New Feminist Economics*. London: Macmillan.

Warren, K. 1970. *The British Iron and Steel Sheet Industry since 1840: An Economic Geography*. London: Bell & Sons.

Williams, R. 1960. *Border Country*. London: Chatto & Windus.

Willis, P. 1977. *Learning to Labour: How Working Class Kids Get Working Class Jobs*. Lexington, MA: Lexington Books.

Wise, M. 1949. On the Evolution of the Jewellery and Gun Quarters in Birmingham. *Transactions of the Institute of British Geographers* 15: 57–72.

Wolff, R.P. 1982. Piero Sraffa and the Rehabilitation of Classical Political Economy. *Social Research* 49: 209–238.

Chapter 5

Theory and Theories in Economic Geography

5.1   Introduction

Part I to this point has provided initial strokes toward painting the critical overall picture of economic geography to which this book aspires. In doing so, it has frequently brushed up against, but has not colored in any detail a subject to which we now explicitly turn: the matter of theory. All of the dominant contemporary narratives of economic geography identified in Chapter 2 both draw upon and develop, to varying degrees, theoretical arguments; the history of economic geography sketched in Chapter 3 is a history, in part, of theoretical developments and debates; and the traffic to and from other social sciences examined in Chapter 4 has typically been theoretically freighted traffic. Now is the time, then, to bring theory and theoretical questions specifically to the fore.

One of our main objectives in this chapter is to demystify. The idea of "theory" often scares people. It can appear abstract, in the sense of intangible or, worse, irrelevant; it can appear overwhelming, in respect of its scope; it can appear just too plain complicated. And while theory can indeed be some or even all of these things, it does not have to be and arguably should not be. We want to help make theory approachable, but without caricaturing or dumbing down.

The chapter has four sections. The first considers theory's definition and the varying meanings and roles imputed to "theory" in economic geography. Theory often means different things to different practitioners and is expected by them to do different types of work.

Economic Geography: A Critical Introduction, First Edition. Trevor J. Barnes and Brett Christophers.
© 2018 John Wiley & Sons Ltd. Published 2018 by John Wiley & Sons Ltd.

Notwithstanding this variance, we argue in the second section for the importance *of* theory, and especially of theory with a critical edge. We suggest that theoretical engagement and development is essential to economic geography's health and dynamism.

Many of the theories that have most influenced economic geographers and geography have been taken from elsewhere, and we explore the range and types of these theories in the third section (cf. Chapter 4). As Sunley (2008, p.16) says, economic geography historically has shown a "proclivity continually to import theories and models from other disciplines." Indeed it has.

But to leave things there would be to do a disservice to theoretically innovative and generative work *within* economic geography. To recognize such work is, in turn, to raise a set of other crucial questions. What is the theoretical status of economic geography itself? Is there such a thing as "economic geographic theory" and, if so, what does it look like? These questions – concerning economic geography not as importer-user of theory but *as* theory – are the subject of the fourth and final section, which concludes in critical-evaluative mode by taking the pulse of today's economic geography specifically in theoretical terms. If, as we suggest, economic geography must be theoretically oriented to thrive, what kind of shape is it currently in? The answer, we suggest, is mixed, but mostly positive.

5.2   Meanings and Roles of Theory in Economic Geography

A principal difficulty of discussing theory and theories in economic geography is that "theory," like economic geography itself (Chapter 2), means different things to different people, both within economic geography and more widely. There is a basic definitional issue that makes the relation between theory and economic geography an inherently complicated one. Economic geography is not only characterized, as we will see, by a range of different theories; it is also home to a range of different understandings *of* "theory," with different theories themselves often construing "theory" in different ways. Before we can consider the particular theories that have shaped economic geography, or economic geography's own theoretical content and status, we first therefore have to consider some of the main approaches to "theory" definition circulating in economic geography, and some of the main reasons for this variance.

5.2.1   Theory as explanation

As in social science in general, theory in economic geography has traditionally meant explanation. As David Harvey submitted in his influential theoretical treatise *Explanation in Geography* (1969, p.172), theorizing entails "developing general explanatory statements of considerable reliability." Harvey's succinct definition also highlights two other important aspects of theory as conventionally figured. One is generality: the explanation offered by a theory should not only apply in one place or time. It should be generalizable. The other, closely related, is reliability: a good theory consistently works as an explanation; its explanatory capacity will not often let us down, although occasionally it might.

While the structure of theoretical explanation can vary greatly, it typically entails an assertion of causality: in other words, theories that profess to general explanation do so by positing causal relations between outcomes and underlying causative factors. A causes B. The capitalist exploitation of workers causes social inequality of incomes. Often, but not always, adherents to this definition of social-scientific theory take things a step further. Theory, they say, should not only identify the causes of the world as we see it. Theory, given certain starting assumptions, should also be able to predict the world. More on that in due course.

But if theory, conventionally figured, is centrally about causal explanation, what does it actually look like? How can we recognize it *as* theory when we happen upon it? The literary critic Jonathan Culler (1997, pp.2–3) offers some helpful pointers. Pared down to essentials, theory is an "established set of propositions": A causes B; C causes D; the combination of A and C causes E. In fact, usually a theory will contain even more linked propositions than this. Theory, Culler says, "can't be obvious; it involves complex relations of a systemic kind among a number of factors; and it is not easily confirmed or disproved."

Perhaps more than anything else, however, causal explanatory theory is *abstract*. What do we mean by this? We don't mean abstract as in vague, or intangible, or difficult. Rather, abstract means abstracted from, and thus absent reference to, the empirical particulars of individual cases or situations. Kieran Healy (2017, pp.122–123), a sociologist, usefully glosses: "We begin with a variety of different things or events – objects, people, countries – and by *ignoring* how they differ, we produce some abstract concept like 'furniture', or 'honor killing', or 'social-democratic welfare state', or 'white privilege'." A theoretical explanation is thus an account of the causal relation between at least two abstract entities. For example, within mainstream economics we might state the causal relation between the abstract entity consumption and the abstract entity marginal utility: the increasing consumption of a given good causes the marginal utility of that good to fall.

This traditional causal-explanatory construal of "theory" flourishes in much of contemporary economic geography. Consider two of the dominant narratives or "schools" of economic geography identified in Chapter 2: "Geographical Economics" and the political economic tradition of "Geographies of Capitalism." The former, says one proponent (Overman 2004, pp.512–513), considers it "most important to explain the rules" underlying "the fact that economic activity is unequally distributed across space." Those working in this field use formal mathematical models (Chapters 4 and 6) to express these alleged rules because they believe that they "allow us to see what factors are crucial." Economic geography's political economists, meanwhile, are no less wedded to theory in the form of causal explanation, aiming to generate "powerful conceptualizations of the processes that [generate] geographies of economies" (Hudson 2001, p.16). We will encounter some of those conceptualizations later in the chapter. While the two schools have major differences, they share a penchant for explanatory theory and especially for abstraction. Each argues that to explain the world we need to abstract from individual cases to isolate general "rules" or "laws."

5.2.2 Theory as something else, or more?

Yet by no means everyone in the economic geography tent subscribes to the conventional causal-explanatory understanding of theory. Some vigorously contest it. Why?

The sociologist Charles Tilly's book *Why?* (Tilly 2006) suggests one reason. Concerned with the different modes and forms of explanation people use when answering that eternal question (in its myriad guises), Tilly argues that social context is all-important: different contexts call for *different* explanations; explanation cannot be context-independent.

We can extend this insight to theory in economic geography and in social science more generally. As the philosopher Simon Critchley (2015) notes, Max Weber famously argued that some phenomena require causal explanation whereas others call for a different *type* of theory, more akin to "clarification" or "elucidation – richer, more expressive descriptions." Critchley agrees. And he thinks that as a society we are not very good at recognizing "when we need a causal explanation and when we need a further description, clarification or elucidation. We tend to get muddled and imagine that one kind of explanation (usually the causal one) is appropriate in all occasions when it is not."

Part of the reason for the circulation of different understandings of "theory" in social science in general and economic geography in particular is because some scholars believe that causal explanation is not always necessarily the right "theoretical" response to an object of study. As an example, take Christian Berndt and Marc Boeckler's (2011) work on the economic geographies of commodity chains (see also Chapter 4). They argue that because the particular parts of the chains that they study are normally excluded from academic and political discourse they need a theory that helps elucidate (as much as, or more than, explain) the heterogeneous networks and associations that make up these chains. This is why they turn to actor-network theory (ANT) (Chapter 7), which offers something like the ethnographic method of "thick description" championed by the anthropologist Clifford Geertz (1973). ANT may not emphasize causal explanation, but for its users it *is* a theory, and one appropriate to the content and context of the study in question.

Other economic geographic critics of the conventional approach to theory, like J.K. Gibson-Graham (2008; 2014; see also Chapters 3 and 4), suggest that the problem is with explanation per se. They therefore argue for a conception of theory that is released altogether more forcefully from its traditional rooting in abstract, causal, explanatory power.

Why would they do this? The rationale is partly "academic." Gibson-Graham maintains that generalizable and reliable explanation of the complex, messy, shifting terrain of actually existing economic geographies is too much to ask. It is unrealistic, if not arrogant and elitist, betraying a masculine "God complex," the conceit that the scholar is the cognitive master of all he surveys. If not skeptical of theory-as-explanation in its entirety, Gibson-Graham and others question theory's ultra-confident and expansive causal claims: A definitively causes B, always and everywhere.

The rationale for this challenge to conventional "theory" is also political. Is explanation, critics argue, not overly ivory-towerish? Marx said that the ultimate point is not to understand the world but to change it. Gibson-Graham sees this as the main objective of her work, and theory-as-explanation can only get us so far in this endeavor. The world needs, she says, more non-capitalist or more-than-capitalist economic formations such as "community economies" encouraging ethical living (see also Chapter 3). If our goal is indeed the "proliferation of different economies, we may need to adopt a different orientation towards theory," one which will "help us see openings, to provide a space of freedom and possibility" (Gibson-Graham 2008, pp.618–619).

Alongside and sometimes reinforcing these different perspectives on theory-as-explanation are frequently deep differences in underlying philosophies of (social) science, which also help account for the varying understandings of "theory" within economic geography. Exponents of theory as traditionally figured often claim social science should strive to be *epistemological* in the way natural science is – designed, as Bent Flyvbjerg (2001, p.3) writes, to "produce explanatory and predictive, that is, epistemic, theory." Pivotal to such a view is the premise that our theory should accurately mirror an external and independent reality, and can be adjudicated as such. For critics like Flyvbjerg, however, this philosophy is inherently flawed. "After more than 200 years of attempts" to ape the natural-science paradigm, he observes (p.32), "one could reasonably expect that there would exist at least a sign that social science has moved in the desired direction, that is, toward predictive theory. It has not."

But not all theory that strives for explanation is rigidly epistemological in Flyvbjerg's sense. Two economic geographic traditions that we aligned with one another earlier – "Geographical Economics" on the one hand and "Geographies of Capitalism" on the other – part company here, as do their respective understandings of theory. The former tradition, where prediction *is* a core ambition, is, much like mainstream economics itself, an epistemological exemplar. The latter, while no less committed to theory that accurately mirrors reality, shakes off some of the epistemological baggage. Much of this divergence comes down to underlying scientific philosophy.

"Geographical Economics" is largely philosophically *positivist*: it assumes that we can access reality through simple sensory perception (i.e., observation) that is theory- and value-neutral. Political economy in the "Geographies of Capitalism" vein is not positivist. For one thing it accepts, as Harvey (1972, p.4) wrote in another seminal intervention on theory in (economic) geography, that "social science formulates concepts, categories, relationships, and methods, which are not independent of the existing social relationships which exist in society. Thus the concepts used are themselves the product of the very phenomena they are designed to describe." For another, political economy disputes the idea that anything truly significant about social reality can be simply, unproblematically, observed. Society and economy's core dynamics are hidden behind appearances, in underlying social structures. Our theoretical task therefore should be to generalize not about the relations among empirical objects, but about the structures that produce them. Political economy is not positivist but *structuralist*. And it is wary of prediction, for all Marx's own famous expectation of revolutionary futures.

All this begins to help to explain the critical observation with which we began: that economic geography is home to a range of different understandings of "theory," or what Jamie Peck (2015) refers to as its "theory-cultures." There is neither one economic geographic theory, nor agreement about what *constitutes* "theory" in economic geography.

It is not possible to provide a full inventory of the various ways in which theory is understood and invoked in economic geography. It probably wouldn't be very helpful anyway. We argue, though, that there are two broad theory-cultures or orientations *to* theory. One is the conventional construal of theory centered on explanation; "Geographies of Capitalism" and "Geographical Economics" for all their differences both belong here. The second "culture" is effectively everything else, a variegated culture perhaps best defined by its skepticism *of* conventional theory and its explanatory focus and ambition – by what it is not rather than what it is.

Other geographers have previously discussed these two broad theory-cultures (albeit not always in relation specifically to economic geography), and rather than suggesting a new terminology for them we would refer readers to these existing classifications, three of which we parse in Figure 5.1.

	Gibson-Graham (2014)	Katz (1996)	Peck (2015)
Conventional construal of "theory"	**"Strong theory"** ■ "Powerful discourses that organize events into understandable and seemingly predictable trajectories" ... "theorizes strong connections between certain practices ... while ignoring others [and] draws on a select set of motivations said to animate economic change – such protean forces as individual self-interest, competition, efficiency, freedom, innovative entrepreneurship, exploitation, and the pursuit of private gain"	**"Major theory"** ■ "One of mastery; a way of dealing with knowledge in a progressive, linear, and commanding way that garners respect for those who play by its rules" ... "the theory or theories that are dominant in a particular historical geography under a specific set of conditions" ..."Majoritarianism is secured not in being the most numerous, but in being the unmarked 'subject of enunciation' produced as 'a constant and homogeneous system' that sees all but renders itself transparent"	**"Lumping"** ■ "Concerned with the refinement and reconstruction of generalized categories of analysis" ... "finding patterns and connections across diversity" ... "drawn towards bigger, connective categories [like] capitalism, financialization, or neoliberalism"
Alternative construal of "theory"	**"Weak theory"** ■ "Attends to nuance, diversity, and overdetermined interaction. Weak theory does not elaborate and confirm what we already know; it observes, interprets, and yields to emerging knowledge" ... "does not assume there is any one direction for economic change but is alert to the ways in which crisis and stability are experienced differentially across the heterogeneous economic practices that constitute an 'economy'"	**"Minor theory"** ■ "Strives to change theory and practice simultaneously" ... "tears at the confines of major theory" ... "relentlessly transformative and inextricably relational" ... "Minor theory is not about mastery. Although its politics are rooted in a yeasty notion of theory making – lively and playful – full of possibilities, its intent is to mark and produce alternative subjectivities, spatialities, and temporalities"	**"Splitting"** ■ "Wont to identify significant exceptions to these and other overarching formula(tion)s, to demand recognition for alternative configurations and visions of the economic, to pull the practically neglected and ostensibly marginal into the spectrum of the theoretically and politically visible, and to value exuberant difference over tendential singularity"

Figure 5.1   Economic geography's theory-cultures: three different cuts.

We do not want or need to say too much about these classifications. They largely speak for themselves. But one clarification is important. It concerns the meaning of "weak" and "minor" as used by Gibson-Graham and Cindi Katz, respectively. These terms could easily be misunderstood. So, to be clear: Gibson-Graham's "weak" theory is not weak per se, and especially not in terms of its professed ability to conceive and help bring into being different life-worlds such as diverse, non-capitalist economies. Similarly, Katz's "minor" theory is not minor in terms of its importance. Rather, Gibson-Graham and Katz use the labels "weak" and "minor" to refer principally to the scope of the explanatory enterprise and claims of the theory in question.

Weak/minor theory means in short more humble theory. Explanation may not even be its primary goal. Following the literary theorist Amanda Anderson (2005, pp.3–4), such theory may privilege "ways of living" rather than understanding, thus highlighting practical and existential dimensions and "subjective effects and enactment." Its underlying philosophy of social science is typically more *hermeneutic* than epistemological (Flyvbjerg 2001; Barnes 2001). According to this philosophy there *is* no single explanatory answer (Critchley 2015). And hermeneutic science is not only about understanding the world, but about understanding *understandings* of the world, including that of the author her- or himself. Switching from an epistemology to hermeneutics does not mean merely exchanging one theory for another. Rather, the very idea of theory is transformed. Hermeneutic theory neither possesses a single form nor holds an exclusive truth. It is necessarily messier, sprawling, and indeterminate.

5.3   For (Critical) Theory

In providing our critical introduction to economic geography, and in this chapter specifically to theory in/and economic geography, we are agnostic both about how "theory" is itself understood and mobilized, and about which particular theory is privileged. In terms of theory cultures or philosophies, there is clearly a place in economic geography for both Peck's (2015) "lumpers" and his "splitters" (Figure 5.1). If a particular theory successfully demonstrates, or helps an economic geographer successfully to demonstrate, the substantive implication of geography in economic processes, then we would likely assess it positively. This does not mean that we think all theories are equally robust, useful, or valuable (or that we disavow theoretical preferences in our own work – we do not). It does mean openness to theoretical innovation and diversity in economic geography at large, however.

Yet we are not agnostic about the importance to economic geography of theory per se or especially critically oriented theory. To be and to remain relevant, insightful, and constructive, economic geography must be both theoretically engaged *and* theoretically critical. Let's take these two dimensions of "good" theoretical practice in economic geography in turn.

5.3.1   Theory as sustainable generalization

First, we want to make a forceful case *for* theory. Economic geography should *be* "theoretical," which is to say engaged with existing theory and concerned not just to use but to (re)build it. To make this case we suggest a very minimalist working definition of theory, one that can potentially embrace the range of theory-cultures evoked in the previous section. Neil Smith (1987, p.67) called theory "sustainable generalization." Such generalization *may* be explanatory, and very often will be, but it need not be. For example, a theory's usefulness may lie in its political valence, in its rhetorical power, or in providing a mandate and guide for different kinds of economic action and being, rather than in its explanatory power.

Why should economic geography be invested in generalization? Because, otherwise, it lurches toward what Smith (1987, p.60) identified as the "abyss of empiricism," where everything is unique, particular, different, and disconnected from everything else. This is not to suggest that empirical studies are redundant. Of course, they are not. But as Smith (p.62) rightly insisted: "It is what we make of the empirical data that counts." If it stands on its own, expected to tell its own story, this data has limited value. The kind of "shared theoretical perspective" advocated by Smith – though not necessarily the specific (Marxist) perspective he favored – brings that data to life and in turn is enlivened by it. This is why, *pace* Smith, theory "must be an ever active ingredient; theory cannot function simply as a backdrop to an unfolding empirical play but must be a co-star on the front of the stage."

Note the "co-" qualifier. Smith was well aware that "the abyss of abstract theory" is no less problematic than the abyss of empiricism. As the sociologist Richard Swedberg (2014, pp.14–15) observes, abstract, explanation-oriented theory sometimes does not connect to empirical data at all, and even where it does it is frequently only to "illustrate" the theory. Theory, in such cases, comes first. But this was not the relation of empirics to theory championed by Smith, and nor is it the one we recommend.

Instead, we think the very best theoretically engaged economic geography is that which ploughs the "middle ground" figured by Smith (1987, p.60), attempting "to walk a knife edge path" between the abysses. The question of how exactly to walk this walk is one that has exercised some of the most fertile minds in economic geography. We don't have a textbook answer. Andrew Sayer (1982, p.72) argued that empirical studies should be designed to "check the adequacy" of "established conceptualizations." We agree. Ray Hudson (2001, p.28) called for the construction of "middle-level theoretical bridges" to link theoretical abstractions to "empirically observable forms of production organization and its geographies." We think this is a good approach too. But we would also emphasize, with Doreen Massey (1995a, p.304), that the generalization involved in theorizing should "not imply propositions about necessary empirical outcomes." We share Massey's "commitment to uncertainty" (1995a, p.311), although it was criticized, including by Smith.

5.3.2   Theory as critique

In addition, the best economic geography is theoretically critical. "Good" theory not only adopts a critical perspective but is critical in its very nature. Good theory *is* "critique." For at its best, says Culler (1997, p.4), social theory constitutes "a pugnacious critique of common-sense notions, and further, an attempt to show that what we take for granted as 'common sense' is in fact an historical construction, a particular theory that has come to seem so natural to us that we don't even see it as a theory." For Culler, critique of the taken-for-granted is what theory is all about. Theory therefore involves questioning and, where necessary, reconceptualizing both that which we recognize as received theoretical wisdom *and* that which habituation or propaganda has converted from theory into presumed fact.

The history of theory, therefore, is a history of critique: a never-ending "process of rigorous conceptualization, itself drawing on [and questioning and reformulating] previously-achieved understandings of the phenomena in question" (Massey 1995a, p.304). Perhaps the most famous example of economic theory in this mold, Marx's *Capital*, also happens to be the most felicitous example because the label Marx gave his theory was not "political economy" but a *critique* of political economy. Marx did not theorize capitalism in a conceptual vacuum or out of thin air. He did so by taking existing political economic theory (especially that of Adam Smith and that of David Ricardo), identifying its contradictions, silences, and vested interests, and radically reworking it. Some of the most important and influential economic geographic theory, as we will see later in the chapter, has in turn been produced by doing much the same thing with Marx's own theory. Such theory is quintessentially "critical."

To think about theory as a form of relentless but respectful critique is to appreciate that equating "theory" with the vanity of "mastery" (Katz 1996) ultimately misses the mark. To be sure, some theorists may aspire to mastery, and some students certainly yearn to master theory. But when conceived as critique, as Culler (1997, p.16) recognizes, theory "makes mastery impossible, not only because there is always more to know, but, more specifically and more painfully, because theory is itself the questioning of presumed results and the assumptions on which they are based." To choose to be theoretical, as we encourage readers of this book to do, is in this sense to accept humility: "The nature of theory is to undo, through a contesting of premises and postulates, what you thought you knew," and so "to admit the importance of theory is to make an open-ended commitment, to leave yourself in a position where there are always important things you don't know" (p.16).

But does "pugnacious" critique of existing theory represent the limit to critical theory, the most critical that theory in economic geography and elsewhere can be? Absolutely not. Many scholars, including prominent economic geographers, would insist that while such critique is necessary, it is clearly not sufficient in order for theory to be truly, progressively critical. Harvey (1972), for example, thinks geography needs not "critical" but "revolutionary" theory. This label may bring to mind crude images of revolutionary political slogans and calls to mount the barricades. Harvey's argument is sophisticated, though. It merits consideration for signaling what theory/critique in economic geography can potentially be and do when it is pushed.

As it happens, there is an irony in Harvey using the term "revolutionary" rather than "critical" theory because the type of theory he has in mind is remarkably close to the approach to theory developed by the so-called Frankfurt School of philosophy, sociology and political economy, which *is* often referred to precisely as "critical theory" (see Box 5.1: Critical Theory). That School has important lessons for economic geographers, especially in its recognition of the significance of three particular types of critique, which we will discuss alongside Harvey's own depiction of "revolutionary" theory.

First, critical theory in the Frankfurt tradition is self-critical. It insists on confronting the perspective and power relations embedded in one's *own* theory/critique. "Theory is always *for* someone and *for* some purpose," in the words of the political scientist (and critical theorist) Robert Cox. "All theories have a perspective" (Cox 1981, p.128). That is partly because all theory, as another admirer of the Frankfurt School, the geographer Neil Brenner (2009, p.202), says, is enabled by "specific historical conditions and contexts"; no theorist can "stand 'outside' of the contextually specific time/space of history."

The first step in being theoretically self-critical is to appreciate we are "using" theory even when we think we are not. After all, every observation, however innocent and unprejudiced, is theory-laden in the sense of being "refracted through a pre-existing frame of meaning" (Gibson-Graham 2014, p.148). The positivist ideal of rooting scientific explanation in theory-neutral observation is just that: an ideal. Theory is "never absent, nor can it be bypassed" (p.148).

But recognizing theory's ubiquitous presence is not enough. We also need to ask questions, often uncomfortable ones, about whose perspectives our theories privilege and the types of social relations that make them conceivable. Do economic geographers always ask such questions? Hardly.

Second, "critical theory" à la Frankfurt is critique of the prevailing social order and especially of the relations of power and dominance that scaffold it. It looks hard, says Cox (1981, p.129), at "the prevailing order of the world and asks how that order came about." How did the status quo become such, and whose interests does it serve? More challengingly, it theorizes alternatives to the prevailing order and therefore represents an existential threat to it. Harvey's "revolutionary" theory is in this respect precisely of a piece. He has no time for scholarship that merely maps "more evidence of man's patent inhumanity to man" such as "the daily injustices to the populace of the ghetto." Harvey lambasts this as "counter-revolutionary" theory and as "moral masturbation" (1972, p.10). Gibson-Graham has no time for it either. What is the goal of (critical) economic geographic theory? Not "to extend knowledge by confirming what we already know, that the world is a place of domination and oppression"; it is, rather, to "help us see openings, to provide a space of freedom and possibility" (2008, p.619).

Box 5.1   Critical Theory

What is "critical theory"? There are two main answers to that question. The first is generic: critical theory is any theory that is in some respect critical – whether of capitalism, of political oppression, of other theories, of empiricism that eschews theory, or of something else. The second is much more specific: critical theory is the particular tradition of theory associated with the Institute for Social Research in Frankfurt, with the leading theorists based at or affiliated with that institute from the 1930s through 1950s, and with those theorists' most prominent intellectual successors. Critical theory in this particular tradition is theory concerned not merely with understanding society but with critiquing and transforming it.

Initially inspired by Marx, the Frankfurt Institute was founded in 1923. Its early years were relatively unexceptional, but this began to change from 1930 when it came under the leadership of Max Horkheimer, a German Jewish philosopher and sociologist. Against the historical backdrop of a gradual decline in the influence of the German workers' movement, the institute, guided by Horkheimer, began a long-term process of shedding Marxian orthodoxy and becoming a much more heterodox intellectual establishment.

Alongside Horkheimer, arguably the two most influential Frankfurt theorists were his long-term friend and collaborator Theodor Adorno, whom he had met as a student in the early 1920s, and Herbert Marcuse. Both Adorno and Marcuse were also Institute members. Walter Benjamin was not, but he was closely associated with it and occasionally funded by it, and his thought is very much part of the "critical theory" canon.

Critical theory Frankfurt-style ranged so widely that it defies summary, but perhaps the key book was Adorno and Horkheimer's *Dialectic of Enlightenment* (1944). It makes for pessimistic reading, arguing that while intellectual enlightenment certainly holds out the potential for human emancipation, it tends, in the same moment that it enlightens, to become the very basis for various forms of domination (of other humans, and of nature) and alienation. This was the "dialectic," the symbiosis of "good" and "bad." At the end of the day, enlightenment was a "mass deception," propagated in no small part by the so-called culture industry.

While the direct influence of these famous Frankfurt theorists waned somewhat after the 1950s, it was kept alive by key intellectual descendants, most notably Jürgen Habermas.

And third, critical theory in the Frankfurt tradition, like Harvey's revolutionary theory, is a critique of theories that explicitly justify and buttress the existing social order or that "pose no threat to the existing order since they are constructed with the requirements of that existing order broadly in mind" (Harvey 1972, p.5). The most famous statement of this critical imperative is found in the essays making up Max Horkheimer's *Critique of Instrumental Reason* (1974). Instrumental reason represents a means-oriented, "problem-solving" rationality that simply accepts the ends toward which those means are targeted. Rather than questioning the status quo, it is concerned with making it more, say, "efficient." Harvey (1972) identifies mainstream urban economics as an exemplar of such instrumental reason or, in his terminology, counter-revolutionary theory. Revolutionary theory provides its critique.

5.4   Theories in Economic Geography

Whatever its particular theoretical disposition, economic geography has long maintained theoretical vitality by importing theory from elsewhere, from scholarly communities not "doing" economic geography in the sense in which we have defined it (Chapters 2 and 4). This theory inflow, which is a crucial component of the interdisciplinary exchange discussed in Chapter 4, makes economic geography a vibrant intellectual space, buzzing with ideas and arguments of varied form, objective, and origin. It also makes it a confusing space, far harder to grasp, at least in theoretical terms, than those disciplinary spaces characterized by theoretical singularity. Indeed, there is practically no limit to the types of theory that economic geography has sought to mobilize in coming to terms with geographies of economies.

There seems little point is trying to offer a sweeping overview of the range of imported theories one finds in economic geography (and partly this has been done already in Chapter 4). Not only would such an overview take up inordinate space, it would be of little practical interest beyond the encyclopedic. Nor is a historical narrative in order. There has not been a neat temporal progression, for instance, from economic geography importing theories of type X toward a reliance on theories of type Y.

What is potentially instructive, however, is to consider the principal *categories* of theory that economic geography draws upon. Doing so can help students and practitioners alike better orient themselves amidst the swirling currents of theoretical movement and countermovement.

By "category" we have something quite specific in mind. We refer to the degree of explicitness and centrality with which the "economic" features in the theory in question. The interesting point is that while economic geography uses theories that are explicitly and centrally *about the economy*, it also draws liberally on theories where "the economic" either is not nominally a concern at all, or is a concern but not necessarily the only or central one. To be useful for economic geography, then, the imported theory does not necessarily have to be *economic* theory.

To this end, we distinguish between three theoretical categories (Figure 5.2). The first, we refer to simply as "economic theory." Not surprisingly, much of the theory that animates economic geography *is* economic theory, theorization about specifically economic processes, structures, and outcomes. Such theory can and does vary widely, ranging from resource allocation and price determination in mainstream economics to the production process in Marxian economics. But in all its variants, the economy is the remit.

The second category goes under the ungainly label "theory pertaining to the economy." This is theory that exists at the next conceptual remove. It is concerned with the economy, but the theorization is not derived from "economic" theory. And if that sounds

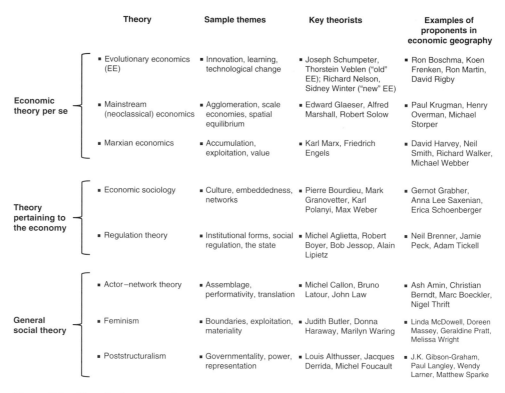

Theory	Sample themes	Key theorists	Examples of proponents in economic geography
Economic theory per se			
▪ Evolutionary economics (EE)	▪ Innovation, learning, technological change	▪ Joseph Schumpeter, Thorstein Veblen ("old" EE); Richard Nelson, Sidney Winter ("new" EE)	▪ Ron Boschma, Koen Frenken, Ron Martin, David Rigby
▪ Mainstream (neoclassical) economics	▪ Agglomeration, scale economies, spatial equilibrium	▪ Edward Glaeser, Alfred Marshall, Robert Solow	▪ Paul Krugman, Henry Overman, Michael Storper
▪ Marxian economics	▪ Accumulation, exploitation, value	▪ Karl Marx, Friedrich Engels	▪ David Harvey, Neil Smith, Richard Walker, Michael Webber
Theory pertaining to the economy			
▪ Economic sociology	▪ Culture, embeddedness, networks	▪ Pierre Bourdieu, Mark Granovetter, Karl Polanyi, Max Weber	▪ Gernot Grabher, Anna Lee Saxenian, Erica Schoenberger
▪ Regulation theory	▪ Institutional forms, social regulation, the state	▪ Michel Aglietta, Robert Boyer, Bob Jessop, Alain Lipietz	▪ Neil Brenner, Jamie Peck, Adam Tickell
General social theory			
▪ Actor–network theory	▪ Assemblage, performativity, translation	▪ Michel Callon, Bruno Latour, John Law	▪ Ash Amin, Christian Berndt, Marc Boeckler, Nigel Thrift
▪ Feminism	▪ Boundaries, exploitation, materiality	▪ Judith Butler, Donna Haraway, Marilyn Waring	▪ Linda McDowell, Doreen Massey, Geraldine Pratt, Melissa Wright
▪ Poststructuralism	▪ Governmentality, power, representation	▪ Louis Althusser, Jacques Derrida, Michel Foucault	▪ J.K. Gibson-Graham, Paul Langley, Wendy Larner, Matthew Sparke

Figure 5.2   Theories in economic geography.

wrongheaded – how can we theorize the economy *except* by using economic theory? – then consider the central shared premise of theories that come under this header: that to understand the economy, we need to understand *more* than just the economy. We need to understand also at the very least states that regulate the economy, laws that police it, sociocultural institutions in which it is embedded, and so on. "Theory pertaining to the economy" theorizes the economy from more-than-economic perspectives, rejecting the economy's arbitrary conceptual separation from the political, the cultural, the ecological, and the like.

Finally, there is "general social theory." By this, we mean theory about the social world in the very broadest sense; often overlapping with philosophy, such theory can extend to such weighty questions as the nature of social reality, the nature of our knowledge about it, and the nature of the human subject. While the link between theory of this kind and questions of economic geography may not be obvious, it does not have to be, and often is not. Economic geography can and has drawn upon such theories in myriad creative ways. Its borrowing from feminist theory is a prime example, with nothing about "traditional" economic geography – research methods, subjects and objects of study, core propositions – remaining untouched by the encounter (Chapter 4).

For each of these three broad categories, Figure 5.2 provides a sampling of imported theories that have proven influential in contemporary economic geography. The word sampling is important. The table is not comprehensive; it is, in all respects, illustrative and partial. Needless to say, there is also in reality plenty of overlap that the table does not reveal: there is not necessarily anything to stop a proponent of, say, evolutionary economics from simultaneously incorporating a feminist theoretical perspective; and while some economic geographers have stuck to singular theoretical guns throughout their careers (as did Neil Smith with Marxism), others (e.g., Ash Amin) have hopped around to such an extent that they could be aligned with at least half of the theories indicated. In short, the aim of the table is not to "fix" either theories or individuals. It is to serve as an aid to navigation.

Also helpful, we think, is to provide a concrete example from each category: an example of one of the theories being drawn upon to positive effect in empirical economic geographic study. The reason for providing these examples is that because the nature of "theory" differs (in the ways identified above) between the categories, so does economic geography's relation *to* the theory. Since in each category theory has different ends and means, it has to be put to work in different ways. Encouraged to engage with theory, students often wonder what to do with it, how to use it. The answer, of course, depends on the nature of the theory one is reading and thinking about "using." Our three examples, which are purely illustrative, show how different types of imported theory can be variously mobilized in and for economic geography.

i. *Evolutionary economics, technological variety, and regional growth*: In a series of studies (e.g., Rigby and Essletzbichler 2006), Jürgen Essletzbichler and David Rigby have made a significant contribution to understanding the varying pace in recent decades of economic change and growth in different regions of the United States. The theory to which they have turned to help them crack this puzzle, or at least one element of it, is a heterodox economic theory: evolutionary economics, which, rejecting the steady-state equilibria so often assumed by mainstream economics, attempts to conceptualize the principal dynamics of economic change or "evolution" (Chapter 4). Essletzbichler and Rigby take from evolutionary economic theory *the alleged centrality of technological variety to differential trajectories of economic development*, and explicitly spatialize this relation. Manufacturing production techniques, they find, vary significantly by US region, even within individual industries. This variance, moreover, endures over time (as a result of what evolutionary economists call "path dependence"). Consequently, the regional economic imbalance this creates similarly endures. Essletzbichler and Rigby's work is an excellent example of economic theory being given a specifically geographical twist, turning theoretical economics into applied economic geography.

ii. *Regulation theory, neoliberalism, and uneven development*: In analysis also laid out in a series of studies (e.g., Tickell and Peck 1995), Jamie Peck and Adam Tickell have provided a very different perspective on a nominally similar problematic, the persistence of uneven regional development in the United Kingdom since the beginning of the 1980s. The difference in perspective derives from a difference in theoretical influence. Peck and Tickell draw not on an economic theory but on "regulation theory," which theorizes economic dynamics specifically in and through their regulatory setting (Chapter 4). Peck and Tickell take from this theory its premise that when economies run into periodic trouble, as they invariably do, they require an "institutional fix": *the prevailing regulatory mix of social, political, and legal institutions is no longer fit for purpose and must be overhauled to put the economy back on track.* Peck and Tickell draw on this theory to suggest that when the UK economy hit the buffers in the 1970s, the attempted institutional fix – neoliberalism, Thatcher-style – failed. The fix did not work (Chapters 1 and 4). Thus, rather than enabling, as regulation theory's *successful* fixes do, a renewed period of stable growth, Thatcherite neoliberalism instigated a renewed period of sustained crisis, albeit one experienced very unevenly in different regions. In short, the contradictions seeded by the failed regulatory experiment of neoliberalism include stark geographical ones.

iii. *Feminism, dualism, and hi-tech*: Our third example concerns the use in economic geography of feminism: a theory, or bundle of theories, that conceptualizes social rather than strictly economic relations, but which has clear relevance to the economy insofar as social relations both shape, and are in turn shaped by, economic processes and practices (see also Chapter 4). The example is taken from the work of Doreen Massey, one of the pioneers of feminist approaches in economic geography from the 1980s (Chapter 3). In the 1990s, Massey carried out collaborative research on hi-tech science parks in the United Kingdom. She was interested in the influence of gendered dualisms – "dualisms" being concepts organized around presumed polar opposites, such as "right" and "wrong" or "up" and "down" – in and on the lives of hi-tech workers and their families, and especially on the geographies of these lives. She took from feminist theory the thesis that *abstract dualisms such as, in her study, "reason" and "non-reason" are not just gendered, aligned in this case with "male" and "female"* (the smart male scientist and his wife or partner), *but lived.* While they are representations, they also actively structure real-life social relations. Massey's (1995b) study demonstrated this active structuring in practice; explored the significance of the public/work and private/homely spaces attached to the respective sides of the reason/non-reason dualism; and argued that even when men tried to resist the dominant version of hi-tech masculinity, it tended paradoxically to reinforce the dualism.

5.5   Economic Geographic Theory

Economic geography is not *only* an importer of theory. Its modus operandi is not merely to take theory developed elsewhere and apply it to empirical economic geographic questions, although to be sure, that often happens. Rather, economic geography is itself a locus of theory development, with such development invariably occurring in conversation with existing theory/theories, whether imported or indigenous or both. What then does theory developed within economic geography look like? Is there such a thing as a distinctively "economic geographic theory"? If so, what are its defining characteristics? Or to phrase the question differently, how might we understand economic geography *as theory*?

Of course, the existence and attributes of economic geographic theory can only be addressed once a prior question has been answered: What is economic geography? If we proceed on the basis of our own favored definition of the latter (Chapter 2), economic geographic theory would be *that which explicitly theorizes* the substantive implication of space, place, scale, landscape, or environment in economic processes. So, does theory of this type exist?

5.5.1 The problem of place?

Economic geographic theory of the indicated type does exist, yes. But economic geography, like human geography more generally, has long wrestled in an existential struggle with its theoretical status. And one issue in particular has caused more headaches than any other: the meaning and role of place.

The reason is that place is generally construed as the locus of the particular, the unique, the contextual. In contrast, theory aspires to generalization. Indeed, insofar as it aspires to abstraction, theory *rids* itself of the particular. The specific is seemingly anathema to theory; the latter, surely, has no room for "place." As Flyvbjerg (2001, p.47) says of the conventional wisdom, "a theory must be free of context." Why? Because, "otherwise it is not a general theory."

Unsurprisingly, the one strand of economic geography never displaying the slightest self-doubt about its status *as* "theoretical" is the strand that consciously and deliberately ignores place, even disdains it. This is "Geographical Economics." Other economic geographers frequently criticize geographical economists for their neglect of "real places." "Real communities in real historical, social and cultural settings with real people, going about the 'ordinary business of life' (as Marshall once described economics)," complains Ron Martin (1999, p.77), "are completely bypassed." They are, but to leave things there is to miss the crucial point: that geographical economists regard this bypassing not as a weakness but as a source of strength. It is an essential part of what *makes* "Geographical Economics" "theoretical." Accordingly, some such economists counter that the attention other strands of economic geography pay to "real places" is not something to celebrate; rather it is their Achilles heel, because "treating each situation as something unique and each idiosyncrasy as something crucial teaches us nothing" (Overman 2004, pp.512–513). It is, in short, *a*theoretical; one cannot see the wood for the trees.

In reality, Overman's picture of economic geography as an atheoretical discipline privileging the unique through its focus on place, and at the expense of theoretical generalization, is mostly a caricature (we will return to the "mostly" qualifier in due course). This is not to say that economic geography was never like that. It was. Prior to the 1960s, as we saw in Chapter 3, economic geography was very much focused on place, and could scarcely have been less interested in abstraction or generalization: ideographic and descriptive, it was "short on theory and long on facts," in the words of one economic geographer (Ballabon 1957, p.218) of the era. As Harvey (2000, p.76) later recalled:

Traditionally, geographical knowledge had been extremely fragmented, leading to a strong emphasis on what was called its "exceptionalism". The established doctrine was that the knowledge yielded by geographical enquiry is different from any other kind. You can't generalize about it, you can't be systematic about it. There are no geographical laws; there are no general principles to which you can appeal – all you can do is go off and study, say, the dry zone in Sri Lanka, and spend your life understanding that.

But that was half a century or more ago. And although human (including economic) geography's struggle with the implications of its concern with place for the possibility and nature of "geographical theory" has never entirely gone away, subsequent generations of geographers, with economic geographers notably to the fore, have made enormous progress in transcending the place–theory dichotomy. Harvey himself was in the vanguard. "I wanted to do battle with this conception of geography," he recalled about his work in the 1960s when faced with geography's entrenched, atheoretical, place-bound exceptionalism, "by insisting on the need to understand geographical knowledge in some more systematic way" (2000, p.76). He wanted to respect place *and* to be theoretical.

In the 1960s, the answer, as Harvey and others saw it, was positivist spatial science (Chapter 3). From the early 1970s, as Harvey and others argued, the answer increasingly was Marxism. Both spatial science and Marxism were and are in Gibson-Graham's terms "strong" theoretical approaches; but both, at least as Harvey understood them and their application to economic geography, allowed for theory to be reconciled with the importance of place. Thus, in *Explanation in Geography* (1969, pp.172–173), published before his turn to Marxism, Harvey insisted that while the "ultimate purpose of geographic analysis may be to understand individual cases," this did not mean that "a separate map or theory must be created for every instance … Above all, we cannot conclude, as many appear to do, that because geographers are very much concerned with particular cases, there is no possibility for formulating laws which can be applied to explain those particular instances."

These were valiant and no doubt ingenuous words, yet there were and are many economic geographers for whom Harvey's own distinctive brand of "Geographies of Capitalism" (to which we will return shortly), much like his earlier spatial science, does not go nearly far enough in terms of respect for and understanding of place, just as, relatedly, it is judged to veer too close to Smith's abyss of abstract theory. Barnes (1996), for example, argued that the use of both Marxian *and* (in the style of "Geographical Economics") mainstream neoclassical economic theories in economic geography generates an emaciated view of place because these theories are "essentialist," and called for economic geography to engage instead with a range of more "contextual" economic theories.

One of the most vocal critics in economic geography of Harvey's treatment of place has been Massey, and in a series of articles (e.g., Massey 1991) she lays out an approach that allows a far more comfortable and effective rapprochement of place with generalizing theory than Harvey arguably has ever achieved. Contra conventional understandings, Massey argues that place is not the static realm of the merely empirical and descriptive, nor, crucially, of just the "local." You *can* theorize it, specifically in terms of the dynamic interaction of wider and perhaps even global socioeconomic relations. Place is a process. This is not to say that each place is not unique. It is. But this uniqueness reflects the particular outcome of the general (and thus theorizable) process that is the interaction of extra-local relations (see the discussion of Massey's spatial divisions of labor thesis in Chapter 3). If ever there was a categorical debunking of the shallow Overman-type critique of economic geography that says because of its focus on place the discipline cannot be theoretical, Massey's was it.

5.5.2   Theorizing in economic geography

Aside from an axiomatic concern with the geographies of the economic, is there anything especially distinctive about economic geographic theory and theorizing that meaningfully distinguishes them from theory and theorizing in other social-scientific fields? Or are such theory and theorizing much of a muchness with, say, anthropological theory and theorizing, or economic theory and theorizing, or sociological theory and theorizing? We want to highlight two distinctive features that are important and connected to one another.

First, partly by dint of their relative newness, theory and theorizing in economic geography constitute a more open and expansive field than in neighbor disciplines. Here we think the comparison with sociology is instructive (see the discussion on sociology in Chapter 4). Economic geography remained a resolutely atheoretical domain until as recently as the end of the 1950s, only "opening up" to theory and attempting to develop theoretical accounts from the 1960s (Chapter 3). By this time, theoretical sociology had existed as a formal and relatively circumscribed field of inquiry for more than half a century.

One result of sociology's historical institutionalization has been an expectation that practitioners align or position themselves in relation to one of sociology's three "founding fathers": Karl Marx, Emile Durkheim, and Max Weber. We argue that this makes for an inherently closed and restrictive mode of theorizing. Even (especially?) some sociologists think the same. Pierre Bourdieu (1990, p.27) said that sociology's "are you a Marxist or Weberian or …" question was "almost always a way of reducing, or destroying, you." Bourdieu hoped to expunge this question from sociological practice.

Economic geography, by contrast, has never had its own Webers or Durkheims (or even, arguably, its Bourdieus). Theorizing within economic geography, there *are* no fixed stars by which one navigates. The first generation of economic geographic theorists in the 1960s was entirely free to theorize, unconstrained by fixed historical lodestones. And for all the theoretical development and redevelopment in the decades since, the same remains broadly true today. The license to theorize continues to be granted extremely broadly. Some observers, of course, might consider this feature a sign of weakness, interpreting the lack of ineluctable reference points as a lack of powerful and persuasive theory. We do not. We think it is a sign of strength, reflecting a collective, enduring commitment to keep theoretical debate in economic geography open and plural.

Perhaps the closest economic geography comes to having a founding theoretical "father" – and it does happen to be a father – is David Harvey. Nigel Thrift (2006, p.223), for example, argues that Harvey's work is indeed "canonical," not just in economic geography but in human geography more widely. "Harvey," Thrift observes, "often seems to be one of the primary points of reference in many of the discipline's contemporary takes of itself: his writings have conditioned much of what has been subsequently said."

This brings us to the second distinctive feature of economic geographic theory and theorizing. While Harvey's and others' Marxian-influenced economic geographic theorizing is far from a *necessary* reference point for contemporary economic geographers – and there is certainly nothing remotely like an "Are you a Marxist/Harveyite?" question – some of this theorizing *is* to one degree or another baked into economic geography's collective theoretical cake. Peck (2015), following Eric Sheppard (2011), refers to this as economic geography's "critical common sense" or "unity-within-difference."

What Peck and Sheppard (and we) mean is that despite the multitudinous differences among them, the majority of economic geographers would likely subscribe to a core set of foundational theoretical premises that like the common sense discussed by Culler (1997) have been absorbed so thoroughly that their status *as* theoretical is often overlooked. These premises originally emerged out of geography's encounter with Marxism, even if the link is no longer always explicit or necessary. What are the main ones? That conflict, instability, and crisis are endemic to capitalism; and that capitalism generates, and is in turn powered by, socio-spatial inequality in the form of uneven geographical development (Chapter 8).

While many economic geographers have contributed to the production of this critical theoretical common sense, Harvey's significance has indeed been profound. His pivotal theoretical intervention was without doubt *The Limits to Capital* (1982). He laid out there many of his most influential, and emphatically economic geographic, theoretical ideas such as "spatial fix" (see Box 5.2: Harvey's "Spatial Fix").

Box 5.2   Harvey's "Spatial Fix"

When Harvey encountered Marxism in the 1970s he found most appealing and valuable its theorization of capitalist crisis (see also the section on Harvey in Chapter 3). He saw the signs of crisis all around him in the world and became appalled by the powerlessness of existing (positivist) geographic theory "to say anything really meaningful about events as they unfold around us ... It is the emerging objective social conditions and our patent inability to cope with them which essentially explains the necessity for a revolution in geographic thought" (1972, p.6). In theorizing crisis, Marxism seemed to him to hold the keys to a successful revolution.

On delving further into Marx, Harvey discovered that his theorization of crisis was limited. *Capital* offered, Harvey argued in *Limits*, only a "first cut" of a theory of crisis formation. This was centered on Marx's famous theory of the tendency of the rate of profit to fall as capitalism expanded. Furthermore, Harvey did not even find this first cut particularly persuasive. (He still doesn't.)

Some of the most innovative sections of *Limits*, therefore, were dedicated to trying to piece together a more comprehensive and workable (but still Marxian) theory of crisis formation. Specifically, Harvey offered second and third "cuts" at such a theory. The second cut was largely concerned with the role of the credit system. By enabling capitalists to defer or delay the onset of crisis tendencies, albeit without eradicating those tendencies altogether, the credit system provides capitalism with what Harvey called a "temporal fix" to its underlying contradictions. The third "cut," meanwhile, dealt primarily with geography rather than history, space rather than time, and was encapsulated in the "spatial fix" concept.

What was this spatial fix? Just as credit-time, or the passage thereof, can defer capitalism's crisis tendencies, so too, Harvey claimed, can *the utilization or production of new economic geographic configurations* displace those tendencies. Again, it does not eradicate them; but by using "space" in creative ways (e.g., selling unsold commodities in untapped geographic markets, or ploughing idle cash into rebuilding the urban built environment), capitalists can effectively kick the metaphorical can down the road – making sure crisis does not occur here and now but later in a different form, and probably elsewhere. This is a broadly functionalist argument: space, in one way or another, performs a useful function *for* capitalism by enabling the displacement of otherwise critical crisis tendencies. In the years since Harvey coined the term, economic geographers have filled his accommodating "spatial fix" conceptual vessel with all manner of different content, effectively using it to refer to any economic geographic transformation that serves to mitigate capitalism's propensity to crisis.

Harvey's overall objective in this book is well captured by its title, which signals the two different sets of limits that Harvey sought to identify and understand. First, capital is understood after Marx as a process of circulation. Money is used to fund production, generating a surplus/profit that then funds expanded production. That circulation, however, is subject to all manner of limits that give rise to capitalism's crisis tendencies. Second, *Capital*, Marx's original theory, like all theories, also has limits. Harvey argues that one crucial set of such limits concerns the relative weakness of Marx's theorization of capitalism's geography. In writing *Limits*, therefore, Harvey sought not only to put Marx into geography (the discipline), but to put geography into Marx. He succeeded on both counts.

For all its importance, including for his own subsequent thinking and writing, Harvey's opus is not the final word for economic geographic theory. Far from it. There are wide swathes of economic geographic theory that do not engage with Harvey's work or the issues he theorizes. And, exemplifying the idea of theory-as-critique, there are also wide swathes of economic geographic theory formulated partly through critical engagement *with* Harvey's work. Some criticize it, as Massey's alternative theorization of capitalism's uneven geographical development, *Spatial Divisions of Labour*, implicitly did, for being too "strong," for looking at the world "as if it were merely the pre-determined product of a set of laws and tendencies" (1995a, p.6). Some criticize it for being too masculinist (Massey 1991). Some criticize it for being too Western (Gidwani 2008). But if, as we have suggested, large parts of contemporary economic geography do share a critical theoretical common sense, Harvey was instrumental to its development.

5.5.3 Economic geography and/as theory: a (critical) stock-taking

One aspect of writing a "critical" introduction to economic geography that we emphasized in Chapter 1 was that of being evaluative: weighing pros and cons, judging relative merits, taking a position. But we have not really done that yet in this chapter. We have identified different perspectives on theory and different theories. What we have not done is critically assess the current condition of economic geography *as* a theoretical field. Stepping back from the fray and from the detailed theoretical ebb and flow, how do things look? Is the body of knowledge we define as economic geographic generally in a good state theoretically, a bad state, or somewhere in between?

One reason for posing this question is that voices of doom are in the air, and it is important to consider whether they are correct. At the broadest scale, Swedberg (2014, p.14) reckons that "theory is currently not in very good shape in social science." Where economic geography in particular is concerned, Overman (2004), among others, concurs. Aside, naturally, from his own strand of economic geography ("Geographical Economics"), Overman eviscerates the field for being theoretically vague and sloppy. He implies that Harvey's (1969, p.134) critique of geographic theorizing of the late 1960s as "inexplicit and fuzzy" continues to apply to most economic geographic theorizing half a century later.

Do we agree? For the most part, no. The past two to three decades have seen economic geography develop theoretically in highly creative and productive ways in terms of both "using" imported theory and, even more importantly, generating original economic geographic theory. Furthermore, the best of such economic geographic theory has not only transformed the field itself but traveled and impacted far and wide, both inside academia and occasionally outside of it. And this is true across the board: whichever core narrative/ version of contemporary economic geography one takes (Chapter 2), one finds numerous examples of innovative and influential exported economic geographic theory. Here are four examples, one from each of the narratives we identified:

i.   *Decolonizing development*: An example of theoretically generative work in the "Geographies of Capitalism" vein is Wainwright's (2008) study. It sutures histories of colonialism and development in Belize with political economic and postcolonial theoretical concepts to advance a critical theory of what Wainwright calls "capitalism qua development."

ii. *Regional advantage*: Fitting squarely in the "Geographies of Business" segment of economic geography, Saxenian's (1996) study of the markedly contrasting fortunes of two US industrial regions, California's Silicon Valley and Route 128 in Massachusetts, developed a theory of "regional advantage" that explained the former's competitive success in terms of the network advantages associated with the horizontal, decentralized, and cooperative structuring of its tech sector (Chapter 12).

iii. *New trade theory*: A centerpiece of "Geographical Economics," new trade theory (e.g., Krugman 1995) seeks to explain in formal terms why international trade occurs and why it assumes the patterns it does. It is "new" insofar as it loosens many of the core assumptions of "old" trade theory, like perfect competition and constant returns to scale (Chapter 3).

iv. *Community economies*: In contrast to the preceding examples, Gibson-Graham, Cameron, and Healy's (2013) theory of "community economies" generalizes not for the purposes of explanation but instead for *living*. They provide a handbook of sorts for an ethical existence. It is an example of practical economic geographic theory, connecting peoples, places, and cases through shared ethical and economic concerns, showing the difference that can result.

And yet we do not believe that where theory is concerned the story of modern-day economic geography is an unremittingly positive one. Two developments strike us as cautionary, even potentially problematic.

The first concerns the "cultural turn" that parts of economic geography have experienced since the mid-1990s (Chapter 4). Though this turn comprised numerous components, including an enhanced awareness of the cultural and gendered constitution of geographic economies, it was arguably "first and foremost … a turn to 'context'" (Pickles 2012, p.539). As such, it raises the (perennial) problem of how to reconcile theoreticism with a close interest in particularity and place.

Recall Flyvbjerg's dictum that "a theory must be free of context." In sophisticated hands, such as Massey's, this dictum is clearly misleading. She demonstrated one can *theorize* context, or place, and its production. But Pickles (2012, p.539) cautioned that "the turn to 'context' and its contingencies must be thought through carefully to avoid slipping back into naïve empiricism or loose descriptivism." He said that presumably because those who effected the cultural turn in economic geography had in his view *not* always thought things through carefully, or shown Massey's sophistication, and *had* slipped back into the empiricism/descriptivism of atheoretical place studies bemoaned by Overman. Indeed, even Henry Yeung (2012, p.26) admits, with the benefit of hindsight, that the turn's "protagonists (myself included) may have suffered from an excessive emphasis on the micro," noting that "when broader generalizations about capitalisms are made from these cultural turn-inspired geographical studies, they tend to be ad hoc and haphazard rather than systematic and structural." Back, in other words, to the "inexplicit and fuzzy" geographic theorizing that Harvey dissected 50 years ago.

The second development about which we should be cautious involves the rise to prominence, and arguably dominance, in economic geography, broadly contemporaneously with the cultural turn, of the so-called global production network (GPN) approach (e.g., Henderson et al. 2002). GPN represents an attempt to develop an explicitly non-methodologically nationalist understanding of economic globalization and the interconnected world it creates. Particularly concerned with the growth and increasing influence of multinational corporations, and applied to questions of regional and national as well as international economic development, it defines a "network" as "the nexus of interconnected functions and operations through which goods and services are produced, distributed and consumed" (Henderson et al. 2002, p.445). Those researchers involved in developing and using this approach have taken an especially keen interest in questions of value creation and capture, as well as the organization and governance of global industries. They have demonstrated the multi-actor and multi-scalar attributes of numerous different transnational production systems.

Why should we be cautious? Our concern is not with the GPN approach per se; rather, we deem its ascendancy during the past decade as potentially problematic in view of what it suggests about the state and status of theory in contemporary economic geography at large. Its rise is *symptomatic*. But of what?

We believe the appeal of GPN is indicative of a recent drift away from theory, or of a devaluation *of* theory. We say this for a simple reason: the GPN approach is, or for the bulk of its existence has been, atheoretical, notwithstanding assertions to the contrary (e.g., Hess and Yeung 2006). We will return to the nature of this atheoreticism in a moment, but our main argument is that GPN would not have made the inroads it has into the fabric of economic geography if the latter were as theoretically committed as we have maintained it needs to be.

A theory, recall, is a proposition or set of propositions. It allows space for, and to one extent or another refutes, alternative propositions. Theory is exclusive. Cutting through the noise, it claims that class matters, or power matters, or geography does.

But the GPN approach has generally gravitated in the exact opposite direction. It has tried to be all things to all people by enumerating all of the potentially material factors and relations that shape production networks and that researchers should "check off" when analyzing industry formations. Did we forget to include finance on this list? Okay, then let's throw that into the pot too (Coe, Lai, and Wójcik 2014).

In a scholarly world widely obsessed with "nuance" (Healy 2017), this hedging of bets is certainly not unique to GPN. "People complain," Healy (pp.123–124) says, that "some level or dimension has been left out, and they demand it be brought back in." In this climate, there is an all too understandable tendency "to fall back on assertions of multi-dimensionality, or worry that one has to 'account for' everything at once." But this tendency toward "unconstrained additive complexity," as Healy (pp. 123–124) explains, is the very opposite of good theorizing:

> By calling for a theory to be more comprehensive, or for an explanation to include additional dimensions, or a concept to become more flexible and multifaceted, we paradoxically end up with *less* clarity. We lose information by adding detail.

Expanding a proposition to accommodate all angles and integrate all factors dilutes it and renders it less theory-like. If everything potentially matters, it is impossible to tell what *really* matters; a theory of everything is in the end a theory of nothing.

The crucial question is therefore the one posed by Bernstein and Campling (2006, p.435) in examining the equally wide range of socioeconomic relations invoked as potentially material by GPN's close cousin, the "global commodity chain" (GCC) approach: "How might we start to unravel and order those relations so as [to] investigate and explain them, rather than simply relish listing them?"

It would be fair to say that Gibson-Graham's theory of diverse economies betrays comparable listing-relish. What social relations bear on economic practices? "[T]rust, care, sharing, reciprocity, cooperation, divestiture, future orientation, collective agreement, coercion, bondage, thrift, guilt, love, community pressure, equity, self-exploitation, solidarity, distributive justice, stewardship, spiritual connection, and environmental and social justice … to name just some"! (Gibson-Graham 2014, p.151). Surveying this list alongside the GPN approach, it is hard not to sympathize with Overman's (2004, p.508) claim that much of contemporary economic geography is unable systematically to distinguish the trivial from the fundamental: "richer theories (in the sense of being more complex) … are [not] necessarily better theories (in the sense of helping us to better understand the real world)." This is precisely Healy's and Bernstein and Campling's point.

It is therefore interesting to observe that the most recent significant intervention in the GPN tradition has seen two of its leading proponents (Coe and Yeung 2015) row back some distance, changing the prevailing direction of travel. Instead of adding further complexity and greater nuance, they have tried to cut it out, to do some of the "unraveling and ordering" desired by Bernstein and Campling. They try to craft a considerably more parsimonious proposition concerning the core factors that generally drive and differentiate global production networks. This looks much more like theory, specifically in the "Geographies of Business" mode. (The "structures and dynamics of class inequality and social inequality more broadly" that Bernstein and Campling (2006, p.440) find missing from the GCC approach are still largely absent here too.) Economic geography's response to this new-generation GPN model will therefore be instructive. It will be indicative of the field's current relation to theory.

5.6   Conclusion

Although of a very different hue, *Explanation in Geography* (1969) was just as much of a call to arms for economic geographers as any of David Harvey's later, much more famous, and much more widely read books. It was a call to *theoretical* arms. Surveying the history of a field – geography more broadly – that had only recently begun to take theory seriously, and which was still experiencing the birth pangs associated with the empirical-to-theoretical revolution that he and others precipitated, Harvey concluded that theory would be indispensable to future disciplinary strength. "Perhaps the slogan we should pin up upon our study walls for the 1970s," he wrote, "ought to read: 'By our theories you shall know us'" (p.486).

In the decades since then, there has been an enormous amount of theoretical debate and development in economic geography. This chapter has tried to delineate some of the more significant aspects of economic geography's always-evolving relationship to theory. As we have seen, economic geography has for a long time now been a major importer of theories from other fields. But, increasingly, economic geography has also become a place of origination and export: a field where existing ideas are mulched, reseeded, and harvested in new and distinctive forms. For the most part, "theory" in economic geography appears to be in good health.

One way in which that health is manifest is in economic geography's evident openness to, and indeed active development of, different understandings *of* theory. In other words, as well as being open to different theories, it is open to different views on what theory is, what purposes it might serve, and where its powers should be directed. If we have one concern, however, it is that the historic turn away from singular (and singularly "strong") conceptualizations of theory has recently been accompanied to some degree, in some quarters, by a turn away from theory per se. This is not so positive. In current scholarly and wider socioeconomic environments, it is difficult to imagine that economic geographic scholarship will remain compelling to those not versed in its vernaculars (and even to some of those who are) unless it retains and recovers theoretical vitality.

The point of concluding with this observation is to insist, recalling Harvey's 1969 call, that work remains to be done. The case *for* theory, and for economic geography to be both theoretically embedded and theoretically generative, has been not yet conclusively and consistently made. We shouldn't take down Harvey's slogan just yet. And if we are to be known or to *make* ourselves known by our theories, we cannot escape the linked question of methodology and the methods by which, like it or not, we shall also be known. It is methodology and methods we look at next.

References

Adorno, T., and Horkheimer, M. 1979 [1944]. *Dialectic of Enlightenment*. London: Verso.

Anderson, A. 2009. *The Way We Argue Now: A Study in the Cultures of Theory*. Princeton, NJ: Princeton University Press.

Ballabon, M.B. 1957. Putting the "Economic" into Economic Geography. *Economic Geography* 33: 217–223.

Barnes, T.J. 1996. *Logics of Dislocation: Models, Metaphors, and Meanings of Economic Space*. New York: Guilford Press.

Barnes, T.J. 2001. Retheorizing Economic Geography: From the Quantitative Revolution to the "Cultural Turn." *Annals of the Association of American Geographers* 91: 546–565.

Berndt, C., and Boeckler, M. 2011. Performative Regional (Dis)integration: Transnational Markets, Mobile Commodities, and Bordered North–South Differences. *Environment and Planning A* 43: 1057–1078.

Bernstein, H., and Campling, L. 2006. Commodity Studies and Commodity Fetishism II: 'Profits with Principles'? *Journal of Agrarian Change* 6: 414–447.

Bourdieu, P. 1990. *In Other Words: Essays Towards a Reflexive Sociology*. Stanford: Stanford University Press.

Brenner, N. 2009. What Is Critical Urban Theory? *City* 13: 198–207.

Coe, N.M., Lai, K.P., and Wójcik, D. 2014. Integrating Finance into Global Production Networks. *Regional Studies* 48: 761–777.

Coe, N.M., and Yeung, H.W.C. 2015. *Global Production Networks: Theorizing Economic Development in an Interconnected World*. Oxford: Oxford University Press.

Cox, R. 1981. Social Forces, States and World Orders: Beyond International Theory. *Millennium* 10:126–155.

Critchley, S. 2015. There Is No Theory of Everything. http://opinionator.blogs.nytimes.com/2015/09/12/there-is-no-theory-of-everything/?_r=0 (accessed June 20, 2017).

Culler, J. 1997. *Literary Theory: A Very Short Introduction*. Oxford: Oxford University Press.

Flyvbjerg, B. 2001. *Making Social Science Matter: Why Social Inquiry Fails and How it Can Succeed Again*. Cambridge: Cambridge University Press.

Geertz, C. 1973. *The Interpretation of Cultures: Selected Essays*. New York: Basic Books.

Gibson-Graham, J.K. 2008. Diverse Economies: Performative Practices for Other Worlds. *Progress in Human Geography* 32: 613–632.

Gibson-Graham, J.K. 2014. Rethinking the Economy with Thick Description and Weak Theory. *Current Anthropology* 55: 147–153.

Gibson-Graham, J.K., Cameron, J., and Healy, S. 2013. *Take Back the Economy: An Ethical Guide for Transforming Our Communities*. Minneapolis: University of Minnesota Press.

Gidwani, V. 2008. *Capital Interrupted: Agrarian Development and the Politics of Work in India*. Minneapolis: University of Minnesota Press.

Harvey, D. 1969. *Explanation in Geography*. London: Edward Arnold.

Harvey, D. 1972. Revolutionary and Counter Revolutionary Theory in Geography and the Problem of Ghetto Formation. *Antipode* 4: 1–13.

Harvey, D. 1982. *The Limits to Capital*. Oxford: Blackwell.

Harvey, D. 2000. Reinventing Geography. *New Left Review* 4: 75–97.

Healy, K. 2017. Fuck Nuance. *Sociological Theory* 35: 118–127.

Henderson, J., Dicken, P., Hess, M. et al. 2002. Global Production Networks and the Analysis of Economic Development. *Review of International Political Economy* 9: 436–464.

Hess, M., and Yeung, H.W.C. 2006. Whither Global Production Networks in Economic Geography? Past, Present and Future. *Environment and Planning A* 38: 1193–1204.

Horkheimer, M. 1974. *Critique of Instrumental Reason: Lectures and Essays Since the End of World War II*. New York: Seabury Press.

Hudson, R. 2001. *Producing Places*. New York: Guilford Press.

Katz, C. 1996. Towards Minor Theory. *Environment and Planning D* 14: 487–499.

Krugman, P. 1995. Increasing Returns, Imperfect Competition and the Positive Theory of International Trade. In G. Grossman and K. Rogoff (eds), *Handbook of International Economics, Volume 3*. Amsterdam: Elsevier North-Holland, pp.1243–1277.

Martin, R. 1999. The "New Economic Geography": Challenge or Irrelevance? *Transactions of the Institute of British Geographers* 24: 387–391.

Massey, D. 1991. Flexible Sexism. *Environment and Planning D* 9: 31–57.

Massey, D. 1995a. *Spatial Divisions of Labour. Social Structures and the Geography of Production*, 2nd edn. New York: Routledge.

Massey, D. 1995b. Masculinity, Dualisms and High Technology. *Transactions of the Institute of British Geographers* 20: 487–499.

Overman, H.G. 2004. Can We Learn Anything from Economic Geography Proper? *Journal of Economic Geography* 4: 501–516.

Peck, J. 2015. Navigating Economic Geographies. http://blogs.ubc.ca/peck/files/2016/03/Navigating-economic-geographies3.0.pdf (accessed June 20, 2017).

Pickles, J. 2012. The Cultural Turn and the Conjunctural Economy: Economic Geography, Anthropology, and Cultural Studies. In T.J. Barnes, J. Peck, and E. Sheppard (eds), *The Wiley-Blackwell Companion to Economic Geography*. Oxford: Wiley-Blackwell, pp.537–551.

Rigby, D.L., and Essletzbichler, J. 2006. Technological Variety, Technological Change and a Geography of Production Techniques. *Journal of Economic Geography* 6: 45–70.

Saxenian, A. 1996. *Regional Advantage: Culture and Competition in Silicon Valley and Route 128*. Cambridge, MA: Harvard University Press.

Sayer, A. 1982. Explanation in Economic Geography: Abstraction versus Generalization. *Progress in Human Geography* 6: 65–85.

Sheppard, E. 2011. Geographical Political Economy. *Journal of Economic Geography* 11: 319–331.

Smith, N. 1987. Dangers of the Empirical Turn: Some Comments on the CURS Initiative. *Antipode* 19: 59–68.

Sunley, P. 2008. Relational Economic Geography: A Partial Understanding or a New Paradigm? *Economic Geography* 84: 1–26.

Swedberg, R. 2014. *The Art of Social Theory*. Princeton, NJ: Princeton University Press.

Thrift, N. 2006. David Harvey: A Rock in a Hard Place. In N. Castree and D. Gregory (eds), *David Harvey: A Critical Reader*. Oxford: Blackwell, pp.223–233.

Tickell, A., and Peck, J. 1995. Social Regulation after Fordism: Regulation Theory, Neoliberalism and the Global-Local Nexus. *Economy and Society* 24: 357–386.

Tilly, C. 2006. *Why?* Princeton, NJ: Princeton University Press.

Wainwright, J. 2008. *Decolonizing Development: Colonial Power and the Maya*. Oxford: Blackwell.

Yeung, H.W.C. 2012. East Asian Capitalisms and Economic Geographies. In T.J. Barnes, J. Peck, and E. Sheppard (eds), *The Wiley-Blackwell Companion to Economic Geography*. Oxford: Wiley-Blackwell, pp.116–129.

Chapter 6

Method and Methodology in Economic Geography

6.1   Introduction

On the surface, method and methodology appear as one of the duller topics that books such as this one are obliged to cover. Learning about methods and methodologies can seem like doing manual labor, trench digging, or laying PVC water pipe. They have to be done. No one doubts their importance. They are undoubtedly worthy pursuits, absolutely necessary for accomplishing other tasks. But as activities in and of themselves, they are not very exciting.

One of the ways economic geography distinguishes itself from the other social sciences it borders is generally to avoid talk of method and methodology. Whether that is because of a low threshold for boredom, or due to something else, is not quite clear. But over the last 40 years, when it comes to method and methodology there appears to be a tacit agreement within economic geography of "don't ask, don't tell." This was the official US policy between 2004 and 2011 for gays and lesbians who enlisted in American military service. Legally, gays and lesbians were banned from serving in the US military, but in order to square this legal disqualification with the reality of the many queer men and women who served in various branches of the American armed forces, don't ask, don't tell was enacted. The situation is similar in economic geography. While it is clear that some kind of method and methodology are deployed by economic geographers to obtain their results – how could there not be? – there is an unspoken disciplinary agreement that they are under no obligation to reveal them ("don't tell"), and that the audience (at least if they are economic geographers) should not press for details ("don't ask").

Economic Geography: A Critical Introduction, First Edition. Trevor J. Barnes and Brett Christophers.
© 2018 John Wiley & Sons Ltd. Published 2018 by John Wiley & Sons Ltd.

Don't ask, don't tell is not found in other social science disciplines. Quite the reverse. Many of economic geography's borderland disciplines are fastidious, energetic, and even enthusiastic about airing their method's and methodology's assumptions and practices. Their policy is to ask and to tell. Not to lay bare one's methodological soul is treated by several social sciences as almost a crime, punishable by potential expulsion, or at least by a sharp rap on the knuckles. In some subjects, method and methodology come close to defining the very discipline. They are its woof and weave. Ethnography, for example, *is* anthropology. Other disciplines might do ethnography as well, such as sociology or geography, but if you are an anthropologist *you must do* ethnography. If you do not, then you have not done anthropology. Mainstream economics may be even more unforgiving. Following Paul Krugman (Chapters 4 and 5), if it is not a proper mathematical model with Greek letters, it is not economics. Even if it looks like a proper mathematical model with Greek letters, it still might not be good enough. As discussed in Chapter 4, that was the case for spatial scientists. Their math was just not up to snuff. It required card-carrying economists, that is, those who knew how to construct a proper mathematical model, to show economic geographers how to do the job right.

Don't ask, don't tell was rescinded by the Obama administration in 2011, bringing US military policy on the recruitment of gays and lesbians into the twenty-first century. In this chapter, we would like also to bring economic geography into the twenty-first century by asking questions about both its method and its methodology. We intend to open its hitherto closed closets by telling stories about the methods and methodologies economic geographers have used.

The chapter is divided into three parts: first, we distinguish between method and methodology, and begin to answer the question of why economic geographers have been so keen to hide them; second, we sketch out briefly a history of methods and methodologies in economic geography; and finally, we unpack several specific contemporary methods and methodologies used in economic geography, focusing on their critical potential. Methods and methodologies are not only a means to represent the world in research, but they can also be a means to expose its flaws, to criticize, to change it for the better.

6.2 Method and Methodology

6.2.1 Method, methodology, epistemology, and ontology

We begin by distinguishing between method and methodology. A method is the immediate, hands-on technique used to prepare and analyze data and information to address a given research question. A method could be, for example, a set of inferential statistical tools. Or it could be an ethnography. Or it could be the preparation and administration of a questionnaire. Each method is different, but each requires following certain procedural rules; the time to study and to learn those rules; and frequent practice over the course of which one becomes better at using the method. In addition, and certainly the case for these three examples, there are often also tacit, unstated rules to learn, which makes the deployment of the method an art as well as a science, with some people seemingly having greater aptitude in using a given method than others. Further, different methods trade in different kinds of data, with each type requiring its own skill in manipulation. The data might be banks of "objective" numbers (objective because the information is, at least in theory, independent of the subjective opinions of the person collecting and recording the data); or detailed written field notes littered with subjective impressions; or questionnaire answers that are scored by the researcher and analyzed electronically using a pre-packaged software program.

In contrast, a methodology is less tightly bounded than a method. It is not a specific technique or a tool, but rather a general set of rules, principles, and assumptions that determine the particular research method(s) chosen for a given research problem. A methodology, as it were, lies above methods, limiting which ones might be appropriately selected. For example, if one of the assumptions of the methodology is that the only reliable data are numerical then the set of available methods for selection becomes necessarily restricted. Methods that do not produce numbers, say, ethnography, are ruled out from the start.

In discussing methodology, it is also important to be clear about the epistemological and ontological assumptions inseparably bound with it. An epistemology is a theory of knowledge defining what count as true statements about the world. Take the epistemology known as logical positivism, for example. It says that true knowledge comes in only two forms. First, there is knowledge that is true by definition – for example, mathematical statements such as a mathematical proof – and derived from an initial set of assumptions or axioms. Once the latter are accepted, what is deduced is necessarily true by definition. Second, there is empirical knowledge that can be checked (verified) against the real world. For example, the statement "the cat sat on the mat" is empirically verifiable. Either the cat is on the mat, in which case it is true – you saw it; there are witnesses; you have a photograph – or the cat is not on the mat, in which case it is false – you were there the whole time and there was no cat; no one else saw it; the photo was of only the mat, but no cat. Under logical positivism's epistemology, any other kind of statement that you might make – say, the moral assertion "you are a *good* person" or the aesthetic judgment "you are *beautiful*" – is senseless, even nonsense. They are statements without truth value because they are neither true by definition, nor are they empirically verifiable, the only two criteria for logical positivism that possess truth value. Epistemological assumptions contained within any methodology, say, those of logical positivism, will therefore always limit the kind of research questions that can be asked. Moral and aesthetic questions, for example, are immediately ruled out in logical positivism. Likewise, those same epistemological assumptions will also restrict the kind of methods that can be used. Logical positivism would have no truck with, say, ethnography because subjective judgments contained within it produce only senselessness and nonsense, not truth.

An ontological theory is a theory about what exists. It is a theory about what count as objects in the world that can be legitimately investigated. For example, one ontological theory within the social sciences is methodological individualism. It says that the only (ontological) objects that justify study are individuals. Collective entities like culture, society, or institutions are not real entities. They do not have a fundamental existence. They may look real, but they are not because they can always be reduced to a more essential bedrock entity, the individual. Such an ontological theory was behind British Prime Minister Margaret Thatcher's famous assertion that "there is no such thing as society." For her, society was simply the sum of the individuals who composed it, making the ontological entity of society redundant. Any methodology that incorporates the ontological assumption of methodological individualism is thus constrained in the kinds of research questions that it addresses. It could not ask research questions about society, or any aggregate entity such as culture.

One final point: methods and methodologies often come with unacknowledged baggage – assumptions, values, and beliefs not openly stated. Sometimes they are not acknowledged because the user doesn't know that they are there. Or they become so ingrained that no one thinks to state them. Or again, they might really be hidden, buried, requiring excavation to be found. For this reason, there is often a need to undertake critical analysis when using methods and methodologies (see Chapter 1). It is particularly important to extract the epistemological and ontological precepts submerged within any given methodology. Only once those suppositions are brought to the surface, scrubbed and laundered, recognized for what they are, can they be checked against the research questions asked. Are they consistent? Are they in line with the methods proposed, and the type of data used? And if at some point you find an inconsistency, does it matter? What are the consequences of inconsistency? Historically, economic geography's inclination has been to keep dumb in the face of these kinds of questions, hoping that they would go away, and that things would return to normal. That's what the policy of don't ask, don't tell in the US military hoped too. It was, however, ill-considered, a bit cowardly, if not irresponsible. This is also the case in economic geography, made perhaps even more irresponsible recently by the disciplinary explosion in new types of methods and methodologies the use of which calls precisely for these types of questions.

6.2.2   Economic geography's multiple methods, methodologies, and ends

In the past, there were tight disciplinary bounds on what were acceptable methodologies and methods in economic geography. That is no longer the case. It seems now that everything is permitted, and nothing is proscribed. Here, for example, is a list of some of the methods and methodologies presently used within economic geography: intensive case-study research, in-depth interviews, oral histories, ethnographies, participant observation, discourse and textual analysis, experimental reflexive forms of writing, actor-network theory, and action research, as well as the standard portfolio of quantitative-statistical techniques. It makes for an exciting discipline, diverse and vibrant. But keeping up with developments and appreciating the subtleties of what is available, and even more so, checking for consistency across methods and methodologies, is increasingly complex and difficult.

Adding to that complexity and difficulty are also various new demands that some economic geographers put on the use of methods and methodologies. One, exemplified by Gibson-Graham (2006; 2007; see also Chapters 3 and 5), is to use the research method as an instrument to realize political objectives. Gibson-Graham deliberately recruits non-academics as researchers, believing that through that very process the non-academics will become "new political subjects." By becoming involved in the research process, non-academics will change their subjectivity, change their opinions, change their behavior, and transform the world for the better (Gibson-Graham 2007).

A second demand is to meet certain ethical standards. Traditionally, moral questions in research only revolved around the treatment of the research subject. For example, it was considered unethical to dupe or to lie to research subjects, and especially to cause them harm.[1] Recently, however, ethical concerns have shifted and now bear on the research process itself. For example, some researchers aspire to an elevated moral standard in the very conduct of research: in the relationships forged among researchers within a given project and who have differential rank, stature, and power; or in the relevance of the research, making research more than acquiring knowledge merely for knowledge's sake; or in the source of research funding – while some sources of money are deemed "dirty," not fit to use (e.g., from oil corporations), other money is judged "clean" and appropriate (e.g., from Oxfam).

Yet a third demand is to draw explicitly on multiple methods and methodologies, joining, for example, quantitative-statistical and ethnographic methods. The rationale is that the world is messier than our methods and methodologies for representing it. Consequently, we need to bring more than one resource to bear. Linda McDowell (2013), in the work discussed in Chapter 4 on gendered labor markets, does exactly that. She tries to integrate different forms of data, dissimilar methods, and divergent if not incompatible methodological assumptions. It is an intimidating task, needing a deft hand, as well as creativity and imagination. If it is pulled off successfully, though, it can be riveting.

In sum, there are many potential pitfalls when deploying methods and methodologies, and these are multiplied in economic geography because of their sheer number and diversity. There is no choice, however. One *must* use methods and methodology to carry out research. Without methods and methodology there would be no research. The don't ask, don't tell policy of economic geographers implies doing it, rather than talking about it. That is, to do the job of research at hand without explaining how it is done. We've suggested here that there are good reasons for why there should be more talk and less action, or more talk and better action. What should be our methodology? What methods best address our questions given our methodology? What are our epistemological and ontological commitments? How do they square with our methodology and methods? This is the kind of talk we should be having.

6.3   A History of Methods and Methodology in Economic Geography

6.3.1   Early years – empirical cataloguing

Even the earliest economic geographers were aware of the importance of method for their work (Chapter 3). Chisholm recognized on his opening page that "several [other] works … use methods different from those adopted in the present work" (Chisholm 1889, p.iii). The same with the early American economic geographer J. Russell Smith (1913, p.v), who on his first page wrote, "I believe [my] method permits results that cannot be attained in the same space by the more usual method of presentation." In both cases, they were reacting against the "old" method consisting simply of listing countries, and, without commentary, putting side by side with those lists corollary economic statistics. Both Chisholm and Smith were aspiring to something different using an alternative method. Their books were not meant to be encyclopedias, "general works of reference, … the mere repertory of the where and whence of commodities" (Chisholm 1889, p.iii). Instead, they aimed "to impart an 'intellectual interest' to the study of geographical facts" (p.iii). It was realizing that end that required a specific method.

By method Chisholm and Smith meant a particular form of presentation of empirical information. Both men were dyed-in-the-wool empiricists believing that their task was to report the world as it offered itself as a vast compendium of facts and figures. Empiricism has a long history. As a methodology, it implied both a particular epistemology – that true knowledge took the form of verified facts of the world – and a particular ontology – that the world existed as an immense collection of separate true facts. The facts and figures compiled by Chisholm and Smith were gathered from often various state agencies, but also from private sources such as local chambers of commerce. For example, Chisholm (1889, p.vi) singles out the London *Chamber of Commerce Journal* and industry-specific trade magazines, while Smith (1913, p.ix) mentions the trade journals *Breeder's Gazette* and *Better Fruit*. What Chisholm and Smith did that had never been done before, their method, was to arrange those facts and figures to spark intellectual interest in the geography of commerce. That was the difference their method made. You would no longer see endless lines of numbers. You would be provoked, they hoped, by the very ordering, the cataloguing of their facts and figures, to recognize the larger geographical system of international commerce. The geographical facts and figures would illuminate the global commercial system that generated them.

6.3.2   Regionalism and fieldwork

The methodology of the subsequent regional approach was also empiricist, maybe even more so. By then economic geography was no longer an armchair pursuit, staying at home and collecting and arranging facts and figures from trade magazines and official sources.[2] Instead, it involved going into the field, experiencing the world directly, recording and ordering what one found.

The methodology of regionalism also came with a distinct ontology. It asserted that the basic ontological unit of the geographical world was the region. Geography came to us as regional blocks such as the Canadian Prairies, or the Upper Mid-West, or the Yorkshire Dales. To use Richard Hartshorne's (1939, pp.428–431) vocabulary, each region presented itself as an "element complex," that is, as an integrated whole made up of interlinked separate component parts such as farmhouses, fields, stone walls, irrigation ditches, and so on. The task of the economic geographer was to go into the field to describe those component elements, their relationship, and how as a whole they formed the economy of a given region. The economic geographer would use broad generic (typological) categories to describe all the different regional components they saw (Chapter 3). They would use categories such as "field crops," "irrigation," "livestock holdings," and "building types" to construct a giant catalogue that under each category would contain all relevant observed facts that bore on a given economic region. The exact combination of those generic components would be unique to each region, however. To use Hartshorne's (p.393) economic geographic example, there are common generic elements within the American Corn Belt, the Po Plain, and the Middle Danube Plain. In all of them one would find field crops, irrigation schemes, livestock, and farm buildings, but the precise mix, the exact proportion, in each region and their peculiar relationship to one another would be different. As Hartshorne (p.393) wrote: "the combination of all of them [the generic elements] in the actual mixture as actually found, occurs but once in the earth." So while there were generic elements common to each of the three regions, the precise mixture was unique.

The uniqueness was critical for Hartshorne. It meant that the methods economic geographers used to study regions could only ever be descriptive rather than explanatory. In natural science, phenomena are explained by general principles because they are the same the world over. Indeed, they are the same the whole universe over. Gravity in Vancouver, Canada, is exactly the same as gravity in Uppsala, Sweden, which is exactly the same as found on the furthest known planet in the universe, OGLE-TR-56b, 5,000 light years away. Consequently, one can come up with a single explanatory principle that explains all three cases: Newton's law of gravity. But that approach will not work for regions, economic geography's basic ontological unit for Hartshorne. Each region is different, and so, rather than general principles, explanation, there is only description (see also Chapters 3 and 5).

6.3.3   The scientific method and statistics

The economic geographer's methodology of descriptive regionalism became increasingly out of joint with the methods used by a growing number of social sciences from the early postwar period, however. There was a general push for, and acceptance by several social sciences of, a single method: the scientific method (Chapters 3 and 5). That method began with observing and numerically measuring phenomena, meticulously recording the results as "objective" data. It was objective because it was (ostensibly) the world as it presented itself, independent of the subjective opinions of the person collecting and recording the data. In turn, by carefully inspecting that data, the researcher could identify potential (hypothetical) relationships. One might observe, for example, that the farther in distance a place was away from another place, the smaller the number of people that traveled between them. To confirm this hypothesis, one would have to examine many different cases, establishing that each conformed to this general relation. If the hypothesized relation survived (it was not disproved), it would then likely be turned into a scientific theory. A scientific theory is a general explanation of an observed relationship, in this case, the relationship between the number of trips taken and the distances of those trips (the farther the distance traveled, the fewer the trips). If that theory proved exceptionally reliable, that is, if it was always confirmed, it is possible that the theory would be elevated into a scientific law – a relationship that holds everywhere and at all times (think of Newton's law of gravity, which holds everywhere throughout the entire universe and has done ever since the universe was created 13.82 billion years ago). The geographical theory that connected the number of trips taken and their distances was in fact propounded as a law in economic geography. Waldo Tobler (1970, p.234) called it the "First Law of Geography": "everything is related to everything else, but near things are more related to distant things."

The social scientists who took up the scientific method in those early postwar years justified their approach because it was: (i) transparent – each step in the scientific method was impeccably clear, with nothing implicit or hidden; (ii) repeatable – different people could follow the same set of procedures and arrive at the same conclusions; and (iii) reliable – if carried out according to the rules, the result was foolproof. Epistemologically, the scientific method relied on empiricism. Any claim to have found truth was based on empirical proof, that is, the rigorous demonstration that the outside world conformed to our theory of it. For example, the First Law of Geography was accepted as true because in its case thousands of empirical studies demonstrated that "near things were more related than distant things." Ontologically, the methodology presumed that the world was ordered in mathematical terms. "Mathematics is nature's own language," as Galileo famously said. During the post war years many social sciences, including, as we will see, economic geography, operated on the ontological presumption that the social world presented itself in a mathematical form. Consequently, to describe and explain that world, economic geographers necessarily deployed mathematics.

The scientific method was utilized in economic geography from the mid-1950s. Economic geographers collected and recorded "objective" data, scrutinizing it for possible relationships that could be further explored and tested. The American economic geographer William Warntz was a pioneer in pursuing that method. From the early 1950s, he collected large quantities of (objective) numerical data from the US Department of Agriculture about the geographical production levels of crops. Using early computers to help him, Warntz strove to identify key relationships within his data that then might be the basis of a theory (and if possible, the ultimate, a law).

The theory that he thought best described the geographical relationships within his data was one already discussed, turning on the relationship between the distances between places and the movements between them (see Chapter 3.4.2). That is, Newton's law of gravity. Newton said that the force of gravity (G) between any two bodies, say, planets, would be greater the greater the size of those bodies (M_1,M_2), their physical mass, but would lessen as the distance (d) between them increased. He described that relationship using the equation already given in Chapter 3:

$$G = M_1.M_2/d^2$$

The force of gravity (G) between any two objects, M_1 and M_2, will be proportional to the size of those two bodies, divided by the square of distance (d^2) separating them.

Newton's gravity equation applied to only two bodies, however. He went on to further develop his work by formulating a method to calculate the force of gravity anywhere within a space containing many objects, later called the potential model. The potential model calculated the variable strength of gravity across an area of space filled with different sized mass objects. The force of gravity, the Newtonian potential, would be strongest, most concentrated, in spaces close to large objects, and weakest in spaces farthest from such objects. Using Newton's original equation, Warntz calculated the potential for different crops grown in the United States. Parallel to Newton's original interpretation, potential would be highest around those States that produced the greatest amount of that crop, and lowest around States that produced low levels of that crop or none at all. Figure 6.1 shows Warntz's diagram of the potential for wheat. To obtain this figure, Warntz calculated the potential for wheat for a large number of points across the contiguous United States and then joined up all points which had the same potential value, creating the lines on the map. Those lines are called isolines, lines of equal magnitude, equivalent to a contour line on a map. From Figure 6.1, the isoline for wheat production in the United States (1940–1949) is highest in Kansas, with a value of 35, while the next isoline, of 30, includes not only Kansas but also Nebraska, North Dakota, Oklahoma, and Colorado. These isolines of potential for wheat show the extreme spatial concentration within the United States of the production of this crop. Warntz obtained this result, at least according to him, because he followed the

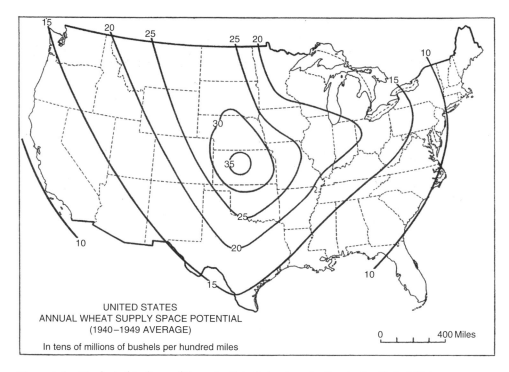

Figure 6.1 Graduated isolines of the potential of wheat production in the United States (1940–1949). Source: Warntz 1959, p.67. Reprinted with permission of the University of Pennsylvania Press.

scientific method: collecting and sifting through reams of (objective) data, looking for significant relationships, and finding them in this case as expressed in the mathematical regularity of the Newtonian potential model. Epistemologically, Warntz's work rested on empiricism. Warntz believed that he could claim the truth because using objective census figures he could show like Newton that the empirical world matched his theory. And ontologically he showed that reality deep down organized itself as an ordered mathematical entity. It came to us in the form of logical equations, in this case as the Newtonian potential formula.

6.3.4 Critical social theory and methodologies and methods

The most recent period in our history of methodologies and methods in economic geography is associated with what we call *critical social theory*. We mean by this term theory that puts society and social relationships at the center of the explanation. It is critical because it questions prevailing economic geographic arrangements and the social distribution of power and resources underlying those arrangements (see Chapter 5). Critical social theory first came into economic geography during the 1970s when Marxism was introduced into the subject. In terms of our earlier periodization, it would include most of the approaches within economic geography that start with radical economic geography and continue up through and include poststructuralism (respectively Chapter 3.5 and 3.6).

The methodologies and methods that burgeoned during this period have most often been qualitative rather than quantitative. Two exceptions were the minority pursuit of "analytical Marxism" (Webber and Rigby 1996; Table 4.1), and more recently, with slightly more proponents, critical geographical information systems (GIS) (discussed further below). Qualitative methods proliferated, however. They have been used to capture aspects of the world that either cannot be expressed in numbers, or, if they are so expressed, lose substantial meaning. The qualitative methods used in economic geography are summarized in Table 6.1. For the purposes of this chapter, perhaps the most important point about them is that when they are used there is often no explicit discussion about them as methodologies and methods. It is don't ask, don't tell.

There are exceptions, however, and we discuss these in detail in the final section of the chapter. Critical realism, which entered economic geography during the early 1980s and was associated with the work of Andrew Sayer, always made its methodology and method explicit. Feminists in economic geography have been another group concerned to make methodology and methods overt. They believe methodological explicitness is a political and moral obligation. And finally, adherents to critical GIS, a quantitative rather than a qualitative method, have been keen also to make their method and methodology transparent.

6.4   Contemporary Methods in Economic Geography

In this final section we discuss in more detail economic geography's most recent methodological period under critical forms of social theory. We first review the relationship between qualitative methods, which have come to ascendency, and quantitative methods, which are now sometimes disparaged, certainly no longer dominant. We suggest that quantitative and qualitative methods should be seen as complementary to the end of critical inquiry, not opposed. Then second we review critical realist, feminist, and critical GIS approaches to economic geography to demonstrate how it is possible to be explicit about method and methodology (to ask and tell) as well as to be critical.

6.4.1   Quantitative and qualitative methods

As already discussed, there was a significant shift in economic geography from quantitative to qualitative methods. Quantitative methods were of the kind championed by William Warntz. They were techniques devised to collect, test, analyze, and display numerical data. For a moment in economic geography's history those techniques were cutting edge, bound up with a view of the discipline as scientific, rigorous, precise, and objective. To be a modern economic geographer was to practice quantitative methods.

From the 1970s, disciplinary beliefs about the virtues of science and modernity, and a concomitant faith in numbers and quantitative methods, were increasingly challenged. Even former revolutionary quantifiers within economic geography became turncoats, criticizing the broader project they helped to initiate. First, they pointed to a yawning gap between the mathematical sophistication of the methods used and the increasing triviality of the substantive studies to which they were applied. David Harvey (1972, p.6), one of

Table 6.1   Qualitative Methods in Economic Geography

Action Research – a method that originated in the 1940s in social psychology concerned with undertaking research explicitly during a period of social change in order to promote and realize that change. Necessarily, such research is normative. It aligns itself wholly with the values of the social change that it studies. Its principal purpose is to guide social action, to fulfill its stated goals. Geraldine Pratt (2012) uses action research in her work with Filipina domestic carers in Vancouver. It is designed to achieve the aims of the Filipina community of workers that are studied.

Archival – an archive is a site at which original records from the past are kept. Traditionally, archives stored paper forms of written documents. Now, though, an archive can be virtual, located in the cloud, storing digital files that can be textual, visual, and aural. While they may appear stuffy, archives are in fact deeply contested political sites. They embody choices that an institution, often the state, has made about what it wants to remember about the past, thus shaping the kind of history about it that can be told in the future. The presentist inclination of economic geography means archival research is rare, but there have been archival studies of corporate history and of the discipline itself (Barnes 2014).

Discourse/Textual Analysis – the close critical reading of a text, usually a written document, but not always. The purpose is to disclose, among other things, hidden assumptions, contradictions, absences, non sequiturs, and dominant metaphors that are there because of various technologies of social power imposed upon the text. The technique originated with Michel Foucault and his painstakingly exact critical analyses of often seemingly minor historical documents. Foucault's discursive analyses showed there was no such thing as a minor document, however. Subject to a discursive analysis, all documents spoke to the relations of power that produced them, signaling a wider significance. Vinay Gidwani's (2008) discursive analysis of interviews he carried out with different caste members in rural Gujarat State, India, brilliantly illuminated the embedded technologies of power at play.

Ethnography/Participant Observation – most associated with anthropology, the ethnographic method involves immersion, close and frequent contact, with the individuals or group one is studying. It entails sustained close interaction, meticulous observation, frequent questions, often joining in, and scrupulous note taking (either during or after contact). Participant observation often goes hand in hand with ethnography in that the researcher fully joins the group that they are studying, carrying out the same activities and practices. In some cases, it is obvious that the researcher is a participant observer. On other occasions, they remain undercover, their identity hidden. The latter was the case in Philip Crang's (1994) exemplary participant observation study of restaurant work in a Mexican-themed restaurant somewhere in South-East England.

Fieldwork – maybe the original economic geographic method. There is the supposition that true, authentic information can be collected only by being "there," wherever "there" is. At its most basic, fieldwork means scrupulous observations of the external geographical sites, which are then written up in various forms (from on-the-spot "jot notes" to later polished entries in a field diary). Exclusive reliance on field observations provides only surface descriptions, however. While useful to the study in providing background color and concrete detail, as well as examples, it needs to be supplemented with other methods such as ethnography, participant observation, or interviews. Ian Cook's (2004) research on papaya cultivation in Jamaica involved him going to that island for a protracted period, making fieldwork observations, but also supplementing by engaging in all three of these other methods as he "followed the thing."

Table 6.1   (Continued)

Focus Group – instead of only one person, a collection of individuals (i.e., the focus group) is interviewed about a specific topic. It is a special kind of interview, though, in that interviewees are allowed to talk to one another, even asking questions to the larger collective. The advantages are that it provides many responses quickly and cheaply to the interviewer, and it allows for the possibility of participants learning from the group. The downsides are that the opinions expressed are not independent but are produced from "group think." That problem is further reinforced by a lack of anonymity, and sometimes by the presence of dominating individuals. Also, the peculiar combination of participants and setting makes the focus group always a "one-off shot," never scientifically replicable. Gibson-Graham (2007) used focus groups in their project on community economies to spread information about both their number and diversity within their study area, and the strategies and techniques used to maintain them.

Interview (see also Box 6.1: The Interview) – a flexible and ubiquitous method found in economic geography. Common forms are the corporate interview; the industry expert interview; the government bureaucrat interview; and the worker interview. More than one person can be interviewed at the same time, and more than one person can do the interviewing. Questions can be closed and narrow, written up and read deadpan (making the interview in effect like a questionnaire). Or questions can be open and broad, made up on the spot. Interviews are never neutral, with interviewer and interviewee equally balanced, but always involve differential social interests and standings. One of the few debates about the interview method within economic geography turned on the issue of gender. Linda McDowell (1992) asked Eric Schoenberger (1991; 1992) what difference it made to her corporate interviews that she was a woman and all her interviewees were men.

Oral History – a special kind of interview where all the questions on a specific topic, for example, around employment, relate to the life experiences of the single individual questioned. What is gained is rich concrete detail, vividness, embodied emotion, and back stories that rarely make it into official histories. Such accounts, though, by definition are partial, relying on fallible human memory that often cannot be checked, and rarely speak either to the abstract or to larger collective historical movements. Oral history is not a common method in economic geography, but Linda McDowell (2013) used it to great effect in her studies of the work histories of different women immigrating to Great Britain beginning from the mid-1940s.

those turncoats, said, "there is a clear disparity between the sophisticated theoretical and methodological framework which we are using and our ability to say anything really meaningful about events as they unfold around us" (see also Chapter 5). Second, in applying quantitative methods economic geographers often violated key internal mathematical assumptions. This was especially true for those who deployed inferential statistics.[3] Peter Gould (1970), not even a turncoat, titled one of his critical articles, "Is *Statistix Inferens* [inferential statistics] the Geographical Name for a Wild Goose Chase?" He thought so. Third, quantifiers were criticized for assuming that the numbers they collected, analyzed, and reported were value-free and objective, rather than socially constructed, impregnated by political interests. For example, critics argued that many numbers used by economic geographers came from the census, reflecting the state's political interests. In fact, the very word statistics is derived from the German word *staatskunde*, state studies. Fourth, quantitative economic geographers were accused of believing that everything should be quantified, and if it couldn't, then it showed only that it was of less value, and shouldn't have been included anyway. Critics asked, though, how much is a human life worth? Democracy? Happiness? Love? All of them have great value, but can any be put into a mathematical equation, subject to the "lore of nicely calculated less or more"? Finally, positivism, the larger philosophical system usually invoked to justify the use of quantitative methods, was criticized (see also Chapter 5). Positivism originated in the nineteenth century. Although it underwent many transformations, it held fast to the notion that only the scientific method and associated quantitative techniques guaranteed Truth. The critique of positivism from the 1960s and 1970s subsequently produced four decades of "theory wars" as a series of other approaches vied to replace it. None of those contenders advocated the use of undiluted quantitative methods.

With this lambasting, quantitative methods became increasingly marginalized from the 1980s. Furthermore, students, both undergraduates and graduates, were no longer trained within that tradition, making it very difficult for them to carry out quantitative work even if they wanted to. Previously, students were mandated to take a quantitative methods course (often more than one) in order to fulfill their degree requirements. At the Geography Department at Ohio State University, graduate students were even required to take two courses in calculus. From the early 1980s those requirements were gradually relaxed at a large number of geography departments, and in some cases, those courses were even excised from the curriculum. Quantitative methods that had been so prominent, within a generation or so became a minority pursuit, and even completely disappeared.

Against the sharp decline of quantitative methods was the steep rise within the discipline of the use of qualitative methods. These were methods used to collect, record, interpret, and present varying qualities of the economic geographic. By qualities we mean attributes or properties or kinds that can be described and represented but not assigned a numerical value. What's your job? What do you do in it? Why did you choose to do that? Typically, qualities are articulated as words and sometimes as images (but not by numbers). They might be spoken in an interview with a researcher, or written down in a document and then interpreted, or shown as images in a video that is watched. There are multifarious forms in which the qualities of economic geographic processes are articulated, and which in turn provide grist for the mill of qualitative methods: one-on-one interview recordings, focus group transcripts, notes, memos, letters, flyers, diaries, brochures, policy documents, grey literature, newspapers, and books, as well as visual and audio sources from YouTube clips to podcasts.

Qualitative methods are the formal means by which to deal with the many forms in which qualitative data is found (Table 6.1). By saying formal, we mean that those methods come with a set of codified rules and procedures just like quantitative methods. Consequently, also like quantitative methods, they can be taught in class, written up in a textbook, and tested on an exam (see, for example, the textbook Gomez and Jones 2010). Qualitative methods are sometimes portrayed as the soft option, where anything goes, where you can make things up as you go along. Not true. Qualitative methods are as rigorous, painstaking, methodical, and subject to scrutiny as quantitative methods. That doesn't mean that qualitative methods are always perfectly deployed. There are shoddy examples of the use of qualitative methods in economic geography, but that's true of the use of quantitative methods as well.

Of all the qualitative methods used in economic geography perhaps the commonest is the interview (Box 6.1: The Interview). It is the go-to method for the discipline. Most immediately it reverses the usual subject position of the academic as expert. It is they who are now in a position of ignorance, and who must ask the interviewee, the expert, for information.

Box 6.1 The Interview

At the basis of the interview is an unequal relationship: one party, the interviewer, is dependent upon the other party, the interviewee. Consequently, the interviewer performs the role of a supplicant, that is, a person who petitions the interviewee by means of a series of questions to grant them necessary information. Sometimes those questions are formally written out in advance, and on occasion given to the interviewee ahead of time. Other times, though, the questions are made up on the fly, formulated on the spot in response to previous answers. There is also variation in how those answers are recorded for later use. Sometimes they are logged electronically by video or digital recorder, allowing access later to the entire interview. Other times, handwritten notes are taken, typically documenting only part of the interview. Yet other times, nothing is taken down, the information stored in the head of the interviewer but then usually written up as field notes afterward. Likewise, how the interviewee data is used can also differ. Very infrequently a whole interview is reported verbatim. Usually, though, only a small portion is used (as selective quotes). Yet other times, nothing is used because, say, it has already been expressed better by someone else, or it is irrelevant to the research theme of the paper, or it is confusingly articulated. There are also different strategies in quoting that interview material. One can use special character marks to signal in the text every pause, grunt, chuckle, ironic inflection, and grammatical peccadillo made by the interviewee. While sometimes this is done in economic geography, usually it is the cleaned-up (non-embarrassing) version that is presented. Now and then, transcripts of the interview are sent back to the interviewee, and they are given an opportunity to change their text, or to correct mistakes, or even to decide whether they really want to allow the text of the interview to appear publicly in print. This last point goes to a larger issue around interview ethics. If the interviews are carried out under the auspices of a university, ethics approval will have to be granted, usually from its office of research services. This involves a university committee deciding on the ethical appropriateness of, among other things, the aims of the project, the methods used, the means by which interviewees were recruited, the questions asked, and the means of knowledge dissemination. It is usually expected that the interviewer will anonymize interviewees in published work, protecting their identity so it is impossible to know their name, where they are from, or for whom they work. There can be exceptions, however, when the research project depends crucially on knowing the name, position, and affiliation of interviewees. One other practical question is: How many interviewees are enough? This is a bit like asking: How long is a piece of string? It will vary with the specific project itself. But a rule of thumb is that you stop interviewing when there is significant repetition of information from the last interviewee recruited. Folklore says this is normally interview number 34. But it is folklore. One last and important issue is that an interview is always a social interaction, modulated by issues of class, gender, and race. Consequently, such encounters are inflected always by relations of power.

That information is recorded, and becomes data for the project. It might take the form of a quotation in the research write-up, or a cited fact, or it might be entered uncited within the larger general narrative, forming another confirmatory instance but never singled out.

While the majority of the research now carried out by economic geographers is by means of qualitative methods, this should not imply that all research should be carried out in this way. To make this assertion is to be as dogmatic and as beset by tunnel vision as were the arch quantifiers of economic geography of the 1960s and 1970s. Just as economic geography has become theoretically polyglot, so it should be methodologically polyglot. There is space in economic geography for both quantitative and qualitative methods. One should not be fixated on one to the exclusion of the other. The inordinate range of material encompassed by economic geography as a field calls for a wide canvas of methodological approaches. Moreover, the criticisms once directed at the use of quantitative methods are now much weaker. There is no longer necessarily a "yawning gap" between the geographical problems and the sophisticated mathematical character of the methods used to resolve them because many problems are now, in part, mathematical problems, a result of our prevailing digital and audit cultures. The world is increasingly struck in a mathematical form. Geographical statistical techniques have also become savvier, neutralizing Peter Gould's criticism. Economic geographers themselves have become savvier, too, recognizing that numbers are not privileged but are no better or worse than any other text, characterized by elision, value judgment, and incoherence. Living with numbers is no different than living with any other text. Finally, economic geographers have long given up the idea that there is a foundational philosophy like positivism that guarantees Truth by quantitative methods. While such methods might not deliver the Truth, they can still be very useful, possibly essential albeit in a non-essential way (Barnes 2009; Wyly 2009).

Methods in economic geography have changed enormously over the past 40 years. Most of the qualitative methodological techniques listed in Table 6.1 would have been viewed as beyond the pale, or at best suspiciously avant-garde, even in the mid-1980s. The problem now is that qualitative methods have become a new orthodoxy. This is regrettable because in the process older quantitative methods have been eclipsed, consigned to the methodological rubbish bin, while the economic geography of the real world now so often comes in the form of numbers (which is not to say that *only* quantitative methods should be used to interpret it). Furthermore, if economic geographers are to speak and to be listened to by other social scientists, as well as those making policy, they must understand quantitative methods. It is their vocabulary. Geographers do not necessarily have to believe in all the qualities attributed to quantitative methods, but they need to cultivate at least conversation-level familiarity. Elvin Wyly (2009) terms the requirement for geographers to speak the language of quantification, "strategic positivism." It will allow economic geographers to talk the talk of quantification but without necessarily committing them to walk the walk.

6.4.2   Three examples: critical methods and methodologies in contemporary economic geography

Let us now turn to three examples of methods and methodologies within economic geography that are both explicit and critical, and hence potential models for the discipline.

Critical realism and economic geography

Critical realism was the first methodological approach to take on systematically economic geography's, and in particular spatial science's, empiricism and positivism. It also provided a comprehensive, fully worked-out methodological alternative. Beginning in the late 1970s, Andrew Sayer, a British economic geographer, set out critical realism's tenets using economic geography as his principal test case. It instantly attracted followers. By the late 1980s it had become the unofficial methodology of the discipline, at least in Britain, for those working on industrial change (*the* topic at that time). Furthermore, as a methodology it was immediately recognizable, with a distinctive vocabulary, syncopation, and style. It could have almost been set to music. First abstraction, then identification of the underlying causal structure, then delineation of contingent and necessary relations, and finally, the denouement, "explanation in economic geography" (Sayer 1982; see also Chapter 5). Sayer's (1984) scrupulously clear prose in his how-to manual, *Method in Social Science*, also meant that everyone was able to follow the instructions on the box.

For our purposes, the virtue of Sayer's critical realism was its self-conscious awareness as a methodology. There was no hiding its imperatives. You knew exactly what you had to do, and what you got, including its epistemological and ontological assumptions (discussed below). While realism as a methodology was fixed and rigid, it was open-minded and pluralist when it came to methods. Here Sayer usefully drew a distinction between extensive and intensive methods. Extensive methods involved the census, large-scale surveys, and questionnaires, seeking broad generalizations, regularities, and patterns. For these ends, quantitative techniques were most useful. Intensive methods were case-study-based, and involved focused study that identified specific causes and mechanisms of change. Qualitative techniques were generally more appropriate in this case. Sayer's larger point was that *both* extensive and intensive methods had their place. Both could be useful. It wasn't either/or.

Critical realism was critical most immediately because it opposed economic geography's long-standing methodology of empiricism, along with the discipline's use of the philosophical justification of positivism. Empiricism contended that the only events that mattered were those that were immediately experienced, and which then became facts. Critical realism, in contrast, insisted that much more important than surface experiences, and the facts they generated, were deep ontological structures you couldn't see or directly experience. They were the real thing. Further, those structures contained causal powers, which inhered in the entities that constituted those underlying structures. This was a very different conception of causation than that of positivism. Positivism said that a cause was simply one thing following another. For example, the spatial scientist William Warntz conceived cause in his work on geographical potential as a constant conjuncture:[4] *if* distance between places increases, *then* fewer people will travel between them. For critical realism, however, causation was found in the very ontological structure that defined an entity's existence. Moreover, whether the causal power was triggered depended upon the presence of contingent circumstances. Sayer's illustrative favorite example was gunpowder. Its surface appearance as fine grey particles was irrelevant. Crucial was only the (ontological) chemical structure of the mixture that by its very constitution necessarily possessed the causal power to explode. For that inherent causal power to be realized, for detonation to occur, the right contingent conditions had to be present. Someone needed a lighted match.

To discern the deep ontological structures in economic geography, critical realism said that researchers engage in abstraction (see also Chapter 5). This meant scraping away surface cover that hid the pure (real) ontological structure beneath. In Sayer's terms, a pure structure was composed of necessary relations. These were relations that must hold for an entity to be what it is. If they didn't, it would be something else. Another of Sayer's favorite examples was the entity "landlord" that necessarily depended on its relation to the second entity "tenant." Landlords could not exist without tenants, and vice versa. They were in a necessary relation. Through abstraction, by forensically cleansing and inspecting, Sayer said we could identify the necessary relations that formed deep ontological structures, such as landlord–tenant, along with their causal powers. Furthermore, not only were those causal powers realized under only specific contingent circumstances (when someone had a lighted match in the first example), but the resulting effects were also contingent, depending on the specific contextual setting in which the cause was realized. For example, the consequences of exploding gunpowder depended on when and where the fuse was lit. At firework displays on November 5 or July 4? Or in 1605 in a cellar under the Houses of Parliament when the Commons was due to sit? Only once abstraction was completed, the ontological structure revealed, and necessary and contingent relations delineated was explanation in economic geography possible.

The methodology of critical realism was used repeatedly over the course of the 1980s and 1990s to explain economic geography. It was deployed first to understand British industrial restructuring and deindustrialization. First, through abstraction, the deep ontological structures of industrial capitalism were identified along with their associated causal powers. With those clarified, economic geographers then used extensive and intensive methods to examine the operation of those ontological structures in the contingent circumstances of specific places, which produced frequent wrenching effects. An example was the late-1980s UK localities project (Chapter 4). Each locality represented a specific set of contingent geographical conditions. The project was to understand how geographical contingency played out both as a causal trigger and as a setting as various deep-seated structural (ontological) causal powers of industrial capitalism were activated. Sayer's critical realism also crossed the Atlantic, and was used to understand a different economic geographic context. Especially associated with the Californian economic geographers Michael Storper and Richard Walker (1990), their version of ontological structure was formulated as "the capitalist imperative." That imperative represented unadulterated power and force, brooking no opposition. It provoked destruction in some places, such as the North American manufacturing belt (which became the rust belt; Chapters 1 and 3). But in other places, and this was Storper and Walker's particular contribution, it incited startling creativity, the erection of factories, warehouses, office towers, whole megacities, where only a short time earlier was only a bare and bereft landscape. Silicon Valley, just outside San Francisco, the enormous complex of high-tech firms that seemingly now rule our every waking minute (Apple, Facebook, Google, Netflix, and Oracle), and collectively worth trillions of dollars, was such a case (see Box 12.3: Silicon Valley).

Feminist methods and economic geography
A second body of work within economic geography that also takes seriously, and makes explicit, methods and methodology comes from feminism. More than anything, feminism is a political project. It aims to raise consciousness about the multiple forms of oppression suffered by women, and to promote beneficial social change that provides women with equal political, economic, cultural, and personal rights. Research in the social sciences

(and economic geography in particular) is one means feminists use to achieve these political goals. Of course, there are other means too, from going on marches to writing op-ed pieces in national newspapers to making decisions about what to wear at work. In each of these pursuits, what's critical is the political project. If it is to be realized by social research, then the methods and methodology must be right. They must be honed to deliver those political objectives. Consequently, it can't be don't ask, don't tell.

Feminist research methods in economic geography have been defined by three broad principles. First, and most basic, that methods, methodology, and research practices were grounded in feminist values and beliefs. This affected, among other things, the topics of investigation, the presentation of the data, the specific methods deployed, and how the very research endeavor was organized. Feminist economic geographers studied topics that spoke to the experiences of women from their standpoint. Those experiences had been typically ignored in the past by an economic geography that was overwhelmingly masculinist (masculinist in that almost all economic geographers were men, and that their primary topics of study were other men doing "manly" things such as heavy labor, operating big machines, and transporting large quantities of material things by lorry, train, and boat; see also Chapter 2). In contrast, the focus of feminist economic geographers was women, and especially their forms of oppression. A central emphasis was women at paid work, but also examined were their roles within the reproduction of the household, and their patterns of movement at all geographical scales. There was no unique set of feminist methods to undertake such investigations, though. Feminists used all the usual techniques for gathering data, from hard-edged inferential statistics to the "softest" qualitative methods. If there was anything distinctive that characterized the research methods of feminist economic geographers it was their willingness to mix and match, and to draw on the full spectrum of techniques from the seemingly most traditional to the most cutting edge. It was always the larger political end that was key, determining the specific methods used, and the presentation of the results. Likewise, feminist values shaped how the research project was organized. Among other things, the research organization was sensitive to power differentials among researchers, to issues of diversity and difference, to ethics, and to the need within the project for open and democratic forms of communication and decision-making.

Second, by definition feminist research was interdisciplinary, found on the borderlands of several social sciences (Chapter 4). This gave feminists working in economic geography a mandate to travel widely, to draw on theorists from many fields, to use different methods, and to incorporate a wide array of subject matter. Geraldine Pratt's (2004; 2012) work on Filipina domestic workers was exemplary. It moved seamlessly from work to home, from class to family, from employer to the state, from the "rational" to the emotional, from economy to culture, and from Vancouver to Manila. Likewise, Pratt deployed a broad assortment of methodologies, methods, and sources of information to realize her political project: interviews, focus groups, informal skits, legal texts, and parliamentary speeches. Everything was grist for the mill and this entailed crossing yet more boundaries in staging the research (literally). Initially written up in standard academic prose, Pratt later transformed her research materials into a dramaturgical script for performance on stage, which subsequently toured in Canada, Germany, and the Philippines (Pratt 2012). The theatrical production was a critical exploration of the culture of work of Filipina domestic workers in Vancouver, and the gendered and racialized assumptions that underpinned it. By showing her "data" to a wider audience, including both Filipina domestic workers and their employers, Pratt aspired to change both the practices of the labor she studied and the cultural assumptions that gave rise to it. Pratt aimed to perform a new world of work into being.

Finally, feminist research methods sought to provide both new knowledge about the economic geographic forms of oppression that women suffered, and novel remedial strategies for improvement. The first feminist work in economic geography in the 1970s drew on surveys and statistical analysis to detail significant inequalities between men and women in urban travel (including commuting to work), in the use of modes of transportation (private versus public transport), and in the forms of their activities (Hanson and Hanson 1980). While men drove a car, making one long journey to work in a day, then engaging in various recreational pursuits, women more likely used public transportation, made frequent short journeys, and when not at paid work carried out unpaid work at home including purchasing provisions for maintaining the household. While this initial research provided new knowledge about gender iniquities and forms of female subjugation, typically it offered little by way of remedy or strategies for social transformation. That changed in the 1990s when feminist economic geographers began using new research methods both to uncover forms of female oppression and, at the very same time, to provide strategies to remedy it. The work of J.K. Gibson-Graham (1996) was pioneering (Chapters 3, 4, and 5). Gibson-Graham's first project involved recruiting female partners of Australian miners working in the Latrobe Valley, Victoria, to assist them in running focus groups, workshops, and interview sessions with other female partners of miners in the same region. Through such interaction, Gibson-Graham aspired to nurturing a different version of what the partner of a miner might be; that is, to cultivate a different subject position, one that would empower them to make change. In effect, Gibson-Graham created a space for feminist politics to alter home life, and, through the effect on the male partner, to alter work life as well.

GIS and economic geography

A final method in which rules of use are explicit is GIS. GIS is a computer-based system designed to store, manipulate, analyze, manage, and visually present all types of spatial data, including economic geographic data. As a technique it first emerged during the 1960s. From the beginning it was associated with geography. Roger Tomlinson, a British Canadian geographer, coined the term geographic information system in 1968. Within the context of geography, GIS was in effect a continuation of the 1950s quantitative revolution by other means. Quantitative revolutionaries such as William Warntz had made early use of the computer, in his case by employing it to make calculations of geographical potential that would have otherwise taken a lifetime to complete by hand. Under GIS the computer continued to make myriad digital calculations at lightning speed, but, and this is what made it different, it also provided maps – geovisualizations – of those numbers.

Just as applying quantitative techniques to an economic geographic problem required scrupulously following a regimen of exact rules, defining the method, so too did applying GIS. One needed to know exactly which steps to follow, and in what order, to produce any results at all. Furthermore, to claim that those results were scientifically valid – as in the quantitative revolution, GIS was conceived as a science[5] – the GIS user was also required to be explicit about how the data met the strict assumptions of the statistical techniques encoded within the computer's GIS software. There could be no don't ask, don't tell. The ability to wring out anything from the machine at all, and to accord it scientific integrity, depended on making one's method crystal clear.

Early applications of GIS to economic geography tended to be within the tradition of spatial science, drawing on standard location theory and techniques such as the gravity model and central place theory (Longley 2015; Chapter 3). That tradition has continued, becoming ever-more sophisticated and able to cope with increasingly large geographical data sets as computer power, that is, the efficiency of the machine to process information, has exponentially improved (see, for example, the work by Ioannides and Overman (2003; 2004) on the economic geography of city systems). That work is conscientious and meticulous, adhering rigorously to GIS's methodological rules. However, one can still strictly abide by those rules and at the same time engage in critical work – to be as critical as critical realism and feminism but to follow a different set of methodological rules than either.

A self-consciously critical GIS first emerged during the 1990s. Since then it has become a vibrant sub-speciality.[6] Part of the critical GIS project has been to demonstrate that GIS methods can be a forceful tool for realizing progressive political ends in the hands of, say, community groups. Knowledge is power. The knowledge provided by GIS can defeat regressive proposals, and make progressive ones winning. Another purpose has been to use critical GIS in a range of geographical sub-fields where it had never been deployed. Mei-Po Kwan's (1999) research using GIS to illuminate the connection among the status of a woman's paid work, their journey to work, and the space-time constraints imposed on them

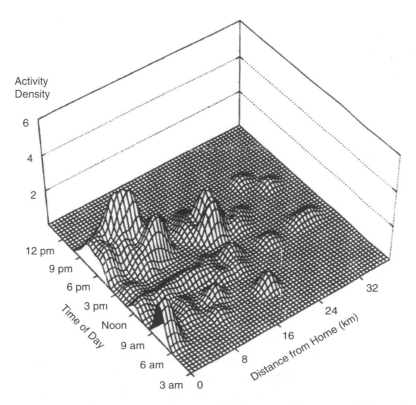

Figure 6.2   A GIS-based 3-D density surface map of daily non-employment activities carried out by women employed full-time. Source: Kwan 1999, p.381. Reprinted by permission of Taylor & Francis Ltd., http://www.tandfonline.com.

by their role in household reproduction, demonstrated a glaring inequality between men and women in their ability to enter high-paid, professional employment. Using GIS-based geovisualization methods, and drawing empirically on daily activity diaries kept by the female workers she studied, Kwan constructed 3-D maps (Figure 6.2). They graphically proved that given where women worked, what they did at work, and most importantly what they did at home, they had no time for anything else. Necessarily, Kwan's GIS method was scrupulously presented, systematic, rigorous, like a methodological open window. The form of her methodology and methods did not prevent her from coming to critical conclusions; in fact, it was the very basis of them.

6.5   Conclusion

Any academic research requires a methodology and a method. Even if you are not aware of them, they must be applied. If they were not, there would be no research results to talk about. Economic geography would grind to a halt. That it has not, and indeed, as a discipline appears robust and animated, shows that methodologies and methods are being used. But they are generally not talked about. The don't ask, don't tell policy in the US military services was repealed in Washington in 2011. Nothing similar has occurred yet within economic geography. Except in a few specific cases such as economic geographic work in critical realism, feminism, and critical GIS, economic geographers still tend to keep their methodological cards close to their chest. Only they, it seems, need to know what they are, usually preferring to fold rather than to show their hand.

Other social sciences are not like economic geography. Some exhibit even an unhealthy obsession. The means of research, that is, methodology and method, become ends in themselves, increasingly divorced from the substantive uses to which they are put. In the 1940s, the economist Paul Samuelson (1947) transformed the method of economics into a hyper-abstract mathematically recondite form. He later reflected, though, that the result was that economists were now like highly trained athletes who had no race to run. The level of the method which they imbibed exceeded the tasks to which it could be applied.

There has been little chance of that in economic geography (although there was the occasional charge during the spatial science period of some of the more high-minded quantitative revolutionaries prosecuting techniques for technique's sake, implicit in Harvey's criticism). In general, though, throughout the history of economic geography economic geographers were concerned mainly to muddle through methodologically, to go with the flow, to be less interested in the how of research than in the what of research. This has brought some complaints from both outside and inside the discipline. The planner and economist Anne Markusen (1999) has criticized the discipline's hopeless "fuzziness" and slapdash approach, while inside there have been grumblings of a lack of rigor, focus, and policy relevance (Martin 2001).

In spite of these methodological foibles, the discipline appears to hold together, and more or less delivers the goods. Paradoxically, it might work in part because it breaks the rules. Some philosophers of science, most famously the maverick Paul Feyerabend, author of *Against Method* (1975), argue that there is no single set of scientific rules to which scientists adhere. Instead, there is only theoretical anarchism. Feyerabend never studied economic geography, but the discipline might be a perfect fit of his thesis.

Notes

1 There have been some famous scientific experiments that deliberately made use of duplicity, lying to the research subjects. The American psychologist Stanley Milgram's 1960s experiments at Yale University were one example. He was interested in studying obedience to authority. Research subjects were told, by someone they believed to be the Head of the Lab, dressed in a white coat, to increase the level of electric shocks that they administered to the "volunteers" in the next room. Subjects could hear but not see the volunteers. The volunteers were really actors, and no electrical charge was delivered. But some research subjects, following orders by the "Head of the Lab," continued to administer charges even though the "volunteer" was audibly screaming in pain and pleading for mercy. More ominously, some of them continued even after the screaming and pleading stopped, when there was only silence.

2 Another early professor of economic geography, Lionel Lyde at University College London, was a chip off the same block as Chisholm and Smith. He relied exclusively, both for his publications and for his lectures, on his massive collection of clippings culled from multifarious sources.

3 The field of statistics is divided into two. Descriptive statistics is about techniques that concisely summarize data such as the mean, the mode, and the median. Inferential statistics is about devising techniques to infer whether properties found in a sample also hold within the larger population. A population (N) is all the potential data points that exist for a given variable. A sample (n) is some smaller subset of that population (n<N). Inferential statistical techniques allow you to take samples from populations and then formally test whether the relationships you find for the sample also hold for the wider population. In order to be confident about the results from inferential statistics a number of strict mathematical assumptions must be met, however.

4 "Constant conjuncture" is the philosophical term for defining a cause as one thing always following another. It was first coined by the Scottish philosopher David Hume (1711–1776).

5 In fact, in 1992 Michael Goodchild renamed the S in GIS "science."

6 See http://criticalgis.blogspot.ca (accessed June 21, 2017); http://manifesto.floatingsheep.org/(accessed June 21, 2017).

References

Barnes, T.J. 2009. "Not Only ... But Also": Critical and Quantitative Geography. *The Professional Geographer* 61, 1542–1554.

Barnes, T.J. 2014. Geo-historiographies. In R. Lee, N. Castree, R. Kitchin et al. (eds), *Sage Handbook of Human Geography*. London: Sage, pp.202–228.

Chisholm, G.G. 1889. *Handbook of Commercial Geography*. London and New York: Longman, Green, and Co.

Cook, P. 2004. Follow the Thing: Papaya. *Antipode* 36: 642–664.

Crang, P. 1994. It's Showtime: On the Workplace Geographies of Display in a Restaurant in Southeast England. *Environment and Planning D* 12: 675–704.

Feyerabend, P. 1975. *Against Method: Outline of an Anarchist Theory of Knowledge*. London: New Left Review Books.

Gibson-Graham, J.K. 1996. *The End of Capitalism (As We Knew It): A Feminist Critique of Political Economy*. Oxford: Blackwell.

Gibson-Graham, J.K. 2006. *A Postcapitalist Politics*. Minneapolis: University of Minnesota Press.

Gibson-Graham, J.K. 2007. Cultivating Subjects for a Community Economy. In A. Tickell, E. Sheppard, J. Peck, and T.J. Barnes (eds), *Politics and Practice in Economic Geography*. London: Sage, pp.106–118.

Gidwani, V. 2008. *Capital Interrupted: Agrarian Development and the Politics of Work in India*. Minneapolis: University of Minnesota Press.

Gomez, B., and Jones, J.P. 2010. *Research Methods in Geography: A Critical Introduction*. Oxford: Wiley-Blackwell.

Gould, P.R. 1970. Is *Statistix Inferens* the Geographical Name for a Wild Goose Chase? *Economic Geography* 46: 439–448.

Hanson, S., and Hanson, P. 1980. Gender and Urban Activity Patterns in Uppsala, Sweden. *Geographical Review* 70: 291–299.

Hartshorne, R. 1939. *The Nature of Geography. A Critical Survey of Current Thought in Light of the Past*. Lancaster, PA: Association of American Geographers.

Harvey, D. 1972. Revolutionary and Counter Revolutionary Theory in Geography and the Problem of Ghetto Formation. *Antipode* 4: 1–13.

Ioannides, Y.M., and Overman, H.G. 2003. Zipf's Law for Cities: An Empirical Examination. *Regional Science and Urban Economics* 33: 127–137.

Ioannides, Y.M., and Overman, H.G. 2004. Spatial Evolution of the US Urban System. *Journal of Economic Geography* 4: 131–156.

Kwan, M-P. 1999. Gender, the Home-Work Link, and Space-Time Patterns of Nonemployment Activities. *Economic Geography* 75: 370–394.

Longley, P. 2015. Analysis Using Geographical Information Systems. In C. Karlsson, M. Andersson, and T. Norma (eds), *Handbook of Research Methods and Applications in Economic Geography*. Cheltenham: Edgar Elgar, pp.119–134.

Markusen, A. 1999. Fuzzy Concepts, Scanty Evidence, Policy Distance: The Case for Rigor and Policy Relevance in Critical Regional Studies. *Regional Studies* 33: 869–884.

Martin, R.L. 2001. Geography and Public Policy: The Case of the Missing Agenda. *Progress in Human Geography* 25: 189–210.

McDowell, L. 1992. Valid Games? A Response to Erica Schoenberger. *Professional Geographer* 44: 212–215.

McDowell, L. 2013. *Working Lives: Gender, Migration and Employment in Britain, 1945–2007*. Oxford: Wiley-Blackwell.

Pratt, G. 2004. *Working Feminism*. Edinburgh: University of Edinburgh Press.

Pratt, G. 2012. *Families Apart: Migrating Mothers and the Conflicts of Labor and Love*. Minneapolis: University of Minnesota Press.

Samuelson, P.A. 1947. *Foundations of Economic Analysis*. Cambridge, MA: Harvard University Press.

Sayer, A. 1982. Explanation in Economic Geography: Abstraction versus Generalization. *Progress in Human Geography* 6: 65–85.

Sayer, A. 1984. *Method in Social Science. A Realist Approach*. London: Hutchinson.

Schoenberger, E. 1991. The Corporate Interview as a Research Method in Economic Geography. *Professional Geographer* 43: 180–189.

Schoenberger, E. 1992. Self-Criticism and Self-Awareness in Research: A Reply to Linda McDowell. *Professional Geographer* 44: 215–218.

Smith, J. R. 1913. *Industrial and Commercial Geography*. New York: Henry Holt and Company.

Tobler, W. 1970. A Computer Movie Simulating Urban Growth in the Detroit Region. *Economic Geography* 46: 234–240.

Walker, R., and Storper, M. 1990. *The Capitalist Imperative: Territory, Technology and Industrial Growth*. Oxford: Blackwell.

Warntz, W. 1959. *Toward a Geography of Price*. Philadelphia: University of Pennsylvania Press.

Webber, M.J., and Rigby, D.L. 1996. *The Golden Age Illusion: Rethinking Postwar Capitalism*. New York: Guilford Press.

Wyly, E. 2009. Strategic Positivism. *The Professional Geographer* 61: 310–322.

Chapter 7

Unboxing Economic Geography

7.1 Introduction

It is not exactly clear when the idea of the black box first arises. Likely it is during the 1920s, associated with research on electronic circuitry. By the 1950s the term is widespread, part of the vocabulary of cybernetics, the postwar science of communication. A black box is an entity in which we see inputs entering, and outputs leaving, but we never see the internal workings where inputs are transformed into outputs (Figure 7.1). Interior processes are hidden, blocked from view, black boxed.

Economic geography as a discipline often appears to be like a black box. It has various visible inputs: jaded undergraduates listlessly filing into a lecture theater for a class on industrial districts; anxious graduate students stammering answers at their thesis defense; and professors seemingly doing nothing but staring at a blank wall but who in fact are trying to think novel thoughts about economic geography. Then there are the outputs: this textbook in economic geography that you are reading; hardbound, thick doctoral dissertations lining university library shelves; and everywhere scholarly papers, found in course readers, electronically on web pages, in between stiff cardboard covers of journals, stacked in a printer's out tray, and shoved in manila folders in one filing cabinet or another.

The sociologist of science Bruno Latour (1987; 1999) believes all sciences present themselves as black boxed. It is a result of their very success, he thinks. "When a machine runs efficiently, when a matter of fact is settled, one need focus only on its inputs and outputs and not on its internal complexity. Thus, paradoxically, the more science and technology succeed, the more opaque and obscure they become" (Latour 1999, p.304).

Economic Geography: A Critical Introduction, First Edition. Trevor J. Barnes and Brett Christophers.
© 2018 John Wiley & Sons Ltd. Published 2018 by John Wiley & Sons Ltd.

INPUTS ⟶ BLACK BOX ⟶ OUTPUTS

Figure 7.1 A black box.

Latour's aim, however, is to make the opaque transparent, the obscure clear. He does it by prizing off the black box's lid, scrutinizing the knotty, messy entanglements beneath.

Here Latour (1987, p.4) makes a useful distinction between "ready-made science" and "science-in-the-making." "Ready-made science" is black boxed science. You deal with only already successfully operating machines, or already established facts – or, in our case, with an already made economic geographic textbook, doctoral dissertation, or journal paper. You don't bother with anything else. Science-in-the-making, however, is all about that anything else, all the otherwise veiled processes going on producing final outputs. Science-in-the-making, or economic-geography-in-the-making, indexes a continually untidy, differentiated, and power-infused process that never finishes. It refers to, among other things, various social entanglements and internalized relations of power, flows of money, bodily travels and potent performances, formal and informal meetings, and all kinds of material bits and pieces from post-seminar beer and chips, to trains, planes, and automobiles, to ancient ivy-clad buildings and brutalist modernist social science towers. All of these elements are hidden from view when black boxed, however (Barnes 2012). The purpose of this chapter is to make visible what is often concealed, to unbox economic geography. Doing so is integral to our book being a "critical" introduction.

Indeed, the justification for unboxing comes from a long-standing body of critical work that over time has taken different forms. It is called variously the sociology of knowledge, the sociology of science, or most recently, science studies. While there are significant differences across the works that fall under these assorted labels, there are at least two common emphases: anti-rationalism and social power.

Rationalism is the view that knowledge derives from pure rationality, and is located within a disembodied mind – a "brain in a vat," to use the philosopher Hilary Putnam's (1981, p.7) arresting image. Under rationalism, mind and body are separated, forming a dualism (known as the "Cartesian divide," following the seventeenth-century French philosopher René Descartes who first recognized it). Because under rationalism only rationality and its home, the mind, matter in producing knowledge, the social and everything else that lie outside the mind are demoted, considered unimportant, secondary, and subordinate. In contrast, the anti-rationalism championed by the sociology of knowledge and science, as well as by science studies (which includes Latour's work), contends just the opposite. The social and everything else are primary, providing an explanation for knowledge. It is rationality that is redundant: it "is a wheel that plays no part in the mechanism," as the philosopher Richard Rorty (1982, p.167) put it.

The second emphasis is on social power, argued to pervade every aspect of the academic production of knowledge, including that of economic geography. The machinations of power are seen immediately the cover of economic geography's black box is removed. It runs the gamut from publishing decisions by journal editors, editorial boards, and university and commercial presses, to judgments of national research funding bodies and hiring committees, to sparkling lectures by high-profile charismatic researchers, to the hegemony of the English language within global academic scholarship, to the changing economic geographic needs of powerful state agencies and private corporations. As Foucault recognized, the exercise of power is productive, enabling things to be done. But it can also be destructive, leaving in its wake constrained, policed disciplines, roads not taken, as well as a trail of individual ruin: truncated careers, emotional trauma, marginality, and feelings of worthlessness and hopelessness.

In 1955, the first ever game show with money prizes was broadcast on British TV. It was called *Take Your Pick!* and shown on the United Kingdom's first commercial channel, ITV, which launched in the same year. After putting participants through their paces, the suitably oily host, Michael Miles, would ask the successful contestant whether they wanted to "take the money" they had already won, or whether instead they wanted to "open the box." The audience always shouted, "open the box!" When opened, it contained either something worthless (a hairnet, a bag of sweets), or, for 1950s Britain, something of amazing value (a washing machine, a week's holiday in Majorca – Oooooh!). This chapter, like the studio audience, wants to open the box. In our case, there is no danger of a booby prize. There is only the promise of a satisfying anti-rationalist rendering of economic geography, making clear how social power, along with everything else, contributes to disciplinary knowledge.

The chapter is divided into three main sections. The first elaborates on the critical project of including within an understanding of a discipline its messy power-laden entanglements – the everything else – which are otherwise neglected or at best marginalized in a rationalist portrayal. The second and longest section focuses on that messiness as it applies to economic geography, specifically examining the role played within the discipline by a disparate group of elements: bodies; conferences, workshops, and summer schools; academic journals and textbooks; machines of all descriptions from the early twentieth-century Brownie box camera to the 2016 iPhone 7; and money. These elements may be disparate, but collectively they are essential to the project of keeping the discipline of economic geography standing. Their sturdiness, resilience, and tight connections form a skeletal frame. Only when something goes wrong, as it sometimes does with our own skeletal system – a bone breaks, or is dislocated, or becomes diseased – is one aware of its importance. Then both we and the body of economic geography slump. To highlight this point, the third section of the chapter briefly discusses an example of just such a disciplinary slump that appears to have been in process for the past decade or so: a weakening of the skeletal frame of institutionalized economic geography in the United Kingdom, where the multifarious elements necessary to keeping the discipline standing no longer seem to be propitiously aligned.

7.2   The Hidden Life of a Discipline

If anyone represented pure rationality, a brain in a vat, it was Sir Isaac Newton (1643–1727), the English scientist. Newton famously could hold an analytical problem in his mind for days, turning it over, trying one logical line of attack, then another, until eventually it succumbed (Gleick 2003). Hence, he was able to invent calculus, the wave theory of light, and the gravity equation. Seemingly, it was all done by sweated brain power. By doggedly following rationality, and only rationality, Truth was laid bare.

It was not that simple, however. Newton's rationality was at best mixed. His achievements were made possible precisely because of his interactions with a larger outside society, with other early scientists, and with a world of things. Newton made astounding scientific breakthroughs. But the majority of his time spent in his two-storied wooden "Elaboratory," as he called it, at the bottom of his Fellow's garden at Trinity College, Cambridge, was not concerned with science. Instead, it involved testing witchcraft recipes to transmute base metals into gold (alchemy), and engaging in frighteningly difficult calculations to reveal God's secrets, which Newton believed were written in arithmetical code within the bible ("Biblical numerology"). In fact, posthumous forensic evidence suggests Newton's later sometimes tenuous grip on life – even his otherwise ardent admirer John Maynard Keynes (1951, p.321) called him "slightly 'gaga'" – was a result of mercury poisoning contracted during those alchemic experiments (Gleick 2003, p.99). If anyone was confronted by the vagaries of a messy life, unpredictable materiality, complex interests, and irrationalities, it was Newton. Furthermore, he was never a lone, detached scientist, but always part of a wider scientific community. One of the earliest elected members of the Royal Society (in 1672), the first learned society for science in Britain, Newton would regularly meet in Gresham College, London, with other scientists, debating them, reading their papers, and witnessing their experiments. That community, though, was shot through with power differentials, rivalry, and jealously. Newton, for example, wrote much of his work in secret code so afraid was he that his ideas would be stolen. Even when Newton was seemingly gracious, recognizing the role of that larger scientific community – he famously said, "If I have seen further it is by standing on the shoulders of Giants" (quoted by Gleick 2003, p.98)[1] – he still used the occasion to put down one of his rivals, Robert Hook, a man of short stature, and no "Giant."

Over the past 50 years or so there has been a revolutionary re-conception of how science operates, leaking into the social sciences, including economic geography. It is an anti-rationalist interpretation. It involves moving away from framing scientists as heroic, isolated, faceless rational automatons, computers on legs, instead recognizing them as sentient, fallible, ordinary human beings, embedded within larger power-permeated social structures and institutions of a given time and place, with specific rules, mandates, and interests. Thomas Kuhn's (1970) *The Structure of Scientific Revolutions* was one key to this anti-rationalist reconceptualization, and was later elaborated by work in science studies. Both Kuhn and science studies provide tools and concepts to understand what we see when we unbox economic geography.

7.2.1   Thomas Kuhn, paradigms, and practice

The Structure of Scientific Revolutions provided a ground-breaking reconceptualization of scientific change, and the practices of scientists. Key was Kuhn's concept of a paradigm (Box 7.1: Paradigm). It denoted the combination of values, assumptions, goals, methods, exemplars, and everyday acts that defined a scientific community. The community was critical. For Kuhn, scientific knowledge was collectively produced, not the result of a lone scientist, however rational, a brain in a vat. Kuhn's idea of a paradigm made that clear. All scientific knowledge came from a paradigm, and a paradigm was shared. From the get-go, scientific knowledge was embedded within a collective set of social relations, agreements, and practices.

Box 7.1   Paradigm

Paradigm was an old English word that meant pattern, exemplar, or model. It was rescued from obscurity by the American philosopher and historian of science Thomas Kuhn (1922–1996) in his short monograph *The Structure of Scientific Revolutions* (1970 [1962]). Reportedly, it was the most cited academic book of the 20th century. The definition of paradigm was contentious, though. One critic counted 21 different meanings in Kuhn's first edition. Roughly, a paradigm was the constellation of norms, rules, techniques, exemplars, and routines shared by a given scientific community, making it what it was. A paradigm shaped what scientists thought about something before they thought it.

Kuhn's use of paradigm was to provide an alternative to the standard rationalist history of science that stressed science's progressive and heroic character. For Kuhn, most science was about puzzle solving, and was carried out within a well-defined frame of reference, a paradigm. Of course, it might take a long time to learn and be proficient within the paradigm, but you didn't have to be a genius. It was 99% perspiration and only 1% inspiration. Consequently, Kuhn thought, the practices of scientists for the most part were run-of-the-mill, not extraordinary. To use Kuhn's term, scientists engaged in "normal science."

Within normal science, however, anomalies sometimes occurred that could not be explained within the prevailing paradigm. Given enough anomalies, a crisis eventuated, precipitating extraordinary research (not just puzzle solving). If successful, in the sense that the new research accommodated the anomalies, a "revolutionary" change occurred in the form of a paradigm shift. The old paradigm was abandoned, and a new one taken up – Kuhn's example was the move from a Ptolemaic (geocentric) to a Copernican (heliocentric) account of the solar system. Kuhn further argued that paradigms by their constitution were not comparable; they were "incommensurable." This was because they contained various value assumptions and practices that could not be made equivalent across different paradigms. Because you could not definitively judge if one paradigm was better than another – they were incommensurable – scientific knowledge moved not as linear progress but as non-comparable, discontinuous shifts.

Within a paradigm, Kuhn thought, scientists undertook routine operations. They did what they were taught to do. This did not make them robotic workers. Their task was to solve puzzles, requiring imagination and creativity. But the solutions to those puzzles always came from within a pre-given set of practical resources that formed a paradigm. Kuhn (1970) later called those resources "exemplars." They were, he said, "concrete problem-solutions that students encounter from the start of their scientific education" (Kuhn 1970, p.196). There are two important points. First, that following a paradigm entails concrete hands-on practices: crunching through equations, drawing maps, interviewing people, taking notes, writing computer code. A paradigm is not simply a good idea with a lightbulb going on.

It requires practice, doing things. It is a worldly achievement. Second, paradigms are intimately connected to the education process. In effect, you come to university to learn a paradigm, to learn exemplars. They can be had by reading a classic textbook; or by doing a set of end-of-chapter problem exercises; or by participating in the re-creation of a famous lab experiment; or by attending a lecture. To make sure students learn the paradigm, they are monitored, disciplined, and tested by exams, quizzes, reports, oral presentations, and classroom observation. It can be ruthless. Those unable to learn the paradigm, to know and practice the exemplars, are weeded out: they flunk the class, they drop the course, they fail their defense, their funding is stopped. A paradigm is serious business.

From Kuhn, then, when the lid of the black box is removed what one sees is a collective enterprise, held together by a paradigm that all scientists (or at least those admitted to the fold) share. Being a scientist is a job much like any other. You follow the rules, using what you already know, the exemplars you learned, to fashion concrete solutions to the problem at hand. The focus is on actual practices of scientists, rather than on what they say they are doing. Indeed, they may not know exactly what they are doing, given that the paradigm they learned as students is so deeply inculcated, drilled into them.

7.2.2   Science studies

Kuhn's work prepared the way for the development of what came next, science studies. That literature is now vast and heterogeneous, beset by vying traditions. It remains united, though, in upholding Kuhn's anti-rationalism, focusing on concrete practices, and asserting that science is a collective project. The fact that many of the researchers within science studies came out of sociology and kindred disciplines also has made them more prone to discuss issues of power and social relations.

The best known approach in science studies is actor-network theory (ANT), developed especially by Latour. The gist is that knowledge is produced by bringing together, "enrolling" in terms of ANT, a heterogeneous set of "actors," both human and non-human. Those actors can include everything from human bodies, to machines, to books, to mathematical equations. Each actor is persuaded to work in concert with other actors, or, in ANT talk, to become an "ally." Only then is a network possible, and knowledge produced.

The linkages among the different actors that make up the network are forged through a process ANT calls "translation." Translation occurs when each different actor realizes that their interest is best served by joining, becoming allies with, other very different actors and participating with them – "performing" is ANT's term – to fulfill a common project. For example, from Chapter 6, the American economic geographer William Warntz, for his project on macro-geography, joined together, that is, orchestrated a performance from, two very different entities: (i) Newton's seventeenth-century theory of gravity, and (ii) statistical measurements of incomes and agricultural prices in the mid-twentieth-century United States collected from the US Census (Warntz 1959). Warntz's achievement was to show that the two were able to work together, that they had similar interests, and could be allies, capable of performing a common project, realized as Figure 6.1.

Specifically, to arrange a successful performance required Warntz to line up a set of heterogeneous actors: numbers collected from the United States Department of Agriculture, reordered and reinscribed on sulphur-yellow pads of paper; an electrical calculator; base maps of the United States; cartographic skills and equipment; Newtonian equations of gravity, mass, and distance; and time and money to allow him to undertake such tasks. The money he received came from the Office of Naval Research (discussed below), with the work he undertook later published in the house journal, *The Geographical Review*, of the organization for which Warntz worked, *The American Geographical Society* in New York.

Warntz's task was hard work, and certainly there was no guarantee of success. In his case, however, bringing together these very different entities was successful, and created both a new piece of knowledge and a new piece of reality. It was new knowledge because hitherto no one had applied the theoretical logic of Newtonian cosmic formulations of gravity potential to terrestrial space. And it was a new piece of reality because no one before had drawn, or even conceived of, such maps of price and income potential (Figure 6.1). They never existed, but because of the performance, they were now out in the world jostling and circulating with other bits of reality.

A performance does not have to be successful, however. Things do not always go right on the night. A performance is a precarious achievement, relying on a myriad of different actors each playing their role and cooperating. The resulting performance might be brilliant, bringing down the house, like producing a vaccination for polio, or the double-helix model of DNA. Or it might never begin, or quickly fizzle out, or finish with an audience's indifferent shrug.

For example, there was a long-promised textbook by William Warntz and his friend, another economic geographer, Bill Bunge, called *Geography: The Innocent Science*. A contract for it was signed with the publisher John Wiley on July 18, 1963 (Warntz 1955–1985, box 5). It was to include even physical props for performing Warntz and Bunge's brand of economic geography. In the end, though, the book never materialized, although there were many drafts and even prop prototypes. In ANT's vocabulary, there was too much "interference" in lining up the different agents and persuading them to work together as allies. The "interferences" were many and frequently involved diverse exercises of power. They varied from Bunge's denial of tenure, which from 1967 left him without a job, to Warntz's battles with his dean at Harvard, which eventually contributed to his departure. The end result was an unfinished book. Its potential to make a difference, to perform, and to create a new reality, was given up in 1971 when Warntz and Bunge declined to accede to their editor's increasingly plaintive pleas to finish the volume (Warntz 1955–1985, box 45).

The point: taking off the lid of the black box reveals a tangle of actors that have temporarily agreed to work with another, to become allies, to complete the project. Those actors put on a collective performance that results, as for Warntz, in new knowledge and new bits of the world. The performance is uncertain, though, and needs various forms of buttressing to continue. If there are some unfortunate power plays by institutions or individuals, or various actors don't step up, or put on a duff show, it could be curtains; the end of economic geography as we knew it. The purpose of the next section is to identify various elements in economic geography's performance that shore up its operations, that keep the disciplinary lifeblood flowing. These elements may not seem like much. But they are the bone structure of the discipline lying beneath the surface. They keep the body of economic geography standing, at least for now.

7.3 Keeping Economic Geography Standing

7.3.1 Bodies

One of the first things we see when we open the box are particular kinds of human bodies. They are normally obscured. For rationalism, only the mind counts. Bodies are secondary, mere vessels, peripheral to knowledge. Science studies contends that the bodies involved in producing knowledge make an enormous difference, however. Different bodies are subject to different regimes of power, enabling and constraining what they can do. Bodies should not be hidden, kept in the dark in a closed box, but rather brought into the light.

Donna Haraway (1997) argues that rationalists treat bodies that produce knowledge as merely "modest witnesses," neutrally recording the world. Such modesty, argues Haraway, is no modesty at all. It is really a front, a ruse, hiding and protecting the wellbeing of just one type of body. For Haraway, the façade of modesty conceals the vested interests of primarily white, Western men. It is their bodies that most benefit from knowledge that is produced. "Modesty pays off … in the coin of epistemological and social power," Haraway (1997, p.23) writes. The rationalist denial of the importance of bodies, hidden underneath the cover of a black box, turns out then to be a strategy to promulgate a particular kind of knowledge that for Haraway is marked by masculinism, racism, and heteronormativity. For this reason, she argues, we urgently need to pay attention to the types of bodies involved in producing knowledge (such as economic geography's), to stare into the black box to see what sort of bodies are at work. Economic geography's knowledge is never disembodied, but always incarnate.

Before 1980, the bodies that practiced economic geography were most often male, and most often white. They were frequently concerned with studying similar kinds of bodies that manufactured tangible goods in heartland industrial regions in Western Europe and North America (Chapter 2). The bodies of economic geographers accumulated "hard" facts, often in numerical form, displaying them in tables and maps, and from the post-war period they undertook statistical analysis including occasionally formal mathematical modeling.

Since 1980, while white male bodies continue to predominate, there has been at least incremental change, with women becoming more numerous as well as non-Caucasians (especially of Asian heritage).[2] Doreen Massey's 1984 *Spatial Divisions of Labour* (discussed in Chapter 3) was a watershed book in two respects. First, it offered an intellectual agenda different from any hitherto proposed. Production and the factory were still prominent, but no longer respectively carried out and inhabited by only men. Moreover, to understand what went on inside the factory, whose workers were also women, Massey argued it was essential also to refer to what went on outside, particularly in the home (the site of "domestic reproduction"). Distinguishing different kinds of bodies and what they did at different sites was therefore critical. The second watershed, of course, was that the book's author was a woman. There were women economic geographers before, but not many. The discipline was another "Boys Town" (Deutsche 1991). But there was never a woman economic geographer before who built into her conceptualization of the disciplinary problematic the absence of bodies, and female bodies in particular, providing a larger theoretical framework to set and resolve that lack. That was one element of Massey's genius.

More recently, the internationalization of the academy, and specifically the globalization of economic geography, has encouraged a wider range of bodies to participate in the discipline. Economic geography now seems less the pursuit of only males of Northern European heritage. Asian economic geographers, in particular, have become more prominent, and not only from former British colonial sites where there was a pre-existing tradition of geography, as in Singapore and Hong Kong, but also from China, Korea, Japan, and Taiwan.

There is at least one more form in which bodies can be significant, also entrenched with relations of power: giving a lecture performance either to students in a college class or to a wider audience whether at a conference or a public event. As part of the creative class, the professoriate is expected to perform. And a performance is always embodied, involving dressing appropriately, knowing one's lines, making emotional connections with the audience, drawing on physical props, and play acting moves, facial expressions, and corporeal gestures. Issues of gender and race, and likely class, spiral through such bodily performances, and their reception. Alpha males, for example, perform, and thereby reproduce, alpha masculinity, with all its attendant fringe benefits and losses.

In all these ways, the bodies of economic geographers are vital. Normally, the closest those bodies come to being recognized is as gender-specific or ethnic-specific names written at the top of a paper or on the spine of a book. Knowing more about those bodies is imperative, however. Minds that come up with and develop economic geographic knowledge cannot be divorced from the bodies that they inhabit.

7.3.2   Interpersonal contact and invisible colleges: conferences, workshops, summer schools, and leaders

You also need to know how those bodies interact. Both Kuhn and science studies suggest that producing knowledge is collective, involving exchange and direct personal contact. Being there makes a difference. It binds you to a collective project, and through interaction triggers new thoughts and novel ventures.

The benefits of being there were first formalized by Diana Crane (1972) in her book *Invisible Colleges*. An invisible college is one means by which those who are part of an academic community remain connected. It facilitates interpersonal ties at particular sites, shores up social solidarity (reproducing the group), and can spark original, fertile lines of inquiry. Crane also emphasizes the importance to the colleges of research leaders. They function variously as organizers, managers, recruiters, muses, and promoters. Necessarily energetic and productive and well cited (citation counts function as an important albeit flawed measure of academic status and worthiness), such leaders, Crane contends, must also be open-minded, encouraging cross-fertilization and the use of outside literatures.

Invisible colleges come down to ground, allowing direct interaction among bodies of individual economic geographers, at sites like conferences, special lectures, workshops, and summer schools. Each venue provides opportunities for strengthening the community, firming up interpersonal connections, trading ideas and research findings, learning and training, debating issues, extending conversations about the discipline, and initiating new projects.

A national conference like the annual meeting of the Association of American Geographers (AAG) held in one major North American city or another every spring is one place where economic geography's invisible college becomes visible. The AAG is massive, though, with upward of 8,000 participants, and 80 concurrent sessions at any given time. Individual papers in economic geography are easily lost. The best chance of being noticed is to be part of a specially designated thematic session organized by individuals interested in that theme, in this case, by the AAG Economic Geography specialty group (one of 65 specialty groups within the AAG organizational structure). Also at that same conference, the journal *Economic Geography* underwrites the annual Roepke lecture given by a supposed luminary in the field. It is the discipline's moment in the spotlight. It is well attended, showcases the discipline, brings a large number of economic geographers into the same room, and the reception afterward of wine and tasty tidbits provides opportunities for community-bolstering, networking, back-slapping, disciplinary gossip, blue-sky thinking, and even on occasion plans for collaboration.

Since 2000 there has been also a specialized international conference for economic geographers, the Global Conference on Economic Geography.[3] That the conference is global – Singapore in 2000, Beijing in 2007, Seoul in 2011, Oxford in 2015, and Cologne in 2018 – speaks to larger processes of globalization, and in particular to the globalization of the academy. The conferences have only gotten larger (the first, in Singapore, was around 250 attendees, the most recent, in Oxford, over 700) and more international (the Oxford conference attracted participants from 50 countries), increasingly drawing in other social scientists. As a site, the Global Conference offers significant opportunities to reinforce economic geography's invisible college. Everyone who is anyone is there.

Compared to conferences, workshops are smaller, more focused, more closely fitting the bill of the original definition of Crane's invisible college, which she expected would have around 100 members. At a workshop, everyone hears everyone else's paper. Workshops may be one-off special events, or an annual series. Their hallmark is a tightly focused theme to which all participants direct their presentation. It might be global production networks (see Chapter 5), sponsored by the Global Production Network Centre at the National University of Singapore, and run by the economic geographers Neil Coe and Henry Yeung.[4] Or it might be on regional innovation systems underwritten by the European Forum for Studies of Policies for Research and Innovation (Eu-SPRI), formally interdisciplinary, but historically connected to Scandinavian economic geographers and their university centers of innovation studies. Those workshops closely integrate participants not only socially and intellectually, but geographically: GPN in economic geography *is* Singapore; innovation studies in economic geography *is* Scandinavia.

A special kind of workshop is a summer school, used often for disciplinary socialization, recruitment, and training. In economic geography, summer schools first emerged during the 1960s as a means to encourage and to teach the techniques and methods of spatial science to a group never before exposed to them. In Britain, the most famous was called the Madingly Hall lectures, organized by Peter Haggett and Dick Chorley. Begun in 1963, those lectures introduced, coached, and drilled sixth-form college geography teachers into the ways of spatial science (discussed further in the next section). The US version was called the Summer Institute, and first held in 1961 at Northwestern University in Evanston, Illinois. Funded by the US National Science Foundation, organized by Edward Taaffe and Brian Berry, it was directed primarily at young assistant professors and graduate students.

Serving as a boot camp for non-quantitative recruits, the Institute functioned multiply as a forum to make converts, to forge disciplinary solidarity, and to extend geographically the new economic geography by participants taking back and practicing their newly acquired knowledge at their home institutions.

A different summer school, although concerned still with economic geographic training, was launched by Jamie Peck (Box 7.2: Jamie Peck) in 2003 in Madison, Wisconsin: the Summer Institute in Economic Geography. The Institute has met for a week more or less every two years ever since (for dates, places, and participants, see Figure 7.2). Unlike the earlier Summer Institute, this one had no particular methodological axe to grind. It was self-consciously "catholic," celebrating all viewpoints, "everything from regional science to feminism and Marxism."[5] Core, though, is its pedagogical goal: "to make a distinctive contribution to the ongoing development and social and intellectual 'reproduction' of the discipline." It does that by training "young researchers around the world" – graduate students, post-docs, newly minted assistant professors – in the art of economic geography. For the reproductive cycle of an academic discipline rests crucially on professors producing offspring, graduate and post-doctoral students, who in turn become professors themselves. For good Darwinian reasons, the more economic geography professors there are, and it is the aim of the Summer Institute to proliferate these, the greater are the chances of disciplinary survival.

Roughly 40 participants take part in the Summer Institute (the gender split is almost even, 45 countries have so far been represented, and there are stipends for those who need financial help). Invited "internationally renowned faculty" lead the group through a crash course of instruction and discussion in everything you need to know to be a high-functioning economic geographer. Topics include publishing, teaching, research methods, and grant writing. Throughout the week there are also myriad opportunities, formal and informal, for "ventilating contemporary controversies and debates" within the field. Apart from conveying hands-on knowledge, practical skills, and concrete exemplars, the Summer Institute provides an instant peer social network, and through the "internationally renowned faculty" bridges to the larger economic geographic community. By the end of the week, participants are on their way to being paid-up, card-carrying members of the discipline.

A final element is leadership. Invisible colleges require movers and shakers, individuals who can organize, spark ideas, bridge sub-groups, and connect and cross-pollinate. Those leaders must be active and articulate, they must have disciplinary status, and they will ideally be fair, judicious, and flexible. Perhaps most important, they need to be less concerned about their immediate self-interest than the interests of the collective, the wider disciplinary community. This is an unusual combination of qualities, but over the history of economic geography, different periods have thrown up just such people. George Chisholm was one, the American regionalist Clarence F. Jones was another, so was the British master of the regional economic survey, Dudley Stamp. During the 1960s, Peter Haggett in Britain and Leslie King in North America played critically important leadership roles. More recently, Doreen Massey, Nigel Thrift, Roger Lee, Gerry Pratt, Linda McDowell, Eric Sheppard, and Jamie Peck have been key leaders (Box 7.2: Jamie Peck). Being a leader does not necessarily mean being the most innovative, or the most cited, or receiving the most research money (although it could). The appropriate skills and qualities are entrepreneurial and managerial, including the ability to see intellectual opportunities and to realize them systematically. The former takes energy and ambition, the latter commitment and many long work hours.

Box 7.2   Jamie Peck

Jamie Peck. Source: Jamie Peck.

Jamie Peck holds the Canada Research Chair in Urban and Regional Political Economy at the Department of Geography, University of British Columbia. He perfectly exemplifies the qualities characterizing a leader of an invisible college. He is exceedingly well published (more than 300 publications) and cited (in the top 1% of all social scientists). The hallmark of Peck's work is connecting theoretically and substantively various outside literatures to the concerns of economic geography. That ability has resulted in the establishment of a series of sub-disciplinary research agendas that other economic geographers have then pursued, but galvanized by Peck's initial explorations. Those agendas include regulation theory, varieties of capitalism, and perhaps most spectacularly, neoliberalism. Peck is not associated with any single ideological position or methodological approach; rather, his intellectual sensibilities are ecumenical, and attested by his role as editor in chief of the *Environment and Planning* series of journals (now five), representing the entire waterfront of human geographical research. His only singular commitment is to the discipline of economic geography. He brings to that task tremendous energy, verve, and dedication, taking the form of: extensive speaking engagements at national and international events; inaugurating and directing the Summer Institute of Economic Geography; serving on editorial boards; organizing workshops, conferences, and speaker series; editing journals and book series; awarding academic prizes; supervising large numbers of graduate students and serving on an even greater number of student committees; and maybe most important, continuing to publish innovative research monographs and journal papers within economic geography.

Figure 7.2   Homepage of the Summer Institute in Economic Geography (http://www.econgeog.net/). Source: Jamie Peck.

Throughout these activities, of course, runs power and privilege. It is clear in the internal hierarchies, insider groups, and pecking order found within the discipline. Many might be called to economic geography, but few are chosen. Those on the inside give plenary talks, or attend invisible college workshops, or have their way paid to teach summer school, or are taken out for very nice lunches and dinners. Then there are the majority of economic geographers who have to pay their own way to conferences, organize their own sessions, have only 15 minutes to present, and go to fast food restaurants for lunch. Partly it is generational, about putting in one's time. Partly it is about social characteristics. And partly it is about formal and informal yardsticks of success as measured by quality and quantity of publications, reliability, and in-person performance.

7.3.3 Immutable mobiles: economic geography's journals and textbooks

If economic geography produces anything tangible it is books and journal articles. On the surface, they appear as neutral media, simply the materialization of economic geographic thought. There is nothing simple about them, however. Books and journals enact complex performances, helping to keep economic geography up and running. Latour coins a special name for them, "immutable mobiles" (Latour 1990). They are mobile in that they easily travel, but are immutable in that the distance they travel does not physically corrupt the inscriptions they contain.

Immutable mobiles contribute to maintaining economic geography in at least three ways. First, they facilitate connections among geographically dispersed members of the discipline, cementing their relation, but also reaching out to non-members, bringing them into the fold, enlarging the community. Since the 1990s, with the globalization of economic geography, the role of immutable mobiles in making distant connections has been ever-more important. Partly it is facilitated by electronic publication. With a few clicks of a mouse or with only one's thumb or index finger, one is able potentially to read almost anything anywhere on whatever device one has at hand – smart phone, tablet, laptop, or stand-alone computer. It is that easy. Well, maybe not that easy. Here being critical is again important. For one thing, the distribution of the internet is highly uneven geographically, centered primarily on North America, Western Europe, Australasia, and parts of Asia. There is also a similar unequal geography of online access to academic journals and books, only possible if a university library subscribes. Moreover, prices of journal subscriptions have climbed enormously over the past two decades ("the serials crisis"[6]). Consequently, one gains access only if the university one is attending is wealthy.

There is also another issue around journals. The vast majority are written in English, reflecting the fact that English has become the *lingua franca* of the global academy. *Lingua franca* was originally a common hybrid language used by traders around the Mediterranean (see also Chapter 4 on trading zones). As the academy has globalized, rather than developing a similar hybrid language, a single pre-existing language has come to dominate, to become hegemonic: English. It means that native speakers have a significant advantage in their ability to publish (the currency of the academy). At the same time, it also creates a marginalized group that doesn't write in English, and necessarily will be on the outside looking in (for discussion of how this plays out in economic geography see Hassink and Gong 2016).

Second, an immutable mobile can become a point of crystallization for a much larger project, shaping the field, bringing people together, providing a disciplinary agenda. Latour (1987, p.159) calls such a text an "obligatory passage point," something to which you must refer in discussing a particular topic because those books or papers have made that topic theirs. For example, if you discuss neoliberalism in economic geography then you must make reference to Peck and Tickell's 2002 paper "Neoliberalizing Space" (it is coming up to 3,500 citations). Or if you want to discuss the economic geography of globalization then you must reference Peter Dicken's (1986–2015) textbook *Global Shift* (now in its 7th edition with collectively over 10,000 citations). If you do not reference these obligatory passage points, you will either not be taken seriously, or your own contribution will never make it to print because it will be rejected by referees (anonymous reviewers contacted by editors to vet submitted manuscripts). Referees will say your work is unsatisfactory because it has not reviewed properly the literature. How can you write a paper about neoliberalism without citing the formative contribution of Peck and Tickell? How can you write about globalization without referencing a founding text of the field, Dicken's?

Finally, immutable mobiles show the ability of the discipline to knit together contributions from outside the field, demonstrating in this case that economic geography is a potential ally to other disciplines, making it more robust and secure. For example, in his classic economic geography textbook *Locational Analysis in Human Geography* (see below) Peter Haggett (1965) was brilliant in making strong alliances with other authors. On two facing pages, pp.118–119 (Figure 7.3), Haggett brought together: empirical studies of the Iowan urban system by Brian Berry; theoretical studies of economic location completed by August Lösch; the statistical techniques of correlation and regression invented in late nineteenth-century Britain by the eugenicist Francis Galton; and elementary packing theory initially devised by the astronomer Johannes Kepler in the seventeenth century. Everyone was brought to the same page, made to work together. It showed the strength of

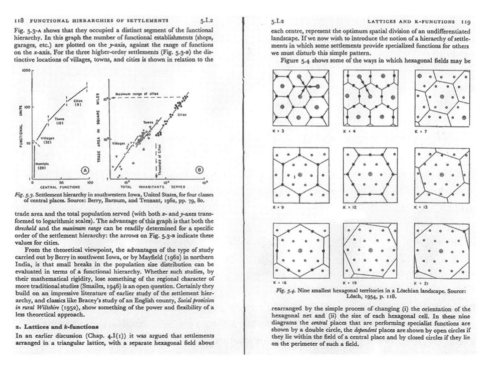

Figure 7.3 Peter Haggett's textbook enrolled the works of both geographers and non-geographers to make his arguments stronger and more persuasive. Haggett 1965, pp.118–119.

the field by the number of connections, but it also meant that if you wanted to argue against Haggett, you needed to take on his allies: Berry, Lösch, Galton, and Kepler. As individuals, each is impressive. As a group, they are daunting.

Journals of economic geography

To legitimate academically a discipline like economic geography it is necessary to have specialized journals. Apart from being a forum in which research in and debates about the field are published and disseminated, they also represent academic turf grabbing. They are a claim on a particular intellectual territory. The launching of a new journal is a declaration that the territory it covers now constitutes an academic field. Don't believe me? Here's the proof: a hardbound version of volume 1 of the journal.

The first journal of economic geography, unsurprisingly called *Economic Geography*, was published in 1925. On its very first page, the journal asserted the distinctiveness of the discipline as an intellectual pursuit, and asserted itself as a necessary outlet for its propagation.

> The need for a full knowledge of the natural resources of the world, and a better understanding of the natural conditions to which man must the more carefully adapt … has only recently made itself felt; it is from this need that the new interest in economic geography has sprung …. Since no journal occuples this field of scientific inquiry in America or abroad, Clark University will publish ECONOMIC GEOGRAPHY.

The journal was claiming a piece of intellectual territory not yet staked. Furthermore, in Kuhnian terms, the journal was an important stabilizing point to anchor everyday practices of the new discipline. It was a journal to read, to publish in, to assign readings from for students, and to cite in one's own work. It was a paradigmatic element, a "concrete exemplar," to reinforce practices of normal science within economic geography.

No further new journals in economic geography were launched until after World War II. When they came, they were associated with a different discipline, regional science (Chapter 3 and Box 4.1: The Rise and Fall of Regional Science). Regional science was the prelude to economic geography's phase of spatial science. Because spatial scientists were in effect shut out of *Economic Geography*, which did not publish quantitative-theoretical papers (Murphy 1979, p.42), analytically minded economic geographers instead placed their work in one of two regional science journals: *Papers and Proceedings of Regional Science* (first published in 1955) and the *Journal of Regional Science* (first published in 1958). By the late 1960s, however, two new journals based wholly within geography were established to publish spatial science, and a lot of economic geography too: the UK-based *Environment and Planning* (later *Environment and Planning A*) and the American-based *Geographical Analysis* edited at Ohio State University, one of the early sites of the quantitative revolution.

It was another long wait, more than 30 years, before the next journal appeared, although this one was exclusively devoted to economic geography, *Journal of Economic Geography* (first published in 2001). Initially jointly edited by an economist, Richard Arnott, and an economic geographer, Neil Wrigley, the idea was to bring economics and economic geography literally on to the same page by fostering dialogue. That larger end has never really been realized, but the current editors, and the editorial board, remain vigorously multidisciplinary, the journal energetic and expansionary (Box 7.3: Two Journals: *Economic Geography* and *Journal of Economic Geography*).

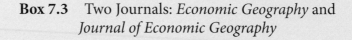

Box 7.3   Two Journals: *Economic Geography* and
Journal of Economic Geography

Two journals: *Economic Geography* and *Journal of Economic Geography*.
Source: Trevor J. Barnes.

Currently two journals act as economic geography's flagships, carrying the banner of disciplinary distinctiveness. The older is the established, more conservative *Economic Geography*. The younger upstart, but which quickly became part of the establishment, is *Journal of Economic Geography*. Both publish works defined as economic geography, although they each have different definitions of the field. Those definitions have themselves changed, though, reshaped in part by papers carried in those two very outlets.

In 1925, the Graduate School of Geography at Clark University, Worcester, Massachusetts, inaugurated the first journal of economic geography, which it continues to own. The then newly hired president, Wallace Atwood (1921–1949), an active geographical educator, accepted the job at Clark on the conditions that a School of Geography be established, and that a new journal, *Economic Geography*, be established with him as editor. Early faculty hired at the School worked on the relationship between primary commodities and the physical environment, determining the journal's initial definition of economic geography (Murphy 1979). Since then, the journal has broadened. But while no longer fixed on natural resources, articles in the journal tend to conform at least to the stylistic template of theoretically informed case studies, which, while they may include quantitative analysis, tend to rely on narrative prose. The journal rarely publishes purely theoretical accounts or technical, mathematical expositions (although there have been exceptions). The journal's website says: "Our long-standing specialization is to publish the best theoretically-based empirical articles that deepen the understanding of significant economic geography issues around the world."

In 2001, a second serial broke the monopoly, *Journal of Economic Geography*. It was a response to the advent of the new economic geography (or "Geographical Economics") pioneered by economists during the 1990s, prominently Paul Krugman (see Chapters 2, 4, and 5). The year before, an important edited collection had appeared, *The Oxford Handbook of Economic Geography* (2000), jointly edited by two economic geographers (Gordon Clark and Meric Gertler) and an economist (Maryann Feldman). That volume paired articles by economic geographers with economists to spur discussion between the two disciplines. The hope was a new version of economic geography where economic geographers and economists were on an equal footing, mutually learning from interaction. The next year, the journal appeared with the same goal: "The aims of *Journal of Economic Geography* are to redefine and reinvigorate the intersection between economics and geography, and to provide a world-class journal in the field" (*Journal of Economic Geography*, homepage, 2017).[7] The journal never achieved the first end. As in the original *Handbook*, economists and economic geographers mostly stayed with their own "tribe." But the journal has become a lively site, less conservative then *Economic Geography*, more interested in pure theory (albeit of the economist's kind), more quantitative, with economists publishing their work at least cheek by jowl if not in direct conversation with economic geographers.

Journals, of course, are not only venues for disseminating academic knowledge, but play a crucial role in the internal sociology of a discipline, with the potential to both make and destroy careers. Most academics, after they are initially hired, will go through a probationary period, called tenure in North America (usually six years). Successful tenure means a university job for life, but tenure denial is the end of the academic road, dismissal, with little prospect of another job (you are damaged goods). Receiving tenure most often depends on publishing ("publish or perish"), and the award of research grants (discussed below). For economic geographers facing tenure, publishing in journals is critical (peer-reviewed papers are the tenure gold standard). Consequently, journal editors, along with the reviewers they choose, are enormously powerful. Further, editors of the most important journals in a field, like *Economic Geography* (a "flagship journal"), are the most powerful of all, given that many departments insist that faculty necessarily publish in them not only for tenure but later for promotion.

Textbooks of economic geography
Economic geography textbooks are another kind of immutable mobile, albeit serving very particular purposes. They introduce and acculturate new students to the field (which is what we are aiming to do as you read these very words); they are repositories of Kuhn's (1970) "concrete exemplars," descriptions and worked-through examples of classic theories, concepts, and case studies on which disciplinary practice, craft, and judgment rest; they are

deposits of disciplinary memory; they are catalysts for creative new projects for the next generation of economic geographers (like some of you); they are objects to give to outsiders to show them what we do, and to persuade them of a field's worth; they are a means to forge allies both inside the discipline and outside, including funding bodies, government bureaucracies, even Royal Commissions. While a large textbook may appear intimidating, its purpose is the reverse, to welcome and encourage, to engage, to persuade, and to gather up.

The first textbook in economic geography in the English language was George G. Chisholm's (1889) *Handbook of Commercial Geography* (see Chapter 3). It created the discipline, giving it substance. With the first ever such textbook you could now hold the new discipline in your hands, and literally pass that discipline around to interested others. From 1896, it reached the briefcases of fresh-eyed university and extension students at Birkbeck College, University of London, where Chisholm first used his *Handbook* as a text for a full-year course of lectures. It was also passed on to education boards, and in the United Kingdom, even to a Royal Commission that determined school and university curricula. It further made its way to other academics, showing that there was a fresh discipline on the block. And it went to professors and students, sometimes far afield, who wanted to practice the new discipline themselves. The American geographer J. Russell Smith used the *Handbook* to argue for the importance of establishing a separate Department of Geography and Industry at the Wharton School. It was created in 1906 with Smith as founding Head. Then in 1913, Smith (1913) used the *Handbook* as a template for his own textbook, *Industrial and Commercial Geography*, which was widely picked up as a text for the new discipline among US colleges offering courses in economic geography (Fellmann 1986). The *Handbook* was a powerful volume.

By the end of World War I, partly as a result of Chisholm's and Smith's textbooks, economic geography was up and running in Anglo-American universities. Already the perspective was shifting, tilting toward the regional perspective, and requiring a different textbook, although a textbook nonetheless. Such textbooks were no longer about imperial global production and the worldwide movement of commodities, but the peculiarities of unique domestic regions, represented using stock typologies into which the economic geographic facts of a region were meticulously sorted. Being an economic geographer now meant learning the typology, and all the facts under each of its categories. Those typologies and facts were found, for example, in Ray Whitbeck and Vernor Finch's (1924) textbook, *Economic Geography*, which used a fourfold classification grid. Clarence Jones's (1935) *Economic Geography* at Clark pushed the boat out farther, providing an expansive eightfold classification. It was Jones's book that caught on, becoming "the standard introductory text on the subject for thousands of college students" (Hudson 1993, p.167).

By the late 1950s something completely different had taken over the discipline, but still requiring a textbook: spatial science. That need was brilliantly filled by Peter Haggett's (1965) primer on geographical theory, modeling, and statistical techniques, *Locational Analysis in Human Geography*. The book was based on "much thumbed and much revised lecture notes" Haggett (1965, p.v) used to teach third-year undergraduate students at Cambridge University on a Saturday morning. Bright and eager, those students would not have been exposed to a theoretical or mathematical geography before Haggett's class. But they were a perfect audience to try out, sharpen, refine, and clarify Haggett's new approach. Like Jones's volume, Haggett's sold like hot cakes, shifting an unprecedented 40,000 copies, becoming an "obligatory passage point," setting out an agenda for the wider discipline. Furthermore, Haggett prepared the ground for his book from the mid-1960s by organizing lectures framing economic geography as spatial science for British sixth-form (late high school) geography school teachers. Consequently, when their students went on to university to take geography they expected to read Haggett. The book was also stuffed with Kuhnian "concrete exemplars" to facilitate the new normal science of economic geography. Most frequently represented by the miniature diagrams that littered its pages (there were162 numbered illustrations in 310 pages of text), each figure showed with great economy and clarity the exemplary virtue of some spatial theory, model, or statistical technique.

Finally, Peter Dicken's extraordinarily successful contemporary textbook *Global Shift* (the publisher reports sales of over 100,000 for the seven editions, from 1986 to 2015) in some ways returned the discipline to Chisholm. Just like the *Handbook*, *Global Shift* showed the relevance of economic geography for understanding the whole world. Hirst and Thompson (1996) have argued that the late nineteenth century experienced globalization comparable to that of the late twentieth century. So, it is not surprising that the two volumes have similarities. They are not identical, however. Dicken eschews Chisholm's environmentalism for an explanatory framework based on the triad of capital (multinational corporations), state regulation (or not), and technological change. *Global Shift*, though, is perhaps even more crammed with facts than the *Handbook*. They leak from the book's pores as lists, tables, diagrams, bar graphs, case studies, text boxes, and maps. It is said that to be a fact is to be a disappointment. But for Dicken (and Chisholm) that is not true for facts about globalization. For him they are vital and dynamic, the fundamental bases for understanding (and explaining why there have been seven editions of *Global Shift* in 29 years, although still a long way shy of Chisholm's 24 editions of the *Handbook* in 93 years). Dicken's genius was to recognize early on both that globalization is *the* story of the late twentieth and early twenty-first centuries, and that economic geographers who had not written on the topic for some time possessed theories, techniques, and tools that were critical to representing and understanding it. Their discipline had special purchase. Globalization and economic geography only needed to be brought together. The person who did that would be a global winner. Dicken did and was. His book became what he wrote about, a global commodity, used from Shanghai to Sheffield, and economic geography – which it expressed and explicated and about which it enthused – did the same.

From their covers, textbooks might seem dull, "chloroform in print" as Mark Twain once quipped. But they perform crucial disciplinary work. That might not appear to be the case. Even once they are pulled from your backpack, sitting flat on the library table, or in your hands while you are propped up in bed, nothing much seems to happen when you open one up. You turn the page, you casually highlight a sentence or two with a yellow magic marker, you scribble a desultory marginal note. As you do these things, though, something is happening. You are being disciplined – disciplined within the discipline of economic geography – and this is the ultimate aim.

Machines and economic geography

Compared to other disciplines, especially those in the natural sciences, economic geography has not been a big user of machines to produce knowledge. Of course, early economic geographers like Chisholm and Smith were deeply concerned with machines, interested not only in individual pieces of equipment, but in the larger genus that by the late nineteenth century created the object of their new study, economic geography. Writing in the late nineteenth and early twentieth centuries, Chisholm and Smith both thought latent Promethean impulses were at last unshackled, breaking loose of old geographical constraints on the production and movement of goods. Innovations in transportation and communications – the railways, the steamship, the internal combustion engine, the telegraph, the telephone – allowed for a hitherto unachieved global integration of commodity production, circulation, and exchange, finally circumscribing nature's historical niggardliness.

In representing machines and their geographical effects, the early economic geographers also used at least some machines themselves. Chisholm's book has no photographs, but Smith's is full of them. Many were gleaned from US government sources, but a significant number were taken by Smith himself with his own camera as he traveled the sites of industrial and commercial geography. Likely he used a Kodak Brownie camera, first introduced in 1900, which for the first time allowed non-professionals to take their own photos. Chisholm's book is notable for its pull-out, two-tone, multi-patterned regional transportation and commercial maps individually glued between folio pages. Those maps were not drawn by Chisholm but by a colleague, F.S. Walker, FRGS (Fellow of the Royal Geographical Society). Accordingly, it was Walker who deployed the machinery of map-making that especially during the second half of the nineteenth century was radically improved as a result of technological change (Pearson 1983 provides a detailed inventory). It was now much easier to print different color tones, incorporate varied map markings, and change fonts and type sizes, all features found in the *Handbook*'s maps.

Machines especially came into their own during the postwar spatial science version of economic geography, with its hallmarks of numbers and calculations. At least initially, the numbers were limited, with calculations relatively simple. A slide rule was sufficient. Better still, though, was a mechanical calculator. There were manual ones such as the Monroe, although increasingly they were supplanted during the 1950s by electrical versions such as the Marchant or the Friden. Dedicated rooms in university buildings were devoted to calculators. The situation wasn't static, though, with the amount of numerical data and the complexity of calculations increasing exponentially. By the late 1950s, even the Marchant and the Friden couldn't cut it any more. More machine help was necessary. Fortunately, precisely at that moment help became available. The first commercially sold computer in North America was the IBM 650, launched in 1954, bought by Columbia University (Figure 7.4). Other universities quickly followed, including the University of Washington and the University of Iowa, which both had geography departments beginning to travel down the road of spatial science. There was no formal training for those early spatial scientists in using the computer, though. It was "bootstrap operations," learning by doing. There was no mouse, no on-screen menu, in fact, no screen, and at first, not even a formal programming language. Economic geographers using early computers programmed them by plugging in and taking out electrodes on a circuit panel. There was no cloud to which to send your work; there was not a hard disk or even a floppy disk where you could save it. Waldo Tobler (1998), a student at the University of Washington at Seattle, which got its first computer in 1955, remembers:

> We had to go up to the attic of the Chemistry building at 2 am so we could run the computer by ourselves. They didn't have any computer operators in those days, and that was before computer languages like FORTRAN. … To cover programming on the 650 you had to pick up two bytes of information on one rotation of the drum. It had a 2K memory which rotated real fast. And if you were clever, you could pick up two pieces of information in one rotation.

Figure 7.4 IBM 650 computer at Texas A&M University, College Station, Texas, likely in 1950s. Source: By Cushing Memorial Library and Archives, Texas A&M (Flickr: IBM Processing Machine) [CC BY 2.0 (http://creativecommons.org/licenses/by/2.0)], via Wikimedia Commons.

In comparison to present computing capacity, these early machines and procedures were like lumbering dinosaurs. But in terms of the culture of the time, they were rocket science, or at least, spatial science. Moreover, machines and operating procedures transformed rapidly, very quickly becoming *de rigueur* within the new culture of economic geography, spatial science. Haggett's (1965, p.248) *Locational Analysis*, for example, extolled the virtues of "high-speed computers," using them to derive specific maps and numerical results in his own text. Those maps and results would never have been produced without those machines. They were not just lifeless instruments (tangles of wires, joining switches, buttons, and flashing lights), but separate and independently contributing actors. How a computer was designed, what it could do, made a difference to the knowledge that spatial scientists generated, thus affecting the very character of economic geography as a discipline. They were powerful shapers of what knowledge could be.

The computer was perhaps the most important of the machines for spatial science but there were others, admittedly less sophisticated, enjoying at least walk-on roles. The lowly slide rule was an early performer, and later the line printer, and the Xerox machine. Key at the University of Washington during the late 1950s was another mechanical device, the duplicator, or spirit duplicator. By typing on specially coated paper, fixing that sheet on a rotating drum, and cranking a handle it was possible to produce multiple "mimeographed" mauve-colored versions of the original each with the distinct aroma of methanol (one of the chemicals used as ink). The Chair of the Washington Geography Department in the mid-1950s, Donald Hudson, gave the graduate students unlimited access to paper and ink, and the unlimited use of a duplicator outside office hours. It was used initially for mimeographing internal discussion papers, but later as the means to reproduce a departmental discussion paper series. Those purple mimeographs were sent around the world, and were instrumental in forging spatial science in other places.

Machines used by economic geographers have multiplied, becoming ever-more sophisticated. Smart phones that we carry around in our pockets are incomparably more powerful and capacious than the IBM 650, which was the size of a large living room. The point, though, is that in both cases we often forget about the machines. We think economic geography consists of only ideas. Machines are seen merely as means, intermediaries, to arrive at those ideas. Consequently, they can be discarded, forgotten, put back into a box and the lid closed. In contrast, our argument is that machines have shaped, and continue to shape, the discipline's knowledge. They are not neutral intermediaries, but active agents. Among other tasks, unboxing economic geography requires examining the discipline from its machines outward.

7.3.4   Money makes the world of economic geography go around

One final element, and bearing on all others, is money. There is an important economic geography standing behind economic geography. Money makes the world go around. The central state has always been a crucial provider of funds, including its military and intelligence arms. On occasion, especially in the United States after World War II, private philanthropies or foundations were also crucial, although on occasion they received some of their money from the state – the CIA laundered large sums of money for university-based area studies programmes through big foundations (Solovey 2013).

It is clear that money talks. The issue is whether it can talk back to power. There is a Russian proverb that says, when money talks, truth is silent. Can research money be used to shape academic knowledge, to make it reflect the political interests of the funders? The historian of American postwar physics Paul Forman thinks it does. In a now classic study, Forman (1989, p.150) showed that between 1940 and 1960 military research funding caused "a qualitative change in [the] purpose and character" of American physics. American physicists changed their research agenda to meet military needs; the kind of knowledge physics as a discipline produced was radically different. Consequently, whenever one takes off the lid of the black box of academic research, as we are doing here, one must always "follow the money."[8]

Both the British Royal Geographical Society (RGS) (est. London, 1830) and the American Geographical Society (AGS) (est. New York, 1851), both of which sponsored early research in geography, drew upon a mixture of private philanthropy and state funds (the RGS was granted a Royal Charter by Queen Victoria in 1859). The research, such as it was, clearly linked into and extended the colonial project by contributing to "expeditions" to the "dark continent," as well as to central Asia and to the Antarctic.

World War II was a particularly important time for the social sciences, especially in America. For the first time, they were systematically integrated and deployed within the military to assist in the war effort. Wartime research was team-based, well-resourced, and directed at one instrumental problem or another. That then became the model for social science research funding after World War II ended, especially in the United States. The US military, through the Office of Naval Research (ONR), created in 1946, funded much social science research through the late 1950s. There was a special ONR Geography Branch that underwrote both physical and human geography, including the spatial science version of economic geography (Chapter 3). The geography departments at both the University of Washington and the University of Iowa, the two earliest centers of spatial science, received crucial ONR funding to support research and graduate students. The other providers of research money in the early postwar period were large private foundations, the three biggest of which were Carnegie, Rockefeller, and Ford. Ford was twice the size of Carnegie and Rockefeller combined, spending significant amounts of research money on academic projects promoting "modernization" of the so-called underdeveloped world. Although formally separate from the state, there is a lot of evidence that these private foundations functioned as extensions of the state, promoting its concerns, backing academic projects including those by economic geographers at the University of Chicago and Northwestern University that produced knowledge that met its interests (Solovey 2013).

In 1958, economic geography became eligible for funding from the US federally financed National Science Foundation (NSF), inaugurated in 1950 but until 1958 restricted to the natural sciences. In the United Kingdom, it wasn't until 1965 that the Social Science Research Council was founded, becoming in 1983 the Economic and Social Research Council (ESRC). Of course, there were other sources of research monies available in both countries for economic geographic research, but the NSF and ESRC represented bulwark research-funding institutions. Other countries have similar national funding agencies providing money for academic research. The larger point: money is never innocent. There is always a politics to be unpacked, a set of power relations to interpret critically, from decisions about individual research applications to government-set priorities for funding.

7.4 Things Fall Apart

It should by now be clear that academic disciplines such as economic geography do not exist and subsist as simple, unitary phenomena. They are made up of multifarious, heterogeneous elements. And for survival to be possible, these elements must continue to interact more or less harmoniously. Take away some of the crucial building blocks, or substantially realign them vis-à-vis one another, and there is a risk of disassembly. Inventing a discipline requires the stars to be aligned, allowing the necessary elements constituting a discipline to come seamlessly together. Under altered conditions, though, things can just as readily fall apart. We discussed just such a falling apart in Box 4.1, the case of regional science.

Something similar may be in the process of happening to economic geography in the United Kingdom, a truly alarming development. The United Kingdom has historically been one of the discipline's strongholds, at least since the 1970s. Much of the most impressive and influential work in economic geography of the past four decades, and which we highlight in this book, has been carried out at UK universities: the work, *inter alia*, of Ash Amin, Peter Dicken, Ray Hudson, Roger Lee, Andrew Leyshon, Doreen Massey, Linda McDowell, Andrew Sayer, and Nigel Thrift. Furthermore, some of the most influential economic geographers now plying their trade elsewhere – David Harvey and Jamie Peck are two – received their early-career training in the United Kingdom.

Today, however, the discipline has become something of an endangered species. Why? Several factors seem to be in play, as ongoing research by James Faulconbridge and colleagues suggests.[9] For some time now, cultural geography has for one reason or another been perceived in the United Kingdom as a more attractive or "sexier" specialization. This perception seems especially pronounced at the stage of the career cycle where a discipline is ultimately reproduced, namely the graduate student phase. To be sure, economic geography had its *own* "cultural turn" (Chapter 4); but the cultural turn of human geography more broadly, and especially in the United Kingdom, has indubitably benefited UK cultural geography and disbenefited UK economic geography.

A striking manifestation of this transformation has been a veritable exodus of trained UK economic geographers, at all levels from the postdoctoral to professorial, from university geography departments to business and management schools. Faulconbridge et al. (2016) have identified 87 individual movers, of whom 92% have shifted to a different discipline since 2000. In the context of a small (sub)discipline, this is a staggering number. Both "push" and "pull" factors are clearly in play. Economic geographers increasingly feel unloved within geography, their specialization often denigrated by cultural geographers. By contrast, business schools have actively recruited disaffected economic geographers, recognizing their merits, and typically offering better pay and work conditions. Those economic geographers who have made the move mostly claim to have retained their identity as economic geographers; but they also say they feel less affinity for the discipline, attend fewer (economic) geography conferences, and increasingly publish in other outlets.

All of this represents an existential threat to the reproducibility of the discipline in the United Kingdom, as the arguments we have made in this chapter should make clear. Who is going to train the United Kingdom's future economic geographers if those "doing" economic geography in the United Kingdom are no longer in geography departments and are, by consequence, doing it – and living it – less and less with the passing of time? Faulconbridge et al.'s research suggests a vicious cycle may already be setting in, with geography departmental heads and university administrators reluctant to create new jobs in economic geography for fear of losing new hires to business schools. And the word from the Summer Institute in Economic Geography is that the number of qualified early-career applicants from the United Kingdom has been in freefall for the past decade. No wonder: with senior role models fleeing to nominally greener pastures, the incentives to become an economic geographer in the first place probably appear much less compelling than they once were.

As economic geography partisans (Chapter 1), we sincerely hope that our concerns for the future of economic geography in the United Kingdom are overstated. And of course, through vision, investment, and energy, this prevailing trend can be reversed. But reversing it requires knowing the various constituent elements of the discipline, those typically hidden within the black box, and knowing how to succor them and begin to stitch them back together. It is not likely to be an easy task.

7.5   Conclusion

The purpose of this chapter was to do what the studio audience bellowed to contestants on *Take Your Pick!* – "Open the box!" In this case, we were concerned to open the black box of economic geography. That the discipline was black boxed to begin with goes to the strong strain of rationalism that infects much Western academic inquiry, and economic geography in particular. As we argued, rationalism has no truck with the body, with power and social institutions, with values and cultural practices, with money, with even material objects in the world. They are all a sideshow, likely to lead people astray, to pervert knowledge. The main event, the only event, for rationalism in the production of real knowledge is logic, reason, and brain power. They are the only bases for correct disciplinary knowledge. Consequently, rationalist accounts of disciplines such as economic geography erase, cover up, and leave out processes of knowledge production that do not fit. This chapter, in contrast, has been about recouping those missing elements, reintegrating them, unboxing economic geography.

The works of Thomas Kuhn and those in science studies, especially Bruno Latour, are an important antidote to the rationalist (black box) rendering of academic disciplines. We used both these bodies of work to identify elements excluded in orthodox rationalist accounts of knowledge, but which under anti-rationalism are vital to include. We focused on bodies, invisible colleges, immutable mobiles, machines, and money, stressing both the productive and the destructive effects of power as it flows though, among, and over these elements. Of course, this is only a start, only a partial list. Even so, it begins to show what might be included. There is a famous line in Shakespeare's *Hamlet*, "There are more things in heaven and earth, Horatio, than are dreamt of in your philosophy." Shakespeare was probably not thinking of rationalism when he wrote that line (although he might have been). But it does seem a good opening position, a recognition that disciplines are complex entities, the product of multiplicity rather than singularity, throughout which power continually courses. That's why we wanted to open the box, to see all those things in heaven and earth that are relevant to realizing the dreams of economic geography.

Notes

1 "Standing on the shoulders of Giants" is inscribed in tiny letters around the edge of the British £2 coin. That inscription celebrates Newton both as the United Kingdom's most famous scientist and as a holder, in later life, of the position of Master of the Royal Mint that issues Britain's bank notes and coins.

2 Katz and Monk (1993) and Maddrell (2007) discuss historic and contemporary discriminatory practices bearing on female academics in geography, producing a glass ceiling that continues to constrain salaries and opportunities for promotion.

3 http://www.gceg2018.com/home.html (accessed July 5, 2017).

4 http://gpn.nus.edu.sg/conference2017.html (accessed July 5, 2017).

5 All quotations in this paragraph and the next are taken from the website for the Summer Institute in Economic Geography; http://www.econgeog.net/(accessed July 5, 2017).

6 There is even a Wikipedia entry, "Serials crisis"; https://en.wikipedia.org/wiki/Serials_crisis (accessed July 5, 2017).

7 https://academic.oup.com/joeg/pages/About (accessed July 5, 2017).

8 "Follow the money" was the famous imperative of "Deep Throat," who anonymously gave advice to the two *Washington Post* reporters Carl Bernstein and Bob Woodward as they took the lid off what became known as the Watergate scandal leading to the 1974 resignation of President Richard M. Nixon. "Deep Throat" was eventually revealed as Mark Felt, a special FBI agent based in Washington, DC.

9 We rely primarily on a recent conference presentation: Faulconbridge et al. (2016).

References

Barnes, T.J. 2012. A Brief Cultural History of Economic Geography: Bodies, Books, Machines and Places. In B. Warf (ed.), *Encounters and Engagements between Economic and Cultural Geography*. Dordrecht: Springer, pp.19–37.

Chisholm, G.G. 1889. *Handbook of Commercial Geography*. London and New York: Longman, Green, and Co.

Clark, G.L., Feldman, M., and Gertler, M.S. 2000. *The Oxford Handbook of Economic Geography*. Oxford: Oxford University Press.

Crane, D. 1972. *Invisible Colleges: Diffusion of Knowledge in Scientific Communities*. Chicago: Chicago University Press.

Deutsche, R. 1991. Boys Town. *Environment and Planning D* 9: 5–30.

Dicken, P. 1986–2015. *Global Shift*. 7 editions. New York: Guilford Press.

Faulconbridge, J., James, A., Bradshaw, M., and Coe, N. 2016. In the Business of Economic Geography: Trends and Implications of the Movement of Economic Geographers to Business and Management Schools in the UK. Paper presented at the RGS-IBG Annual International Conference, London, August.

Fellmann, J.D. 1986. Myth and Reality in the Origin of American Economic Geography. *Annals of the Association of American Geographers* 76: 313–330.

Forman, P. 1989. Behind Quantum Electronics: National Security as Basis for Physical Research in the United States, 1940–60. *Historical Studies in the Physical Sciences* 18: 149–229.

Gleick, J. 2003. *Isaac Newton*. New York: Pantheon.

Haggett, P. 1965. *Locational Analysis in Human Geography*. London: Edward Arnold.

Haraway, D.J. 1997. *Modest_Witness@Second_Millenium.Femaleman© Meets_Oncomouse™*. London: Routledge.

Hassink, R., and Gong, H. 2016. Towards an Integrative Paradigm of Economic Geography. Unpublished paper, University of Kiel.

Hirst, P., and Thompson, G. 1996. *Globalization in Question: The International Economy and the Possibilities of Governance*. Cambridge: Polity.

Hudson, J.C. 1993. In Memoriam: Clarence Fielden Jones, 1893–1991. *Annals of the Association of American Geographers* 83: 167–172.

Jones, C.F. 1935. *Economic Geography*. New York: Henry Holt and Company.

Katz, C., and Monk, J. 1993. *Full Circles: Geographies of Women over the Life Course*. London: Routledge.

Keynes, J.M. 1951. *Essays in Biography*. New York: Horizon Press.

Kuhn, T.S. 1970 (1962). *The Structure of Scientific Revolutions*, 2nd edn. Chicago: University of Chicago Press.

Latour, B. 1987. *Science in Action: How to Follow Scientists and Engineers Through Society*. Cambridge, MA: Harvard University Press.

Latour, B. 1990. Drawing Things Together. In M. Lynch and S. Woolgar (eds), *Representation in Scientific Practice*. Cambridge: Cambridge University Press, pp.19–68.

Latour, B. 1999. *Pandora's Hope: Essays on the Reality of Science Studies*. Cambridge, MA: Harvard University Press.

Maddrell, M. 2007. *Complex Locations: Women's Geographic Work in the UK, 1850–1970*. Oxford: Wiley-Blackwell.

Massey, D. 1984. *Spatial Divisions of Labour*. London: Macmillan.

Murphy, R.E. 1979. *Economic Geography* and Clark University. *Annals of the Association of American Geographers* 69: 39–42.

Peck, J., and Tickell, A. 2002. Neoliberalizing Space. *Antipode* 34: 380–404.

Pearson, K.S. 1983. Mechanization and the Area Symbol/Cartographic Techniques in 19th Century Geographic Journals. *Cartographica* 20: 1–34.

Putnam, H. 1981. *Reason, Truth, and History*. Cambridge: Cambridge University Press.

Rorty, R. 1982. *The Consequences of Pragmatism (Essays 1972–1980)*. Minneapolis: University of Minnesota Press.

Smith, J. Russell. 1913. *Industrial and Commercial Geography*. New York: Henry Holt and Company.

Solovey, M. 2013. *Shaky Foundations: The Politics-Patronage-Social Science Nexus in Cold War America*. New Brunswick, NJ: Rutgers University Press.

Tobler, W. 1998. Interview with Trevor J. Barnes, Santa Barbara, CA, March.

Warntz, W. 1955–1985. Papers, catalogue # 4392, 54 boxes, Kroch Library of Rare Books and Manuscript Collections, Cornell University, Ithaca, NY.

Warntz, W. 1959. *The Geography of Price*. Philadelphia: University of Pennsylvania Press.

Whitbeck, R.H., and Finch, V.C. 1924. *Economic Geography*. New York: McGraw-Hill.

Part II

Doing Critical Economic Geography

Chapter 8

Globalization and Uneven Development

8.1   Introduction

Is there a more obviously economic geographic phenomenon than globalization and the uneven development that characterizes it? Arguably not, and that is why we start Part II with it. Although it clearly has cultural and political dimensions too, at heart globalization is, on most understandings, a matter of the changing geographies of the (global) economy. This perforce makes it a central concern of economic geography. One need not accord it the privileged status it has sometimes been afforded – Daniel MacKinnon and Andrew Cumbers's textbook *An Introduction to Economic Geography* (2007), for example, is subtitled *Globalization, Uneven Development and Place* – to recognize that the economic processes denoted by the term "globalization" are irreducibly geographical and thus demand a geographical approach.

One of the main aims of this chapter is to explicate these geographies and thus to demonstrate the importance of "thinking" globalization geographically. The other principal aim is to argue for the benefits of a *critical* geographical approach. But why explicitly "critical"? There are two reasons.

First, globalization is not some neutral, free-floating, ahistorical phenomenon that occurred by happenstance and is reproduced supernaturally. It is and always has been something *made*, and in its making it is, as geographer Gill Hart (2002, p.12) emphasizes, "deeply infused with the exercise of power," whether this is the power of individuals, governments, corporations, or other powerful actors. Furthermore, globalization *remakes* the power relations that frame and underwrite it, and invariably in socially consequential ways. Critical perspectives on its origins, forms, and effects are therefore essential.

Economic Geography: A Critical Introduction, First Edition. Trevor J. Barnes and Brett Christophers.
© 2018 John Wiley & Sons Ltd. Published 2018 by John Wiley & Sons Ltd.

Second, we need to think about globalization critically because of the significance *to* globalization of existing narratives about it. As innumerable observers have noted, globalization is both a material reality – a set of economic (and cultural and political) processes that have profoundly changed the map of the world – and a "discourse" or narrative – a relatively coherent set of arguments bound together as a story about what globalization "is" or should be. The reality and the discourse of globalization do not necessarily match, however. We will suggest that the dominant narratives of globalization are not just in large part wrong (including about its geography) or, in their normative forms, contestable, but they have socially pernicious effects. That is, discourses of globalization, as Hart (2002, p.12) writes, are pivotal in "actively shaping the very processes they purport to describe" (cf. Herod 2009). This performative aspect – that narratives of globalization help produce globalization – also makes taking a critical approach vital.

The chapter begins with a brief high-level overview of the main historical economic developments to which the term "globalization" typically refers. It discusses how these developments distinguish the recent form of globalization "proper" – normally presumed to begin after the 1970s – from earlier waves of globalization. For globalization is not entirely new (see Chapter 1). It is much older than industrial capitalism, for example, "going on," as David Harvey (2001, pp.24–25) remarks, "since at least 1492 if not before." We even suggested (although slightly tongue in cheek) it might have started with the Ancient Greeks. But industrial capitalism, insofar as it is, in Harvey's words, "addicted to geographical expansion as much as it is addicted to technological change and endless expansion through economic growth," gave globalization its head, especially in the post-World War II era.

In the second section we turn to the question of how globalization, and especially its geographical dimensions, is typically represented, and to the effects of dominant discourses. The third section then takes these conventional discourses to task by drawing on a rich and critical economic geographic literature on globalization to present a very different figuration of its geographies. Finally, in the last section of the chapter, we raise what we earlier (Chapter 5) identified as a crucial component of any critically oriented economic geography: theory development. Given its economic geographic characteristics, how might globalization be productively *theorized*? Among the most powerful possibilities, we argue, are those contained in a literature that many observers believe closely fits globalization, on uneven geographical development.

8.2   Economic Globalization: The Basics

There is no single identifiable "thing" or bundle of things called globalization, still less any agreement on what that thing might be. The concept of globalization is used to invoke all manner of different dynamics and developments. These different usages, moreover, frequently abrade and sometimes even contradict one another. And yet, we suggest, part of the reason why the term has persisted since it forcefully entered the academic, political, and popular lexicon in the 1980s is that recent decades *have* seen a core set of global-scale economic geographic transformations that meaningfully distinguish them from prior eras.

What are the core developments? Economic globalization is ultimately about economic integration and interconnectivity at expanding geographic scales (Chapter 1). Both elements of this basic formulation are key. Not only have different places and their economic attributes and fortunes become more intimately bound up with one another, but the scale at which such integration has occurred has increased. These processes of so-called globalization have proceeded in the postwar era to the extent that more than a decade ago, Doreen Massey (2002, p.295) felt in a position to assert that "the whole planet is, in one way or another, implicated in the daily lives of each of us." This in a nutshell is globalization. And here, more expansively, but alluding to the same basic point, is another geographer of globalization, Philip Kelly (2000, p.3):

> [Globalization] is found in the flows of capital between the world's financial centres ("hot money"), and in flows of foreign direct investment. It is found in the transfer of goods and services across global space – from Coca-Cola to coconuts and from microchips to missiles. It is found in the passage of people between places – tourists, refugees, migrants, and contract workers. It is found in the ideas and information that pass freely, or sometimes not so freely, across space…

Precisely because understandings and uses of the term "globalization" begin to diverge almost as soon as one ventures beyond such elementary formulations, and because such divergences will be discussed later, we resist digging much deeper here. Three brief observations are worth making, however.

First, if globalization has a single trademark feature that is associated with the term (and the era it denotes) more than anything else, that feature is extensive international trade and the processes constituting it. Such processes are not only quintessentially globalizing but also, in economic geographer Eric Sheppard's (2012, p.45) words, "quintessentially geographical." A vast range of statistics can be and have been called upon to demonstrate the exponential growth in international trade brought about by postwar globalization. Agnew and Corbridge (1995, p.171), for instance, cite a stunning fifteenfold increase in trade between high-income countries between 1950 and 1990. Figure 8.1 charts overall

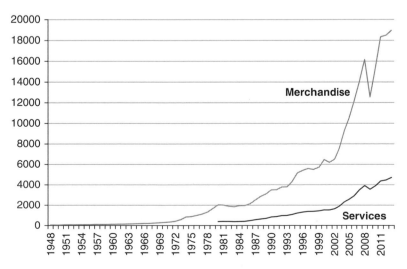

Figure 8.1   Global merchandise and services exports ($US billions, current prices). Source: World Trade Organization.

postwar growth in the value of global exports of merchandise (i.e., goods) and, from 1980, services. It also highlights the crucial fact that such growth occurred from an extremely low base. During the Great Depression of the 1930s and then during World War II, global trade had largely dried up as protectionism took hold and formerly trading nations retreated into their domestic economic shells. Subsequent fifteenfold increases in trading volumes and the like must be understood within this historic context.

And then there is the historic story about trade in services hinted at by Figure 8.1. That no services trade data are available before 1980 is not incidental. Until then, most economists did not even consider services to be "tradable" phenomena; how can you export a haircut, or a carwash? Hence, no trade data. But that view began to change from the late 1970s as trade in financial services in particular boomed. For example, a bank in New York might provide currency exchange support to a company in Taiwan. Further, when what is arguably globalization's trademark institution, the World Trade Organization (WTO), was founded in 1995, it incorporated a new treaty on services trade (the General Agreement on Trade in Services, or GATS) alongside an existing multilateral treaty on goods trade (the General Agreement on Tariffs and Trade, or GATT), which had originally been negotiated in 1947 (Box 8.1: World Trade Organization).

Second, if globalization's trademark public institution is the WTO, its trademark generic private institution is undoubtedly the multinational or transnational corporation (hereafter "MNC"). Organizations that own or control the production of goods or services in one or more countries in addition to their "home country," MNCs are globalization's economic agents and activators par excellence. They are also the focus of easily the most influential book on economic globalization (and also arguably the most successful book by an economic geographer), Peter Dicken's *Global Shift* (see Chapter 7). Although in the immediate postwar era such MNCs hailed largely from the United Kingdom and, especially, the United States, they were increasingly joined over time by companies headquartered in Japan, in the major economies of continental Europe, and, latterly, in China. As well as originating much of the voluminous international trade charted in Figure 8.1, MNCs are conduits within and through which much of that trade actually flows. By the early 1990s, approximately 40% of all global trade was estimated to occur not between different companies in different countries but between different national branches of individual MNCs (Agnew and Corbridge 1995, p.169).

Third and finally, globalization in the particular (postwar) era typically labeled as globalized may not be entirely new, but it *is* different, and the MNC is in this respect pivotal. Let us start with some of the historic parallels. After all, even the most protectionist of national economies have never been totally isolated and disconnected, and tight economic integration at the international scale has a very long history, not least in connection with European imperialism and colonialism post 1492. There have long been global or near-global markets for certain commodities, such as silk, tobacco, or cotton. As this commodity trade "globalized" in the seventeenth- and eighteenth-century era of "merchant capitalism" (the capitalism of traders and merchants), albeit on a scale dwarfed by modern commodity trade, systems of money and finance also globalized in lockstep. Indeed, the levels of international financial mobility and integration achieved by the late nineteenth century, when the adoption of the gold standard substantially increased what were already large international financial investment flows, were unparalleled, certainly compared to what had gone before and perhaps even to what has happened since (Lothian 2001).

Box 8.1   World Trade Organization (WTO)

The WTO is an "intergovernmental organization" comprising at the time of this writing 164 member nation-states, and concerned with designing and policing the rules of trade between member nations. These rules take the form of agreements negotiated and signed by the members, which contain commitments to open markets to foreign companies, for instance, by lowering tariffs and other trade barriers. The WTO's members are estimated to account for over 90% of global trade. "Freer" trade is its raison d'être and guiding motif.

The WTO's all-important agreements have been negotiated in a series of historic rounds of trade talks, known by the names of the places where they were held: for instance, the Uruguay Round (1986–1994), which created the WTO from out of the pre-existing GATT (General Agreement on Tariffs and Trade) agreement. Originally signed in Geneva in 1947, GATT, which remains part of the WTO framework, itself went through six rounds of renegotiation prior to Uruguay: in Annecy (1949), Torquay (1950–1951), Geneva (1956), Dillon (1960–1961), Kennedy (1964–1967), and Tokyo (1973–1979).

While the WTO and its rules – which cover intellectual property rights and trade in services, as well as trade in goods – are extremely complex, its core principle is that of *non-discrimination*. This core principle is itself embodied in and realized through two sub-principles that structure the various rules. The first is "most-favored nation," the idea that all trading partners should receive the same "best" treatment at the border as a country's most-favored trading partner. The second is "national treatment," the idea that imported goods and services should receive the same treatment as those produced domestically. These principles are not mere abstract signifiers. Members are contractually obliged to respect the rules containing them. If they do not, the WTO has a dispute settlement mechanism whereby conflicts are resolved.

The WTO has been enormously controversial. Critics see it as a tool of a particular variant of globalization that systematically favors powerful Western nations and their leading companies. Popular protest against WTO-style globalization famously came to a head at the organization's ministerial meeting in Seattle in 1999.

The WTO has also been internally conflict-ridden. The "rounds" of negotiations have gotten longer and longer because it has generally gotten harder for members to agree on sensitive issues. The current round is a case in point. At the time of this writing, the Doha Development Round, launched in 2001, remains without an agreement.

So, "globalized" finance and commodity trade certainly pre-dated the postwar era commonly associated with globalization; but globalized production and labor generally did not. While money and manufactured things circulated widely across international borders, the same could not be said for the actual commercial infrastructure of manufacturing (in the form of the capitalist firm), or for those pulling the levers at the industrial coal face. This is not to say that there were no significant historical exceptions. Where labor was concerned, the most obvious and important was the slave trade – so significant that the historian

Sven Beckert (2014) prefers the term "war capitalism" to "merchant capitalism" for the era in question. But the MNC, and postwar developments in international labor migrancy, nonetheless ushered in a new scheme of things: globalization as we know it. "If global markets in commodity and finance capital were in place for the postwar expansion [of capitalism]," Smith (1997, p.176) confirms, "the same cannot be so readily claimed about production capital, labor, and cultural capital."

8.3   Framing Globalization

8.3.1   Dominant discourses…

If the brief preceding summary represents more or less the bare factual bones of what has come to be known as globalization, what flesh is conventionally added to these bones by those who use the term? Clearly this is an inordinately difficult question to answer: globalization is pictured in myriad ways. Here, however, we are interested specifically in *dominant* framings, by which we mean those not only that predominate but that are most material, which is to say most impactful – partly by virtue *of* predominating, but also by virtue of the identity and influence of those who propagate them.

The framings we have in mind are those of the powerful institutions – public and private – that favor, advocate, and facilitate or produce globalization, and of commentators – in academia or the media, and sometimes both – who subscribe to similar positions. To be sure, even among these dominant framings there exists variation; but the differences between these framings and the depiction of globalization found in the critical economic geographic literature to which we turn in the next section are much, much bigger.

The best place to begin is with the recognition that just as globalization itself is not new, nor are ways of representing it. This matters because today's dominant framings of globalization replicate in significant ways powerful discourses of yesteryear. The latter implicitly or explicitly inhabit the former. One such longstanding discourse is that of "free trade." In fact, if expansive international trade is the trademark feature of globalization and the WTO its trademark institution, then "free trade" is its trademark narrative; the three support and sustain one another.

The history of the idea of "free trade" can be told in many ways, but most histories focus on nineteenth-century Europe, and especially Britain. In the early part of the century, the consensus was in favor of protecting domestic industries against competition from foreign imports, as recommended by the German economist Friedrich List. This is what Britain's Corn Laws did. But the tide gradually turned. The English economist David Ricardo's arguments against protectionism proved crucial. His key concept was "comparative advantage." It is the theory that when countries specialize in goods in which they have a relative cost advantage, gains to trade will be maximized, benefiting all trading nations. This theory holds even when a country has an absolute cost advantage in every good it produces. Free trade is always better than protected trade, or worse, no trade. When the Scottish businessman James Wilson founded *The Economist* in 1843, he did so specifically to campaign for free trade. The discourse of "free trade" in time became a theoretical "common sense" (Chapter 5). Later books like Henry George's *Protection or Free Trade* (1886) only served to sanctify it.

Other longstanding discourses that continue to animate (or haunt?) dominant contemporary framings of globalization include the related narratives of "development" and "modernization." The thrust of these narratives is that "undeveloped," "less-developed," or "pre-modern" economies can only attain development and modernization by following the course previously pursued by already "developed" and "modernized" nations. Globalization is an essential part of the recommended strategy in achieving those ends. Through embracing globalization, development and modernization become attainable. As a concept, "globalization" can in some ways be understood as a cleaner, less ideologically loaded term than "development" or especially "modernization." This is Neil Smith's (1997, p.172) claim. He argues that "modernization theory ultimately failed at least in ideological terms, and had to be reinvented as globalization."

Whether it incorporates or substitutes for discourses like "free trade" or "modernization," "globalization" as dominantly framed shares with them two vital characteristics. One is its situated nature. Such discourses neither emerge from nowhere, nor do they exist independently of people and institutions with voices. Particular people and institutions in particular places produce and reproduce them. The second characteristic is normativity. These discourses do not only describe the world, but also make a case for how it *should* be: modernized, developed, globalized. It is the combination that is crucially important.

What does the ideal-type model of globalization pictured in dominant framings look like, and what are its distinctive geographies? Let us take an exemplar text, *Why Globalization Works* (2004) by the chief economics commentator of the *Financial Times*, Martin Wolf. The core proposition not just of this book but of the dominant discourse of globalization more generally is that "a world integrated through the market should be highly beneficial to the vast majority of the world's inhabitants" (p.xvii). The reason this statement is so important, of course, is its insistence not so much on global integration – that is, globalization *generally* – but on a particular method of integration. The world that will benefit the vast majority of its inhabitants is one integrated specifically "through the market." Why? Because, Wolf claims, the market is "the most powerful institution for raising living standards ever invented: indeed there are no rivals."

"Good" globalization, the type that by this way of thinking we should encourage, constitutes processes whereby people and places are connected explicitly through market-based arrangements. The flip side of the same normative coin is that one institution in particular should be kept out: the state. In the words of the Japanese business guru Kenichi Ohmae (1995, p.123; also cited in Kelly 2000, p.8), "if a country genuinely opens itself up to the global system, prosperity will follow. If it does not, or if it does so only halfheartedly, relying instead on the heavy, guiding hand of central government, its progress will falter." Ohmae is making an exceptionally forceful claim here, and one that is often heard. He says that globalization *will* deliver prosperity, but only if done properly or, in his word, "genuinely." The significance of this claim is that where countries that embrace globalization encounter difficulties, the likes of Ohmae can retort that globalization per se is not the problem. It (globalization) must simply not have been "done" properly, likely because of a meddling state; if it had been, we know that prosperity would have ensued. In short, the claim is that problems associated with globalization never result from the phenomenon itself but always from the failure of institutions – typically states – to embrace and implement it fully. "The problem today," as Wolf (2004, p.xvii) similarly maintained, "is not that there is too much globalization, but that there is far too little."

Doing Critical Economic Geography

Geographers Agnew and Corbridge (1995, p.171) capture the dominant, market-based framing of globalization succinctly. It reflects and reproduces, they say, an "ideology of transnational liberalism that stresses the efficiency and equity of voluntary economic exchanges in the global marketplace." Before turning to the geographies commonly associated with this framing, it is important lastly to linger a moment on the word "liberalism." Why "liberal"? In what sense is the dominant discourse of globalization a specifically "liberal" one? There are two senses. First, Wolf and like-minded commentators argue that globalization depends upon processes of active *liberalization*: liberalization of markets, and of course of international (hence "free") trade. Second, and more fundamentally, market and trade liberalization theoretically engenders the particular attribute that liberals (in the European rather than US sense) cherish above all else: individual freedom. To live in and benefit from a globalized world free from impediments such as intrusive regulatory restrictions on market mechanisms or barriers to cross-border trade is, literally, to *be free*. A commitment to globalization therefore is at the same time a commitment to defend market-based freedoms from those (such as anybody who questions globalization) that threaten them. "Social democrats, classical liberals and democratic conservatives," implores Wolf (2004, p.4), "should unite to preserve and improve the liberal global economy against the enemies mustering both outside and inside the gates." This liberal global economy has in the past decade increasingly been called a *neo*liberal one. In this context, though, "liberal" and "neoliberal" mean much the same thing (cf. Chapter 1).

8.3.2   … and their spatial imaginaries

As we will see, critical economic geographic work on globalization casts these nominal freedoms in an altogether different light (as, interestingly, does the recent populist backlash against the perceived downsides of "liberal" globalization for white working classes in the Global North, witnessed most emphatically in the 2016 votes for Brexit and Donald Trump; see Chapter 13). This critical work also directly challenges the geographies that dominant framings impute to globalization. These strongly emphasize spatial *fluidity*. Globalization comprises – or, once again, for its acolytes *should* ideally comprise – "friction-free flows of commodities, capital, corporations, communication, and consumers all over the world" (Luke and Tuathail 1997, p.76). The geography conjured is hence a space of rapid, unhindered, often bewildering movement. This movement not only underwrites tight economic integration between places, even those that are a long distance from one another, it also unhinges individual places from traditional place-based restraints, "eroding away fixed in-state places into fluid un-stated places" (p.76).

Of Luke and Tuathail's five C's (commodities, capital, corporations, communication, and consumers), globalization orthodoxy privileges two in particular. One is the commodity. Charged with ensuring its global fluidity is the WTO. The other key C is capital, or money. For evangelists of globalization, capital has nothing short of a natural right to roam the world, flowing freely across national borders in search of investment opportunities, supporting worthy ones and spurning the unworthy. Historically, the International Monetary Fund (IMF), another important intergovernmental organization, has been charged with ensuring that right (Chapter 9).

To the extent that it actually exists, this vaunted, multidimensional fluidity of globalization has been enabled by progressive historical developments in technologies of transportation, information, and communications (itself one of Luke and Tuathail's five C's; Chapter 1). These technologies serve to "shrink" the world or, in Harvey's (1989) influential academic terminology, to engineer "time–space compression": reducing the time taken to traverse geographical space, and thus also squeezing the experience *of* space (Chapter 3). A little over a decade ago, the IMF's acting managing director delivered an address appropriately entitled "Economic Growth in a Shrinking World: The IMF and Globalization" (Krueger 2004). She said that the title was intended to convey a clear message about geography and globalization: "I used the word shrinking in the title of my talk. That is what, in a real sense, globalization means." Marx, much earlier, had called it the "annihilation of space by time."

According to dominant framings, those same globalizing processes that integrate places, unsettle their fixity, and reduce the distance between them also make them more like one another. Globalization standardizes places, effecting their convergence to a single model (Chapter 1). Here, the most striking discursive exemplar is not Wolf's eulogy but an earlier, equally influential intervention by the business economist Theodore Levitt, published in one of the cardinal fonts of globalization discourse, the *Harvard Business Review*. Levitt (1983) began "The Globalization of Markets" with the sentence: "A powerful force drives the world toward a converging commonality, and that force is technology." Approvingly citing Daniel Boorstin's dictum that the chief trait of the new age was "convergence, the tendency for everything to become more like everything else," Levitt went on to claim that the MNC "seeks constantly to drive down prices by standardizing what it sells and how it operates. It treats the world as composed of few standardized markets rather than many customized markets. It actively seeks and vigorously works toward global convergence." In this view there was and is still room for different places in the world economy to specialize in *producing* different products and services, as per the Ricardian theory of comparative advantage. But otherwise, and especially where markets and consumption are concerned, convergence between and among places is the order of the day.

In the standard framing of globalization, the various elements of geographic transformation summarized in the preceding paragraphs are tied in every case to the perceived dominance of a particular geographic scale: for that framing relentlessly privileges "the global." What we mean by this is that globalization is viewed as a set of processes occurring primarily *at* the global scale, and operationalized exclusively by actors with global reach and influence. By contrast, other scales such as the "local," "regional," and even "national," and the places, people, and institutions associated with them, are subordinate and subordinated. They must acquiesce to what Hart (2002, p.13) refers to as globalization's apparently "inexorable market and technological forces." If these forces come from anywhere, Hart elaborates, they "take shape in the core of the global economy and radiate out from there." Globalization is thus envisioned as a centripetal force field whipped up in the Global North and absorbed elsewhere, the "local" being reduced to the status of "a passive, implicitly feminine recipient of [masculine] global forces whose only option is to appear as alluring as possible."

All of this matters. That is why we labor these spatial imaginaries of globalization at some length. Not only do these representations sit uncomfortably with the portrayals of globalization that emerge from grounded empirical research of its actually existing realities (see below), but more importantly, these problematic dominant representations *shape* those unfolding realities. Just because globalization in practice does not correspond to globalization as dominantly framed does not mean that the former is immune from the latter. It is not. As Hart (2002, p.12) emphasizes, common framings of globalization "play a key role in defining and delimiting the terrain of practical action and the formation of political identities."

Kelly's (2000) study of "landscapes of globalization" in the Philippines provides ample testament to such impacts. Globalization rhetoric circulating both internationally and locally matters because it represents, says Kelly, "the discursive context in which politicians and policymakers concerned with economic development currently operate" (p.8). They have to "deal" with it, whatever they make of it. And the rhetoric in question attests both to Smith's argument about "modernization"/"development" bleeding into "globalization" and to the centrality to globalization orthodoxy of market-oriented (neo)liberalism à la Martin Wolf. Kelly describes a "discourse of development that is used to justify neoliberal economic policies in which the state is viewed as a hindrance" (p.11). The result? A "broad process of deregulation, liberalization and decentralization" that "has been facilitated by the constant rhetorical refrain of the necessities of globalization" and that "has left the Philippines with an economy deliberately attuned to the needs of foreign investors" (p.1).

8.4   Geographies of Actually Existing Globalization

8.4.1   Unevenness and unusability

The dominant framing of globalization sketched above is highly problematic, not least in respect of the geographies it specifies. To begin with, globalization does not in reality lead to anything like the geographical standardization and homogenization so often imputed to it. It has, and has always had, geographically differentiated outcomes. Many researchers have demonstrated this. But few make the case as trenchantly as the anthropologist James Ferguson (2005) does in his critique of fellow anthropologist James Scott's (1998, p.8) Levitt-like claim that, today, "global capitalism is perhaps the most powerful force for homogenization." The notion that globalization "works through a fundamental mechanism of homogenization and the gridlike standardization of space" is, Ferguson (2005, p.378) says, simply wrong. He shows this with reference to the experience of Africa, where the levels, forms, and consequences of integration into global constellations of capitalism could scarcely be more geographically uneven both inter- and intra-nationally.

And not only do the processes collectively referred to as "globalization" create geographic unevenness, they actively target and exploit it. One of the most consistent findings of the literature on globalization of and by Western MNCs is that such globalization is heavily concentrated in regions of the world, predominantly in the Global South, where the cost of labor is relatively low. Smith (1997, p.184) pointedly contrasts this "reservoir of cheap labour" with the "reservoir of consumers" anticipated in paeans to globalization like Levitt's. James Meek (2017) has recently written a poignant essay about one example of globalization as the seeking out of cheap labor elsewhere in the world. Between 2007 and 2011, Cadbury progressively shut down production at its chocolate factory in Somerdale near Bristol in England. Most of the 500 jobs were moved to Skarbimierz in Poland, because

Poles could do the same work for less than a fifth of the money. Meek also made an important observation about the nature of the vaunted "integration" of distant places that globalization theoretically engenders. If Cadbury's globalization has connected Somerdale to Skarbimierz, then it "links them, in that strange way of globalization, without doing anything to bring them together."

Further, globalizing Western firms do not only target geographic unevenness in labor costs. *Anything* that makes relocating production activities to one international location more economically attractive than relocating elsewhere is fair game. The world's inherent unevenness is grist to the mill for mobile capitalists. Robert Cox (1992, p.30) summarizes thus: "Global production is able to make use of the territorial divisions of the international economy playing off one territorial jurisdiction against another so as to maximize reductions in costs, savings in taxes, avoidance of anti-pollution regulation, control over labour, and guarantees of political stability and favour."

This, however, is to suggest that everywhere and everyone potentially is folded into the globalization drama, just so long as costs are low enough, conditions are conducive enough, and so forth. But they are not. One of the biggest geographical myths about globalization is that it is, well, *global* – spatially all-encompassing. Yet this is not true. Plenty of places are actively excluded or are simply left behind. Again, Africa, or substantial parts of it, is an exemplar. After William Reno, Ferguson (2005) bluntly distinguishes between "usable" and "unusable" Africa. Usable Africa comprises those places where globalization makes economic sense from the perspective of the globalizing agent; unusable Africa comprises those places that globalization conspicuously bypasses, to which it nods politely, perhaps holding its nose, before moving swiftly on.

From these observations Ferguson maps a striking basic cartography of globalization. Globalization "is 'global' in the sense that it crosses the globe, but it does not encompass or cover contiguous geographic space. … Capital does not 'flow' from London to Cabinda; it hops, neatly skipping over most of what lies between" (2005, p.379). Those in-between, unusable places are the heartlands of what Smith (1997, pp.178–179, 187) no less vividly calls globalization's "satanic geographies." Concentrated in sub-Saharan Africa, these "redundant spaces of accumulation" have been marked out for "disinvestment and abandonment" and thus "denied access to more than a trickle of capital on the global markets, yet condemned to resolve the local consequences of an earlier global adventurism by colonial powers."

8.4.2   Geographies of freedom and power

It is also plainly not the case that globalization *only* breaks down geographical boundaries, which is the implication of dominant representations of fluidity and friction-free flows. For another striking geographical attribute of the "globalized" Africa pictured by Ferguson is that of the freshly created spatial boundaries that crisscross it. A large part of the reason that the forms and effects of globalization have been so unevenly distributed within particular African countries, Ferguson submits, is that new borders have been erected and new spaces carved out at sub-national scales. Many of these are associated with mineral extraction activities. In places like the Sudan, global capital does not engage with the country as a whole but helps fashion territorially delimited enclaves, often secured by private security forces, and from which little or none of the economic benefits accrue to the population at large. Indeed in some cases, such as Angola, the mineral (oil) extraction occurs offshore, using imported equipment and materials, employing foreign workers living in gated compounds, and where "neither the oil nor most of the money it brings in ever touches Angolan soil" (Ferguson 2005, p.378).

More generally, what is increasingly clear is that as well as disassembling certain boundaries, and thus in important ways reducing the distance between places, globalization has led to the formation of new economically meaningful spaces. They are demarcated by borders of varying degrees of porosity at a range of geographic scales. Globalization creates new spaces as well as compressing "space" (distance). This has been one of the main findings of the research on "global production networks" (GPN) discussed in Chapter 5: that globalization has entailed the restructuring of the world economy into a series of linked or "networked" spaces that stand out from a wider plain largely "skipped over" by MNCs. As the historian Manu Goswami (2004) has shown, the production and linking of new economic spaces at the global scale is nothing new. It was the modus operandi of post-1857 British imperialism, or what she calls "territorial colonialism," in the Indian subcontinent. India was fashioned into "a commodified, 'second-order' space embedded within … the broader imperial economy" (p.45).

Equally, where boundaries *are* dismantled, and space compressed or "annihilated" with time, it tends only to be for certain privileged constituencies. Liberal champions of globalization make much play of freedom of movement across geographical borders, and thus of a putatively frictionless world. The reality is that such freedom is not universally enjoyed. Capital may enjoy a "natural" right to flow freely but labor definitely does not. More than a decade ago Massey (2002, pp.293–294) launched a withering attack on the hypocrisy of globalization's cheerleaders in this regard. We will cite it at length because it remains painfully pertinent today. Massey excoriated

> those who argue most strongly for "free trade" as though there were some self-evident right to global mobility; the term "free" immediately implying something good, something to be aimed at; as if it is self-evidently good to be able to roam the world. This is a geographical imagination of a world without borders. Yet come a debate on international migration and the same people will often have recourse to another geographical imagination altogether – equally powerful, equally (apparently) incontrovertible – yet in total contradiction. This is the imagination of defensible place, of the rights of local people to their own local places, of a world divided by difference and the smack of firm boundaries. … And so in this era of the "globalization" of capital, we have people risking their lives in the Channel Tunnel, and boats full of people going down in the Mediterranean.

Indeed, the closer one looks at actually existing globalization, the less one finds of the "freedom" lauded by the likes of Martin Wolf, and the more one finds of its eternal foil: power – and frequently unaccountable power at that. If the power to make and remake the global economy is vested most evidently in vast MNCs and global financial institutions, the role of multilateral institutions such as the IMF, the WTO, and the World Bank cannot be overlooked. In many respects these bodies – outcomes of globalization, but also essential to its perpetuation – are curious phenomena, hard to pin down with traditional categories. Cox (1992, p.30) referred to them as "the official caretakers of the global economy," which is a good way to put it. They reflect and collectively embody what the political scientist Stephen Gill (1992, p.276) describes as "a partial, but still relatively underdeveloped internationalization of political authority."

Globalization has occurred in the context of, and been indelibly shaped by, a complex reciprocal relationship of mutual constitution between these "caretakers" and private-sector capitalist elites. This relationship is shot through with power. On the one hand, corporations have had a major influence, for instance through lobbying, on multilateral institutions and their globalization efforts (Beder 2006). On the other hand, those multilateral institutions, and the consensus formation that occurs in and through them, format the terrain on which corporations operate. And, if globalization has been highly profitable for corporations headquartered in the Global North, then partly this is because the multilateral institutions in question have helped make it so. The World Bank is an exemplar, though not unique. Sociologist Michael Goldman (2005) has shown that inasmuch as it is deeply embedded in the reproduction of elite power networks at the core of the global political economy, the Bank plays a central role in "enlarging the scope for global capital accumulation ... [and] in the making of the highly inequitable global economy" (p.12).

For these reasons, it would be difficult not to see in globalization, and particularly in the matrix of forces shaping it, an attenuation of traditional nation-state-based democracy, or at least a process whereby democracy has come to be redefined in more limited terms. For example, this is the argument of Gill, who associates globalization with what he calls a "new constitutionalism" whereby the economy – and economic globalization – has been increasingly separated from politics. Gill (1992, p.279) defines new constitutionalism as "the efforts by capitalist elites to legally or constitutionally insulate economic institutions and agents from popular scrutiny or political accountability." Globalization fits the bill.

Consider one of Ferguson's (2005, p.380) main observations: that

> the countries that are (in the terms of World Bank-IMF reformers) the biggest "failures" have been among the most successful at developing capital-attracting enclaves. African countries where peace, democracy, and some measure of law obtain have had very mixed records in drawing foreign capital investment. But countries with what are, in conventional normative terms, the "worst" and "most corrupt" states, even those in the midst of civil wars, have often attracted very significant inflows.

On the face of things, this does not make much sense. Certainly, it makes no sense to Ferguson, who calls this pattern "the most surprising finding of the recent Africanist literature." By all "normal" logic (including that of the IMF and World Bank), global capital would gravitate to countries with democratic governments and a transparent rule of law. But if we listen to Gill, the pattern described by Ferguson is perhaps not so surprising after all. Political "failure" can provide precisely the insulation from scrutiny and accountability that Gill says the institutions and agents of globalization prefer. Contra the rhetoric of globalization-as-liberalism, in other words, globalization's "usable" Africa is actually in large part *il*liberal Africa.

None of this is to say that in our globalizing and globalized world of MNCs on the one side and multilateral institutions on the other, the nation-state has necessarily lost its relevance. The national scale and the political authority associated with it may in some respects have become less material in economic affairs, but much of what occurs "globally" is still decided at the national scale – with some nations, naturally, much more influential than others. The internationalization of political authority associated with the likes of the IMF and WTO *is* only "partial" and "relatively underdeveloped" (Gill 1992).

Indeed, the nation-state's continuing significance may in part help explain the weakening of democracy under globalization. The economist Dani Rodrik (2007) argues that "democracy, national sovereignty and global economic integration are mutually incompatible: we can combine any two of the three, but never have all three simultaneously and in full." Retain the nation-state's sovereign role under economic globalization, in other words, and democracy inevitably suffers; democracy and globalization can only co-exist if the nation-state takes a backseat position.

Whatever the truth in Rodrik's theory, it is clearly the case that some of the most prominent and, indeed, inherently globalizing developments of the globalization era have been driven as much by nation-state initiatives as by multilateral institutions, MNCs, or developments in information and communication technologies and the time–space compression they elicit. The best example of this is post-1960s financial globalization (Chapter 9). Both financial institutions and the monetary and financial instruments they handle were enabled to flow much more freely across national boundaries. Technological developments played an important part, but so too did enthusiastic nation-states. Helleiner (1994) emphasized the advocacy and influence of the US and UK governments; Abdelal (2007) focused on the stimulatory activities of German and especially French policymakers. Thus, if Glassman (2012) is right to say that globalization does not bypass nation-states so much as *transform* them – and we think he is – then we can add that this transformation is itself a product in part of nation-state projects.

8.4.3   Glocalization?

The dominant spatial imaginary of globalization is also inaccurate insofar as it envisions "local" (typically sub-national) places and peoples submitting passively to global-scale processes. There is now a significant stock of research demonstrating the flaws in this depiction. For one thing, the local is not some kind of inert stage on which unchanging processes of globalization are unproblematically scripted and performed. People and institutions at the local scale have the capacity, limited though it often is, to shape what forms globalization takes in different local milieux. Globalization therefore is always mediated by and "embedded in place-specific social relations … both as material processes and discursive construction" (Kelly 2000, pp.3, 12). In a study of "Washington's world apple," Jarosz and Qazi (2000) demonstrate just such local mediation/embedding. Globalization of the fresh fruit industry has been powerfully shaped both by representations of this specific locale – "representing the Washington apple as pure and nutritious, emanating directly from a pristine natural landscape" (p.2) – and by its local political economy, in particular the regionally rooted co-optation and exploitation of a transnational labor pool. Globalization does not just take place, it is enmeshed within it.

But to insist that globalization is embedded in and shaped by specific places is still not to go far enough, for it fails to challenge the more fundamental component of globalization's "globalist" imaginary: the notion that the processes experienced as globalization, even if locally mediated, are nonetheless *constituted* at the global scale. This, too, is erroneous.

Not only are the processes we refer to as globalization not unresponsive to local conditions when "touching down" locally, they are not, and never can be, external to the local in the sense of being separable from it. That they *can* be external is the implication of dominant narratives that typically eschew consideration of "marginal" local scales such as the household, the community, and the body except when discussing how the latter absorb globalization's impacts. But researchers including Nagar et al. (2002), Chari (2004), and Bair and Werner (2011) have all shown that globalization is not, and can never be, "out there," somehow beyond the bounds of the local – and especially not separable from those places, peoples, and scales so often neglected in dominant framings. Let us see how they have done so:

i.   *Subsidizing*: In an influential feminist critique of dominant framings of globalization's economic geographies, Richa Nagar and colleagues argue that the economic profits of global capitalism are made possible by the very places and peoples neglected in standard narratives. They emphasize often-overlooked and heavily women-centric "informal economies of production and caring." By "cheapening production in sweat-shops and homework," these female economies "subsidize and constitute global capitalism." In short, the classic globalization-era growth model for countries in the Global South of manufacturing wares for consumption in the Global North is, as economic geographer Marion Werner (2015, p.4) notes, "not simply export-led but also female-led."

ii. *Provincializing*: Geographer Sharad Chari (2004) also analyzes a classic export-led globalization model, in his case through an ethnography of garment production in Tiruppur, South India. This production regime is controlled locally by an emergent capitalist class that Chari styles a *fraternity*, one enabled by precisely the gender-based subsidy identified by Nagar et al. Chari submits that this community of former peasant workers plays a sufficiently significant role in globalization to require us to decenter our understanding of this phenomenon: their achievement being no less than "provincializing the globalization of capital" (p.282). Far from being peripheral to capitalist globalization, Chari's garment producers are active at the very heart of it, directly influencing its wider dynamics.

iii. *Articulating and disarticulating*: Bair and Werner (2011) study the shifting fortunes of commodity producers in north-central Mexico, finding that "global" transformations in industrial production can only be understood in the context of local, "deeply rooted historical processes of struggle, dispossession, and accumulation" that shape localities' articulation *and* "disarticulation" with wider production networks. For this reason, they argue that the influential GPN (Chapter 5) and "global commodity chain" approaches to globalization are inadequate insofar as they privilege the global scale in much the same way as liberal framings do. We need, they maintain, to "turn commodity chains on their head," to emphasize that Mexican localities "are constitutive of, as well as constituted by … configurations of global production." Such places are no more "peripheral" than Tiruppur.

So, is the spatiality of globalization best captured in terms of a simple dialectic between the global and the local, the two co-constituting one another? (It is this that Erik Swyngedouw (1997) calls "glocalization.") Not exactly. Yes, particular local places and processes help constitute the global, but they invariably do so, as Hart's (2002) study of globalization powerfully shows, in connection with *other* local places and peoples. This underlines the importance of attending in globalization research to what Sheppard (2002) describes as a locality's relative situation or "positionality." Analyzing investment by small-scale Taiwanese industrialists in South Africa in the 1980s and 1990s, Hart (2002, p.14) argued that each place – Taiwan and South Africa – was transformed in relation not only "to a larger whole" (aka globalization) but "to one another." As such, it is the *connected* trajectories of change in a vast array of such local places that she says "actively *produce* and drive the process we call 'globalization.'" And this leads her to redefine the geography of globalization, conceived now as "the multiple, divergent but interconnected trajectories of socio-spatial change taking shape in the context of intensified global integration" (p.13).

Based on the very different picture of globalization's geographies drawn in this section compared to the dominant framing outlined in the previous one, we think Hart's reframing is useful. Globalization *is* helpfully thought of as the combination of connected local trajectories. But it is not only that. Hart's definition fails to specify, for instance, the extraordinary unevenness of globalization, or its tendency spatially to hop or skip rather than to flow and universally embrace. We therefore end this section with the rather fuller positing of globalization (or "global space–time") proffered by Goswami (2004, pp.27–28), who speaks of it as: "radically relational (i.e., defined by systemic interdependencies between its various parts), multispatial (i.e., composed of distinct yet entwined spaces and scales), multitemporal (i.e., constituted by different temporalities and time horizons), and endemically uneven (i.e., constituted by the structurally engendered processes of uneven development)." Globalization is indeed all those things.

8.5   Theorizing Globalization as Uneven Geographical Development

8.5.1   Toward a theory of globalization?

Such a complex, multifaceted framing – radically relational, multispatial, multitemporal, and endemically uneven – would seem at first blush to militate against the kind of sustainable generalization that theorization (Chapter 5) of globalization would presumably call for. How, after all, can any sort of reliable generalization be ventured regarding a phenomenon that is at once *not* absolute, *not* even, and *not* (geographically or temporally) singular? In this final section of the chapter, however, we suggest that for all its undoubted complexity and sheer variegation, theorization of globalization that aims to conceptualize its key underlying dynamics is not quite as hopeless a task as one might imagine.

Our initial touchstone in this respect is Neil Smith's (1997, p.182) styling of globalization as "the latest stage of uneven development." Or, more fully (p.183): "The global restructuring of the 1980s and 1990s embodies not so much an evening out of social and economic development levels across the globe as a deepening and reorganization of existing patterns of uneven geographical development." But what has this claim to do specifically with the issue of theorization?

The answer to this question is, a lot. As noted at several points in this chapter, the types of integrative and connective economic processes referred to with the label "globalization" are not entirely new, even if they have taken different forms and exhibited different velocities and patterns since the 1980s. And neither is their inherent unevenness new: processes of international economic integration were deeply uneven under previous rounds of internationalization, just as they are today. The significance of this, in turn, is that prior to the modern globalization era, scholars had expended considerable effort in seeking to understand the core drivers of this economic geographic unevenness; that is, in *theorizing* uneven geographic development at the international (but not only international) scale. Then came "globalization." Thus if, as Smith claims, globalization represents a new iteration of an old phenomenon (uneven development), then surely theories formulated to explain the latter might still, to one degree or another, apply.

As we shall see, Smith was by no means a disinterested bystander vis-à-vis such theoretical possibilities. That globalization could conceivably be theorized as uneven development had significant implications for him and his scholarship because one of the most influential theorizations of uneven development happened to be his own (Smith, 1984). But it was (and is) not the only one.

We turn now then to consider some of the main ways in which uneven development has been theorized, including by Smith. Before that, to set the stage for what follows, two preliminary points are in order. First, we essentially agree with Smith, both that figuring economic globalization as "the latest stage of uneven development" is meaningful and helpful *and* that existing theories of uneven development formulated for the most part before the heyday of globalization have much to tell us. Second, however, it seems equally clear that such theories need to be selectively refined and retooled, in particular to account for some of the other geographical dimensions of globalization documented in the previous section besides its unevenness.

8.5.2 Theories of uneven development

Much of the most influential theoretical work on uneven economic geographic development was produced in the two decades from the late 1960s. But it has a much longer heritage. For instance, Adam Smith and David Ricardo, the two great British political economists of the late eighteenth and early nineteenth centuries, shared a deep concern about the obvious economic disparities between nations, and their respective inquiries into national wealth – its nature and the causes of its growth – were at the same time investigations of uneven international development.

By the early twentieth century, however, the question of uneven development had largely disappeared from the discussions of mainstream economists. Neoclassicism was now ascendant, and neoclassical growth and trade theory predicted convergence (on a developed economic model) in the long term – just as Theodore Levitt would do in his early disquisition on globalization later in the century. This did not mean that theorization of uneven development was abandoned. Now, though, it was taken up and set in dramatically new directions by radicals. The most influential theorization was arguably that of "uneven and combined development" by the Russian Marxist revolutionary Leon Trotsky (Box 8.2: Leon Trotsky and "Uneven and Combined Development").

In the 1960s, persistent "underdevelopment" in the "Third World," including in those regions which emerged from colonial rule, cried out for understanding and theorization, and

Box 8.2   Leon Trotsky and "Uneven and Combined Development"

Today Leon Trotsky is best known as the founder of the Red Army and, via The Stranglers' hit single "No More Heroes," as the man horrifically murdered with an ice pick (in Mexico City in 1940). But he was also a prolific and highly creative thinker, and one of his most influential contributions was his theory of "uneven and combined development."

This theory was very much concerned with the dominant form of globalization of that day and age (the 1920s and 1930s), namely, European imperialism. For example, Trotsky emphasized that the imperial powers and colonial and "semi-colonial" societies were deeply interconnected and that it was important to think in terms of a "world economy."

He was not the first Russian Marxist by any means to talk about "uneven development." Lenin had, seeing it as inevitable under capitalism. But Trotsky offered a more sophisticated treatment.

At heart Trotsky's theory was a theory of revolution, which sought to theorize a political possibility – socialist revolution in primarily agrarian societies – that conventional Marxist uneven development theory appeared to preclude. Such theory suggested that "backward" (e.g., feudal) countries developed more or less in the manner suggested later by "modernization" theorists by reproducing the earlier development trajectory of advanced countries. This implied that socialist revolution was most likely to begin in more developed, industrialized countries: countries had to pass "through" industrial capitalism in order to "get to" socialism.

But this theory jarred with what Trotsky saw (in other countries) and had seen (in Russia). He saw higher levels of working-class consciousness and organization in agrarian, colonized societies than in more developed countries. And he had seen peasant land cultivation techniques in Russia remain at what he described as seventeenth-century levels right up until the revolution. How to explain this?

Trotsky argued that it was not necessary or perhaps even possible for "backward" societies to "catch up" with more developed countries before revolution was possible. Instead, rapid urbanization and industrialization in agrarian societies could create complex "combined" social formations, often featuring "an amalgam of archaic with more contemporary forms" (Trotsky 1977, pp.26–27). This was what had happened in Russia: seventeenth-century cultivation techniques thrived alongside modern capitalist urban industry. Such combined formations, Trotsky argued, were fertile breeding grounds for revolution because they were inherently less stable than the more homogeneous formations in developed societies. This allowed agrarian societies not so much to catch up with developed societies but to "overleap" them, thus achieving socialist revolution before them.

the question of uneven development reemerged with a vengeance. Over the next two decades, all manner of different theories were advanced. They fall into four broad categories:

i.   *Dependency and cognate theory*: A number of Marxian theories revived and reworked Trotsky's ideas. One was so-called dependency theory, closely associated with the economic historian Andre Frank. According to this theory, development and underdevelopment are two sides of the same coin (uneven development). Rich ("core") countries perpetuate a state of dependency in poorer ("peripheral") countries through their economic, military, and cultural policies. Another theory was Arghiri Emmanuel's "unequal exchange." Emmanuel maintained that rather than fostering convergence, international trade exacerbates uneven development because when products are traded between high- and low-waged countries there will always be a transfer of wealth from the latter to the former. Out of both dependency and unequal exchange theories but with critical perspectives on each there then emerged, from the late 1970s, what is known as "world-systems" theory. Proponents such as Samir Amin and Immanuel Wallerstein argued that the unevenly developed world contained not only a core and a periphery but a "semi-periphery." It played a crucial role in the economy by acting as a stabilizing entity between the core and the periphery, also experiencing the most intense restructuring and conflict. The core/semi-periphery/periphery hierarchy was fluid and dynamic, with each country's position influenced by the effects of periodic crises in capital accumulation and resulting periods of stagnation and growth.

ii. *New International Division of Labor (NIDL) theory*: Theorists in this alternative tradition focused on the fact that the division of labor between the Global North and the Global South had seemingly shifted during the postwar decades from one in which the economies of the latter were dominated by agriculture and mineral resource extraction to one in which manufacturing played a much larger role. Theoretically, this "school" diverged from the previous one inasmuch as its explanation for enduring uneven development highlighted not Marxian structural factors but rather the activities and growing influence specifically of MNCs. In the influential work of Stephen Hymer, for example, MNCs were essentially seen to create a world in their own image, where the internal corporate spatial division of labor – paradigmatically, with headquarters in cities of the Global North and production activities increasingly outsourced to "developing" countries – was reproduced in the international division of labor more widely.

iii. *New trade theory*: Developed from around the same time as world-systems and NIDL theories, so-called new trade theory, already discussed briefly in Chapters 4 and 5, is a stream of mainstream economics. It uses the analytical (model-based) techniques of economics and adheres to its formal conventions, even if it revisits certain standard neoclassical assumptions. Paul Krugman was one of the leaders of this new branch of economics. Its key contribution to understandings of uneven development was to throw into question the traditional, neoclassical convictions regarding the long-term (convergence) benefits of "open" international trade. As Autor, Dorn, and Hanson (2016) have recently shown, for example, labor markets do not necessarily respond to trade expansion with effortless adjustment to an agreeable new equilibrium. For example, employment opportunities in US industries exposed to import competition from China have deteriorated and there are no offsetting improvements in other industry sectors. Just ask Donald Trump's legion supporters.

Before turning to the fourth and final category of postwar uneven development theory, it is instructive to pause to identify a feature that the first three "schools," for all their important differences, notably share. The final school differs from the earlier three by *not* sharing the feature in question. It concerns the conceptual status of the phenomenon of uneven development. With the first school, we have a theory of dependency, a theory of unequal exchange, and a theory of world systems; with the second, we have a theory of the spatial division of labor; and with the third, we have a theory of trade. We do not yet have a theory *of uneven development*. In none of the cases is uneven development the essence and basis of the theory, the dynamic spindle on which it turns. Rather, uneven development is the *outcome* in each case of a theory oriented principally to explaining and understanding something else.

This is where the fourth school stands apart. It contains an explicit theory of uneven development. Not only is uneven development the nub of the theory, it is also posited by this theory as the nub of much wider political economic processes. The theory holds that uneven development is integral to and constitutive of the capitalist mode of production. Uneven development is stitched into capitalism's structural fabric; underwrites, drives, and formats its historic (as well as geographic) development; and thus helps make capitalism what it is. Without uneven geographical development capitalism would not be capitalism.

iv.   *Uneven development theory*: While other scholars have also laid out a theory of uneven geographical development along such lines (see especially Harvey 2006), Smith's seminal (1984) *Uneven Development* stands as the most influential statement of such a theory. Smith argues that at the heart of capitalism are contradictory geographical tendencies toward "differentiation" and "equalization." The former occurs both between individual capitals (firms) and between different sectors of the economy. But capital is also paradoxically a "leveler" (cf. Christophers 2016), tending to equalize across space the conditions of production and the level of development of productive forces. For Smith, these two opposed tendencies, and especially the contradiction between them, are what cause uneven development at all geographic scales. This process, he emphasizes, is inherently dynamic: as opposed to getting "stuck," places that are development laggards at one moment may surge ahead later, and highly developed places can regress. Why? The reason, Smith argues, is that in the development of one place, capital creates there conditions that increasingly militate against further *profitable* development – and vice versa in places of relative underdevelopment. As such, capital is apt to oscillate or "seesaw" geographically from developed areas to undeveloped areas and back again. However, certain important factors, such as the level of capital mobility, can limit the extent of such seesawing, particularly at the international scale.

We are now in a position finally to return to where we began this section, with Smith's framing of post-1970s economic globalization as the latest stage of longstanding uneven geographical development under capitalism. Globalization can be understood, he says (Smith 1997, pp.178, 183), in exactly the terms in which he earlier theorized uneven development. On the one hand it "expresses a central tendency in the uneven development of capitalist economies … toward a geographical *equalization* of levels and conditions of social production"; on the other hand it expresses "an equally important countervailing tendency, traceable to the continuous division of labors and capitals, toward the geographical differentiation of global space." Globalization *is* uneven geographical development.

8.5.3   From uneven development to uneven economic globalization

As already indicated, we concur that theories of uneven development do indeed provide a fruitful foundation in the essential task of theorizing deeply uneven economic globalization. With our "critical" hats on, we do not necessarily advocate for any particular one of the four "schools" over the others, and neither do we dismiss any of the four out of hand. All four have their merits, though they serve somewhat different purposes. And all four fit our definition of economic geographic theory (Chapter 5), although they align in different ways with the different "narratives" of economic geography laid out in Chapter 2. Two of the schools (the first and last) represent what we called "Geographies of Capitalism"; the second, on the NIDL, fits "Geographies of Business"; and the third, new trade theory, is clearly a form of "Geographical Economics."

Yet, as we also indicated earlier, none of the theories developed in the 1970s and early 1980s to explain historic uneven development is sufficient in and of itself adequately to theorize modern economic globalization. In ending this chapter, therefore, we will for illustrative purposes highlight and briefly expand upon one of potentially several reasons for this. It relates to one of the signature geographical characteristics of economic globalization discussed in the previous section: the fact that while eroding spatial boundaries of certain kinds globalization simultaneously constructs others, and that it produces new economic spaces, at new geographic scales, in the process. Such a characteristic is not theorized in any of the conceptualizations of uneven development we have considered.

But, since the early 1990s, theoretical work *has* been done to explain this recurrent outcome; and a fuller explanation of globalization and its geographies would need to factor such thinking in. The basic thrust of such work is captured in variations on the concept of "territorialization." On the one hand, globalization *de*-territorializes, making territorial boundaries less significant in light, for example, of time–space compression. But it also *re*-territorializes. It makes new territories, especially new economic territories (recall, for instance, Ferguson's enclaves), and in doing so it repeatedly rewrites the territorial or *scalar* organization of capitalism which refers to the array of principal geographic scales at which capitalist processes of production, exchange, and distribution play out.

Goswami (2004) is one of those to have productively probed the dynamic relation between de- and re-territorialization. Another, notably, was Smith himself (see especially Smith 1995). But perhaps the most important work in this area, insofar as it seeks explicitly to theorize de- and re-territorialization and does so explicitly in relation to globalization, has been Neil Brenner's. In a significant sense Brenner's (1998) argument is analogous to Smith's uneven development thesis. It too is predicated on a geographical contradiction – between not equalization and differentiation but what he calls, after David Harvey, "fixity" and "motion." The former refers to "relatively fixed and immobile territorial infrastructures" (p.461) – nation-states, for example. The latter refers, essentially, to time–space compression. The crux of the argument is that capital, almost despite itself, depends upon a certain degree of fixity *in order to remain in motion*; to break down spatial barriers it has to erect them (cf. Harvey 2001, p.25). And the geographies of globalization evidence this.

For our purposes, the key aspect of Brenner's argument is that it helps explain not just the production of new spaces – such as the "networked spaces" of GPN understandings – under globalization but the fact that these materialize at different interlocking scales: sub-national, national, supranational. In fact the deeper argument is precisely that geographical scale is not set in stone. Social and economic processes – globalization, for one – do not unfold within fixed scales so much as *(re)produce* the scales at which capitalism is constituted and in such a way as to optimize the circulation of capital at any particular point in time. Capitalism is thus "re-scaled."

During one historical phase of globalization the "relatively fixed" spatial forms required to put capital in motion might cohere primarily at nation-state and city-region scales. At another juncture, they may comprise nation-states alongside local enclaves and supra-national economic blocs such as the European Union. We can theorize each such territorial configuration as the particular scalar or "territorial fix" (Christophers 2014) that best suits "each historical round of capital circulation" (Brenner 1998, p.461). A careful melding of the concept of scalar fixes with that of uneven development seems well-suited to the task of theorizing, however provisionally, the complex geographies of globalization.

8.6   Conclusion

Just as scale is never fixed, neither is economic geographic theory, whether formulated to explain scale or any other phenomenon. Theory is always incomplete, a work in progress (Chapter 5). Neil Smith was acutely aware of this. In articulating a theory of uneven geo-graphical development in the early 1980s, he acknowledged that it would have to evolve and mutate over time as capitalism – ever restless, always changing – itself evolved.

How have the dynamics and geographies we associate with capitalist "globalization" compelled such theoretical transformation? One provocation for rethinking that Smith (1995) undertook was the production of new spaces and scales alongside the dissolution of existing ones. But there are other provocations too, many of which evaded his scrutiny. Arguably the most important concerns the "rise" of finance (Krippner 2011) to a position of ever-greater prominence and significance within the capitalist economy during the globalization era.

To mention this is not to highlight a fatal flaw in Smith's (or anybody else's) theorization of uneven development. Certainly, a "complete" theory of uneven geographical develop-ment or of the globalization that represents its "latest stage of development" would need to deal substantively with the role that finance plays. Equally certain is that there exist studies that have sought to integrate finance into explanations of uneven development both theoretically and empirically; Patrick Bond's study of *Uneven Zimbabwe* (1998), which demonstrates how and why financial intensification exacerbates uneven development at particular moments in the accumulation process, is an exemplar.

Our point, rather, is that if we want to understand and explain the modern economic world, we need to understand and explain how capitalist money and finance "work." No "theory" of the contemporary capitalist economy worthy of the name can afford to ignore such an important dimension of economic relations and dynamics, any more than it can afford to ignore globalization and uneven development. And to understand money and finance, an explicitly geographical approach – an analytical attentiveness to, and an explan-atory emphasis upon, the substantive implication of space, place, and scale in economic processes – is no less crucial than it is in understanding globalization. The next chapter shows why.

References

Abdelal, R. 2007. *Capital Rules: The Construction of Global Finance.* Cambridge MA: Harvard University Press.

Agnew, J., and Corbridge, S. 1995. *Mastering Space: Hegemony, Territory and International Political Economy.* London: Routledge.

Autor, D.H., Dorn, D., and Hanson, G.H. 2016. The China Shock: Learning from Labor Market Adjustment to Large Changes in Trade. http://www.nber.org/papers/w21906.pdf (accessed July 5, 2017).

Bair, J., and Werner, M. 2011. The Place of Disarticulations: Global Commodity Production in La Laguna, Mexico. *Environment and Planning A* 43: 998–1015.

Beckert, S. 2014. *Empire of Cotton: A New History of Global Capitalism.* London: Allen Lane.

Beder, S. 2006. *Suiting Themselves: How Corporations Drive the Global Agenda.* New York: Earthscan.

Bond, P. 1998. *Uneven Zimbabwe: A Study of Finance, Development and Underdevelopment.* Trenton, NJ: Africa World Press.

Brenner, N. 1998. Between Fixity and Motion: Accumulation, Territorial Organization and the Historical Geography of Spatial Scales. *Environment and Planning D* 16: 459–481.

Chari, S. 2004. *Fraternal Capital: Peasant-Workers, Self-Made Men, and Globalization in Provincial India.* Stanford: Stanford University Press.

Christophers, B. 2014. The Territorial Fix: Price, Power and Profit in the Geographies of Markets. *Progress in Human Geography* 38: 754–770.

Christophers, B. 2016. *The Great Leveler: Capitalism and Competition in the Court of Law.* Cambridge, MA: Harvard University Press.

Cox, R.W. 1992. Global Perestroika. *Socialist Register* 28: 26–43.

Ferguson, J. 2005. Seeing Like an Oil Company: Space, Security, and Global Capital in Neoliberal Africa. *American Anthropologist* 107: 377–382.

George, H. 1886. *Protection or Free Trade.* New York: National Single Tax League.

Gill, S. 1992. Economic Globalization and the Internationalization of Authority: Limits and Contradictions. *Geoforum* 23: 269–283.

Glassman, J. 2012. The Global Economy. In T.J. Barnes, J. Peck, and E. Sheppard (eds), *The Wiley-Blackwell Companion to Economic Geography.* Oxford: Wiley-Blackwell, pp.170–182.

Goldman, M. 2005. *Imperial Nature: The World Bank and Struggles for Social Justice in the Age of Globalization.* New Haven, CT: Yale University Press.

Goswami, M. 2004. *Producing India: From Colonial Economy to National Space.* Chicago: University of Chicago Press.

Hart, G.P. 2002. *Disabling Globalization: Places of Power in Post-Apartheid South Africa.* Berkeley: University of California Press.

Harvey, D. 1989. *The Condition of Postmodernity.* Oxford: Blackwell.

Harvey, D. 2001. Globalization and the Spatial Fix. *Geographische Revue* 2: 23–31.

Harvey, D. 2006. *Spaces of Global Capitalism.* London: Verso.

Helleiner, E. 1994. *States and the Reemergence of Global Finance: From Bretton Woods to the 1990s.* Ithaca, NY: Cornell University Press.

Herod, A. 2009. *Geographies of Globalization: A Critical Introduction.* Oxford: Wiley-Blackwell.

Jarosz, L., and Qazi, J. 2000. The Geography of Washington's World Apple: Global Expressions in a Local Landscape. *Journal of Rural Studies* 16: 1–11.

Kelly, P.F. 2000. *Landscapes of Globalization: Human Geographies of Economic Change in the Philippines*. New York: Routledge.

Krippner, G.R. 2011. *Capitalizing on Crisis: The Political Origins of the Rise of Finance*. Cambridge, MA: Harvard University Press.

Krueger, A. 2004. Economic Growth in a Shrinking World: The IMF and Globalization. https://www.imf.org/en/News/Articles/2015/09/28/04/53/sp060204 (accessed July 5, 2017).

Levitt, T. 1983. The Globalization of Markets. *Harvard Business Review* 61: 92–102.

Lothian, J. 2001. Financial Integration over the Past Three Centuries. *Bancaria* 9: 82–88.

Luke, T.W., and Tuathail, G.O. 1997. Global Flowmations, Local Fundamentalisms, and Fast Geopolitics: 'America' in an Accelerating World Order. In A. Herod, G.O. Tuathail, and S.M. Roberts (eds), *An Unruly World? Globalisation, Governance and Geography*. London: Routledge, pp.72–94.

Mackinnon, D., and Cumbers, A. 2007. *An Introduction to Economic Geography: Globalization, Uneven Development and Place*. Harlow: Pearson Education.

Massey, D. 2002. Globalisation: What Does It Mean for Geography? *Geography* 87: 293–296.

Meek, J. 2017. Somerdale to Skarbimierz. *London Review of Books*, 20 April. https://www.lrb.co.uk/v39/n08/james-meek/somerdale-to-skarbimierz (accessed July 5, 2107).

Nagar, R., Lawson, V., McDowell, L., and Hanson, S. 2002. Locating Globalization: Feminist (Re)readings of the Subjects and Spaces of Globalization. *Economic Geography* 78: 257–284.

Ohmae, K. 1995. Putting Global Logic First. *Harvard Business Review* 73: 119–124.

Rodrik, D. 2007. The Inescapable Trilemma of the World Economy. http://rodrik.typepad.com/dani_rodriks_weblog/2007/06/the-inescapable.html (accessed July 5, 2017).

Scott, J.C. 1998. *Seeing Like a State: How Certain Schemes to Improve the Human Condition Have Failed*. New Haven, CT: Yale University Press.

Sheppard, E. 2002. The Spaces and Times of Globalization: Place, Scale, Networks, and Positionality. *Economic Geography* 78: 307–330.

Sheppard, E. 2012. Trade, Globalization and Uneven Development: Entanglements of Geographical Political Economy. *Progress in Human Geography* 36: 44–71.

Smith, N. 1984. *Uneven Development: Nature, Capital, and the Production of Space*. Oxford: Blackwell.

Smith, N. 1995. Remaking Scale: Competition and Cooperation in Prenational and Postnational Europe. In M. Eskelinen and F. Snickars (eds), *Competitive European Peripheries*. Berlin: Springer, pp.59–74.

Smith, N. 1997. The Satanic Geographies of Globalization: Uneven Development in the 1990s. *Public Culture* 10: 169–189.

Swyngedouw, E. 1997. Neither Global nor Local: 'Glocalization' and the Politics of Scale. In K.R. Cox (ed.), *Spaces of Globalization: Reasserting the Power of the Local*. New York: Guilford Press, pp.137–166.

Trotsky, L. 1977 [1930]. *The History of the Russian Revolution*. London: Pluto.

Werner, M. 2015. *Global Displacements: The Making of Uneven Development in the Caribbean*. Oxford: Wiley-Blackwell.

Wolf, M. 2004. *Why Globalization Works*. New Haven, CT: Yale University Press.

Chapter 9

Money and Finance

9.1   Introduction

"We Live in Financial Times," declared the London-based international daily business and economic news broadsheet the *Financial Times* in a new slogan launched in April 2007. The timing could not have been more propitious. By the end of the year, with the subprime "credit crunch" – the formative episode of what has since come to be termed the Great Recession – in full swing, the United Kingdom had witnessed its first "run" on a deposit bank (Northern Rock) in over a century and the US stock market had begun a decline that would see it lose more than half its value in a little under 18 months. Financial times indeed.

We also live, we are often told, in placeless/spaceless or "ageographical" times. Two years prior to the launch of the new *Financial Times* slogan the US journalist Thomas Friedman published his paean to globalization (Chapters 1 and 8), *The World is Flat*, in which he argued that differences between places were becoming less important as the world became more interconnected. Friedman's thesis was not entirely new. In 1992, the same year that Francis Fukuyama had published his famous book on "the end of history" (based on an essay first published in 1989), Richard O'Brien had published a book subtitled "the end of geography," presaging many of the ideas about the declining importance of place and space that Friedman would later popularize.

The particular importance of O'Brien's book for our purposes here is that he argued that these two ostensible epochal developments were linked. The world was flat – geography had come to an end – *because* we now lived in financial times, or, more specifically, because deregulation and new information and communication technologies had combined to

Economic Geography: A Critical Introduction, First Edition. Trevor J. Barnes and Brett Christophers.
© 2018 John Wiley & Sons Ltd. Published 2018 by John Wiley & Sons Ltd.

create a world of free financial flows in which location was increasingly immaterial, especially to suppliers and customers of financial services. Thus, no less important than the book's subtitle was its main title: *Global Financial Integration*.

O'Brien's claim should be taken seriously. After all, if we live in financial times – and we patently do – any academic approach to the economic world worth its salt must have something useful to contribute to understanding finance and what is often now referred to as the economy's and society's "financialization" (see Box 9.1: Financialization). But if

Box 9.1 Financialization

Social science is riddled with buzzwords – globalization (Chapter 8) and neoliberalism being two obvious ones – and a particular favorite in recent years has been "financialization." It has become especially common terminological currency in economic sociology and economic geography.

Definitions of financialization abound, and they all say something different: sometimes only slightly different, sometimes very different. As is the case with globalization, there is no accepted definition on which everyone agrees. But more or less everyone who uses the term would say that it refers to a set of ongoing historical transformations dating to the 1970s or 1980s when "finance" broadly understood became increasingly important across all spheres of social life – not just the economic but the political and the cultural.

There are three supposed developments central to the history of financialization. First, scholars assert that finance has become more important in individuals' daily lives, with financial concerns and calculations shaping decisions and behaviors that hitherto reflected non-financial considerations. Second, they assert that finance has gotten more important in the life of corporations. Where corporations used to base their strategies and operations on the needs and values of a diverse range of relevant stakeholders (employees, local citizens, shareholders, customers, perhaps even non-human actors and environments), increasingly all that matters is one stakeholder group – shareholders – and one type of value – financial, shareholder value. And third, scholars assert that capitalism *itself* has become financialized. What they mean by this is that if we conceptualize capitalism in terms of the different types of corporations active in the economy and the different types of income streams they earn, finance appears to have grown in significance: more and more income has been flowing to financial corporations rather than non-financial corporations; and more and more of the income earned by non-financial corporations has itself been of a "financial" type (e.g., interest or dividend payments).

Not everyone thinks "financialization" is necessarily a useful concept for understanding contemporary capitalism and its economic geographies, especially given the promiscuity with which it is increasingly invoked (Christophers 2015). Nonetheless, it has become a commonplace and looks set to stay. Understanding what scholars mean, or might potentially mean, when using the word is essential for today's economic geographers.

geography is at an end, if finance itself has helped bring about this termination, and if such flatness is writ especially large on the financial landscape, what on earth can economic geography bring to the table?

The present chapter answers this question. It does so by showing that O'Brien's claim is extraordinarily wide of the mark. Space and place (still) matter profoundly not only to processes and relations of finance but also to those of finance's twin, in isolation from which finance can itself not be understood: money. Hence to understand our financial times and the centrality of money to them, it is imperative to understand their geographies.

9.2   The Beginning of Geography

9.2.1   Stretching and connecting

If his specific reading of changes in the relationship between finance and geography was awry, O'Brien was nonetheless right to highlight the significance of – and thus the importance of studying and understanding – that relationship per se. Geographical factors are acutely material to both monetary and financial realities; money and finance, in turn, fundamentally shape both worldly manifestations and social experiences of space and place. And, while this chapter is interested primarily in the former set of influences (the implication of the geographical in monetary and financial dimensions of the "economic"), it is hard to appreciate these without first illuminating the latter.

Our argument is that modern finance (and money) does not herald the end of geography so much as its beginning. Or, to put things differently, money and finance constitute a core foundation of the particular (capitalist) experience of geography that most of us take as "normal": they underwrite the spatial infrastructure that we *start* from in our daily lives. How so?

Consider first how most goods and services, the mutual provision of which fuels all social reproduction, are typically exchanged. For example, we – the two authors of this book – both work in universities. The services we *provide* are teaching and researching. The goods and services we *procure* in life, meanwhile, are virtually limitless, ranging from food and clothing to internet provision and air travel. Crucially, the latter are, in a very real sense, *exchanged* for the former. To be sure, the state or family might come to our aid if we were in need. But it is in providing something deemed "valuable" by society that we most reliably procure things that are also deemed valuable. Our key point is that this exchange is rarely immediate in either time or space. The food and air travel "exchanged" for teaching services are not received in the classroom at the time we give lectures, nor are they procured from those we teach. They are supplied at another time, in another place, by someone else.

How is this separability in time and space facilitated? The answer, of course, is money. Capitalist money has several functions (see Box 9.2: Money) of which two are crucial to our example. In serving as both a *medium of exchange* and a *store of value*, money "permits the separation of buying and selling in both space and time" and is thus, David Harvey (1989, p.175) observes, "the great integrator and unifier across the grand diversity of traditional communities and group interests." Receiving money for our work, we do not need to spend it straightaway. We have the freedom within certain important limits to spend it where we like, when we like, on what we like. Such "freedom" is so engrained within us, so seemingly

Box 9.2   Money

What is money? This is probably the wrong question to ask; you will likely soon find yourself in a terrible muddle if you think about it too closely. The better question to ask is: What function (or functions) does what we recognize and accept *as* "money" typically serve? If something reliably fulfills this function (or these functions) then, by definition it *is* money. It can be gold; it can be paper; it can be something else.

In modern capitalism, "money" is generally expected to serve four different but related functions. Some of the most heated debates in scholarship on money turn on the relative significance of these different functions. Which one is most important? Least important? Which one came first? And indeed, *does* something have to provide all four functions to be money? We are going to stay well out of those debates (see Ingham 2004 for a brilliant introduction), simply listing the four generally accepted functions in alphabetical order.

- *Means of payment*: If someone provides us with a good or service, unless it is a gift, we incur a debt. How is that debt settled, or paid for? One method is through barter: providing a good or service in return that is deemed of equivalent value. Another, of course, is money. Unless something can be used as payment to settle a debt, it is not money.
- *Medium of exchange*: Trading or exchanging various goods and services for one another is at the very heart of capitalism. This exchange can be either direct or indirect. Direct exchange is barter, where, as noted, the value of the bartered items is considered equivalent. Where exchange is indirect, something effectively *intermediates*, or comes in between. For example, instead of item A exchanging for item B, A exchanges for X, which exchanges for B. The intermediary, X, is referred to as a medium of exchange. Money fulfills this function by allowing the value of both A and B to be assessed and rendered in *its* terms. Crucially, the exchange facilitated by money is not only indirect, it is potentially multilateral. Bilateral exchange sees A traded for B directly or indirectly; in multilateral exchange, the butcher's wares can exchange for the brewer's … which can exchange for the baker's … which can exchange for the butcher's … or someone else's.
- *Store of value*: Money would not be anything like as powerful and useful as it is if the aforementioned exchanges, direct or indirect, had to occur simultaneously. I would be much less inclined to accept money as payment for a good if I had to spend it immediately. (Imagine the stress.) But I do not. I can hold onto the money and spend it later. That is because money *stores* the value that facilitates exchange; this is its third function. It allows deferral. Of course, money does not store value perfectly. Inflation often eats away at this value. The higher the inflation, the greater the incentive not to defer expenditure, and the greater the stress.
- *Unit of account*: Last but not least (Max Weber actually thought it the most important function of all), money serves as unit of account. If exchange is to occur through a medium, that is, indirectly, it is absolutely necessary for the worth of all goods and services to be capable of being calibrated relative to one another.

> Is a horse worth 10 smart phones or five? In other words, the property that all exchangeable things are said to have that makes them exchangeable – what we call "value" – needs a consistent, comparable *measure*. This is the unit of account, also referred to for obvious reasons as the "measure of value." Under capitalism, it is money that provides this measure.

"natural," that we take it utterly for granted. Yet, were a dependable and trustable medium of exchange and store of value not readily available, our geographical lives would look totally different. It is in this sense that money is a – if not the – beginning of those lives.

Further, finance increasingly serves to reinforce and deepen the connectivity between places and peoples that money, by "stretching" exchange relations, underwrites. The most basic meaning of the verb "finance" is to supply with money, and when individuals or companies in one place finance individuals or companies in another place, those economic actors and the places they inhabit become inextricably connected until such time as the obligations associated with such financing – for example, in the case of a simple monetary loan, to pay interest and repay the "principal" loan amount – have been discharged. If a distant loanee defaults on a substantial loan there are likely to be repercussions for the provider of the loan and for the local economy in which to one degree or another it is embedded.

Historically, such distant, finance-based geographical connections were emphatically not the norm. Most financing was local, among those who knew each other personally. One of the first major capitalist institutions to be funded in significant part from elsewhere was the Bank of England, some 43% of the shares of which were held overseas as early as 1762 (Christophers 2013, p.67). Only once such distant financing became more common did a phenomenon about which we often hear today – international financial "contagion" – become possible. Previously, when money was loaned or invested largely locally, the risk of loans not being repaid or investments going sour was locally circumscribed. The risk might be contagious, with one default triggering others in a domino sequence, but without an extra-local transmission mechanism it could not spread far. Historians often identify the panic of 1847 as "the first spreading international financial crisis. It first occurred in England and France," reports Sell (2001, p.17), "and from there it spread to the Netherlands, to Germany and even to New York." Such contagion was rooted in international financing.

9.2.2 Stitching and blending

Today, the idea that major economic developments in one place will have knock-on effects elsewhere is arguably no less part of our commonplace, baseline experience of geography than the idea that we have spatial freedom in procuring basic goods and services. A major reason for this has been the enormous growth in recent decades in the use of particular modern financial techniques – especially *securitization* – and instruments – especially *derivatives*. People often prefix these with the word "complex," but generally they are not, even though they are often made to *appear* complex both by the financiers who specialize in them and by the scholars who style themselves financial "experts" (Christophers 2009).

Securitization and derivatives are essentially straightforward. Let's briefly consider some of the implications for global economic geographic connectivity of each of these in turn.

To understand financial securitization it is necessary to know what a financial security is. It is a *tradable* financial asset. A non-tradable financial asset is something like a basic bank loan. When a bank provides you with such a loan, the right to receive your interest payments and, at the end of the loan period, your repayment must remain with the bank. It is the bank's asset (and your "liability"). It is not a financial security. If it *were* a financial security, the bank could trade it, by selling it to another company. That company would then hold the asset and receive your future payments. So what is securitization? It is simply the technique for converting the first type of asset – in this case, a non-tradable loan – into the second – here, a tradable loan *security*. That is all. It makes financial assets exchangeable.

Securitization matters to our geographical story because it enables the stitching together of distant financial relationships that were more or less inconceivable pre-securitization. As most people know from films like *The Big Short*, one of the loan types most heavily securitized in recent decades was the residential mortgage. Such securitization was centrally implicated in the 2007–2008 credit crunch and ensuing financial crisis. Famously, it allowed investors like Scandinavian town councils to buy the mortgages of US homebuyers, which had been bundled together and converted into financial securities by US banks. Without securitization, the idea of a US homebuyer (effectively) borrowing from a Scandinavian municipality would have been unimaginable; the necessary relationship and trust was absent. But it was not unimaginable once the clever people on Wall Street had developed the new technique. "In Norway, for example, the municipalities of Rana, Hemnes, Hattjelldal, and Narvik invested some $120 million of their taxpayers' money in [credit assets] secured on American subprime mortgages … In no time at all," quipped Ferguson (2009, p.270), "the risk was floating up a fjord." Norwegian municipal economies became tethered directly to US municipal economies.

Meanwhile, derivatives are themselves a breed of financial security, ones *derived from* underlying assets. The range of such underlying assets is essentially limitless, but derivatives derived from commodities (copper, wheat, iron ore, etc.) and financial assets (company shares, national currencies, etc.) are especially common. Prominent derivative contract types are the forward contract, which denotes an obligation to buy or sell an underlying asset at a particular price at a particular future date, and the option, which denotes an option to do the same. So, an example of a forward would be an agreement with a counterparty to buy a barrel of crude oil from them for $50 in six months' time. A contract that gave me the ability, but *not* the obligation, to sell to a counterparty 100 shares in Microsoft Corporation for $50 apiece in a year's time would be an option.

But what have derivatives got to do with economic connectivity across space? Bryan and Rafferty (2006) help provide an answer with their concept of "blending." What derivatives fundamentally do, according to these authors, is establish pricing relationships that powerfully enable the blending, or *commensuration*, of different types of assets and the places with which they are associated: a Chilean soy bean, the Swedish currency, and so on. They impart spatial continuity to the world, fusing the economy of one place with that of others.

The type of derivative that does this best is the swap, of which the most common type is an interest rate swap. Here is an example. Company A, in the United States, has a $1 billion loan requiring it to pay interest of 6.0% per annum for five years. Company B, in India, has a loan of 100 billion rupees requiring it to pay interest annually, also for five years, at the prevailing rate (6.5% at the time of this writing) set by the Reserve Bank of India, the nation's central bank. A swap essentially allows the two to swap their interest rate obligations: you pay mine and I'll pay yours. Why they would want to do this is not important here. What matters is the effect, which is to bind the US and Indian economies together by blending their interest rate environments, burying the very notion of a "national" monetary space. Think we are exaggerating? Think again. In mid-2015, the estimated value of open interest rate swaps was just under $300 trillion (BIS 2016), an amount far greater than the value of annual global economic output. Finance, like money, structures the geographical connectivity with which we now live, largely unknowingly, on a daily basis.

9.3   Place Matters

As finance orients geography, so also geography orients finance. If geography really did not matter in the world of finance, as O'Brien suggests is the case, then that world would look nothing like it does.

For one thing, one would presumably anticipate a relatively even, perhaps random spatial distribution of firms supplying financial services – banks, insurers, and the like. With money and finance flowing freely across space, and access to them instantaneously, electronically mediated, suppliers could operate from more or less anywhere, or at the very least might be expected to gravitate to low-cost locations.

But of course the map of the financial industry does not conform in any respect to the "end of geography" narrative. Instead it features, *inter alia*, a nested hierarchy of "central places" acting as regional, national, and international financial centers. And this hierarchy varies between different segments of the financial industry, too. Why, for instance, is the international reinsurance industry so heavily concentrated in Switzerland (Zurich) and Germany (Munich)? How and why did London, New York, and Hong Kong historically develop into today's preeminent nodes of the global securities sector? Plainly we cannot come to grips with the dynamics of the reinsurance and securities industries without explaining the particular roles that such central places play and, relatedly, the reasons for their economic centrality.

A great deal of research has been conducted, for instance, into London's pivotal role in global financial circuits. Even as the United Kingdom's wider historic dominance in the global economy declined through the twentieth century, particularly with the rapid dissolution of its empire in the middle decades, London maintained and even strengthened its specifically financial supremacy. As the economic geographer Gordon Clark (2002) has shown, numerous different explanations have been posited for this, including a history of highly favorable local regulation (on which more below) and the distinctive nature and quality of market relationships centered there. Clark himself emphasizes that London's strength is as much about its relative position in international "space–time" as it is about its absolute qualities as a place. As "a switching point between Asia, and Europe and North America," it is "a vital point in time and space for the global financial institutions that manage on a 24-hour basis flows of capital and transactions around the world" (p.450). Major global financial institutions *have* to operate there to be credible; they need the hub capacity that London uniquely provides.

Meanwhile, the financial economy is not a flat, place-neutral world for customers either. Where those who buy financial services are located is often just as material as the location of those who provide those services. For several decades researchers have called attention to processes of so-called redlining whereby customers are treated differently based on where they live. The classic example of redlining concerns mortgage finance: financial institutions discriminating against people in particular locations (often characterized by distinctive racial or ethnic demographic profiles) either by withholding home loans altogether or by charging higher interest rates (e.g., Williams 1978). Yet mortgage finance is certainly not the only domain in which customer location affects service availability and cost. Insurers, for example, routinely categorize different places as representing different risks, and they price – or even withhold – their products accordingly. Recognition of such practices informs a well-established critique of a wider set of so-called geographies of financial exclusion (Leyshon and Thrift 1995).

That the effects of monetary and financial practices are experienced differently in different places is not necessarily always by design. Though geographical differentiation may be built into the very fabric of a product or service it can also be an "innocent," though no less inevitable, by-product. Turning our attention momentarily from finance to money and from private to public institutional practices, Mann (2010) has discussed such uneven outcomes in relation to the somewhat shadowy world of monetary policy. Central banks like Canada's, he shows, use purely aggregate, national-level data in setting national interest rates, thus neglecting regional differences in such critical economic phenomena as inflation, debt levels, and unemployment. "Although formulated as if the nation were homogeneous," Mann notes (p.612), the fact of such (ignored) regional differences means that interest rate changes cannot fail to have "uneven effects" since those differences profoundly influence how the economy responds to interest rate changes.

In sum, place matters across the spectrum of monetary and financial practices, and to all stakeholders. Indeed, O'Brien's claim that the increasingly free-flowing nature of global finance has eroded the significance of place is not just misleading but arguably runs directly *counter* to the reality – which is of place becoming more, not less, material. And if one category of place exemplifies this trend more than any other it is the so-called offshore financial center (OFC), of which the Cayman Islands represent a well-known example.

Consider first those financial service companies that operate in OFCs. Part of their rationale for doing so is that they have clients who – for reasons we shall turn to shortly – want to do business there and who thus call forth willing service suppliers. But this is not the only reason. As Roberts (1994, p.93) notes, OFCs exist, by definition, "beyond the reach of regulation by the originating national economy"; they are offshore precisely because "onshore" regulation does not apply. Historically, the most important example by far is the "Euromarkets" that blossomed in the 1960s and 1970s. The "originating national economy" referenced by Roberts was in this case primarily the United States: a "Eurodollar" was a dollar deposited outside the United States, while a Eurodollar bond was a dollar-denominated bond (a bond being a securitized and hence tradable loan paying a fixed interest rate referred to as the "coupon") issued outside the United States. And the Euromarkets' main offshore center was London. Critically, neither the US nor the UK authorities regulated these (dollar) markets, and nor did anyone else. In a world where money and finance ("capital") flow increasingly freely across national borders (Chapter 8), those in the business of handling such flows were – and still are – strongly attracted to where regulation is lightest.

Why, in turn, are their clients attracted to OFCs? While minimalist regulation is certainly often part of the attraction for them, the more fundamental lure of OFCs turns on taxation. OFCs typically impose a very low – and sometimes even zero – tax burden. Hence, if we combine client with supplier rationales we can say, after Roberts (1994, p.96), that the use of OFCs is collectively about exploiting "the uneven global topography of taxation and regulation to gain advantage."

The most important point for our purposes is that what allow such exploitation are precisely the free financial flows discussed by O'Brien. Had new information and communication technologies and the removal of controls on cross-border capital movements not historically freed up such flows, both those with money to invest and the companies servicing them would have remained considerably more constrained in geographical choice. Far from flattening the globe's topography, in other words, O'Brien's "global financial integration" actually *sharpens* it insofar as it makes the peaks easier to access and the troughs easier to avoid. Geographical differences become *more* material because they are easier for stakeholders to navigate and exploit. The contemporary prominence of OFCs is ultimately a testament to this materiality.

9.4   Spatial Relations of Finance

So far we have mainly been discussing place. For instance, some places are able disproportionately to attract financial company operations, becoming, like London, financial centers, on- or offshore. And all places are affected, albeit in different ways, by monetary and financial operations, whether such differentiation is inherent and intended (as in the case of redlining) or not (as in the case of monetary policy).

While the significance of place arguably represents the most obvious way in which geography influences money and finance, it is not the only one, nor necessarily the most important. Money and finance are also, we now want to suggest, *spatially constituted*, by which we mean two different but equally crucial things.

The first is that the *scale* at which monetary and financial forms and practices are constituted is material to them. Perhaps the best example of this scalar materiality relates to money itself. Most currencies are national currencies, issued by national governments. There are also sub-national, "local" currencies, however. And there are also supra-national currencies, most notably the euro, which is (currently) the legal tender of 19 of the 28 member states of the European Union. The scale of such monies matters enormously to their respective economic, political, and social effects. For now we will park this question of scale before returning to it explicitly in the following section.

Our second understanding of spatial constitution, and the one we focus upon in this section, refers to the fact that financial processes are predicated upon and influenced by *spatial* relations, relations across geographical space. Put another way, different places are connected with one another *through* the operation of financial processes, and those connections between places shape the form and outcome of the processes in question.

We have already encountered an explicit example of the importance of such spatial relations: London's role as an international financial hub and the significance, according to Clark (2002), of its stitching together of international financial space–time. What matters here is not so much London as place but London *in relation* to other places. We have also arguably encountered an implicit example: redlining. This is a bit harder to appreciate. Redlining looks on the face of it to be just about (one) place: the one being financially discriminated against. Yet, if we think about the economics of a lending institution in the round, then it is clear that its overall profitability and viability depends on the performance of mortgages across its whole portfolio. Higher returns in one part of a portfolio can accommodate lower returns in another. It is thus not hard to envision – though it would be much harder to prove – customers in non-redlined neighborhoods being subsidized by punitive interest rates be charged elsewhere. Either way, it is unarguable that for multi-location lenders, lending decisions in one place cannot safely be made independently of decisions and outcomes in others: places are necessarily conjoined in the financing tableau that constitutes the lender's consolidated asset base.

To make the case about the spatial relations of finance more emphatically, it is useful to look at examples from two very different parts of the literature on financial geographies – one pertaining to developments in contemporary China, the other to developments in California almost exactly a century ago.

i. *California branchin'*: Henderson (1998) analyzes the growth of branch banking in the agricultural interior of California in the years between World War I and the Great Depression. By focusing on the spatial relations of finance, Henderson shows why branch-bank networks – and especially the market-leading Bank of Italy, which mushroomed from two branches in 1912 to 24 by 1919 and over 200 by 1927 – were so superior to standalone independent banks. The reason is that the landscape featured many different crops, grown in different places and harvested at different times of the year, implying an extremely lumpy seasonal-spatial profile of financing needs. What appeared as an uncoordinated "jumble of space-time fragments" to individual local banks crystallized as a cumulative, complementary "year-round demand for credit" to "a bank that could channel funds from branch to branch as needed" and thus effectively "coordinate these space-times" (p.107). The effectiveness of financing institutions and of their financing operations depended intimately on their spatial structure – on whether they were rooted in one place, or they were distributed across space and able to realize synergies between those places of operation.

ii. *Intercity China*: Lai (2012) analyzes the roles of the three major financial centers in China today: Shanghai, Beijing, and Hong Kong. In doing so, she seeks to explain why it is that the recent emergence of the first two of these has not led to Hong Kong's decline, as many commentators had predicted it would; rather, all three are flourishing. Like Henderson's, Lai's explanation rests on the spatial relations of finance, or on the relations *between* places. Large foreign banks do not see the three cities as competing locational options, which had been the stock assumption, so much as complementary ones. The cities are a network of financial centers and "it is this network that enables them to capitalize on their respective advantages and perform distinctive roles … within the regional strategies of foreign banks." Specifically, foreign banks' locational decision-making vis-à-vis China "is based on and reinforces the distinctive development of Shanghai as a commercial centre, Beijing as a political centre and Hong Kong as an offshore financial centre" (p.1277). Those banks' operations in China bring together the three cities in distinctive ways, and are themselves shaped by that intercity network in no less distinctive a fashion.

9.5   Geographies of Power

In the remainder of this chapter we want to focus more explicitly on the "critical" dimensions of our introduction to the geographies of money and finance. We think two such dimensions are especially important. One of these is to inquire more closely into the notion of critique specifically in relation to monetary and financial issues. At various points in the book so far we have discussed in more general terms what might be meant by "critical" approaches to economic geography, what "critical theory" potentially entails. But why might a critical perspective on *money and finance* in particular be important? What, if anything, is it about money and finance that makes the adoption of a critical standpoint especially important or worthwhile above and beyond the more generic motivations for being critical?

Second, we aim, in the process, to offer a critical reading of recent work in and around the economic geographies of money and finance – a reading that takes seriously the specificities of monetary and financial phenomena. Here we try to highlight the main ways in which critical approaches to the geographies of money and finance differ from uncritical ones. Although being uncritical is certainly not the only characteristic that potentially renders studies of monetary and financial geographies problematic, we argue that it is an extremely important and also relatively common one.

If there is one word that best encapsulates the reasons for adopting a critical perspective on money and finance, then that word is undoubtedly "power." Power pervades monetary and financial affairs; and critical social science is nothing if not concerned to identify power and interrogate its sources and effects.

Power and money, in fact, go together hand in glove. Money furnishes people and institutions with power, not least power over one another. This, in fact, was one of Marx's key insights: that in capitalist society money is social power objectified since it provides the wherewithal to command living labor – to put waged labor to work in the production process – and thus to create surplus value and realize *more* money. But money does not just confer power; its very viability, which is to say the fact that we take for granted that money has value and can perform reliably the functions we expect of it (Box 9.2: Money), depends intimately *upon* power. This is most emphatically the case in modern societies using so-called fiat money, where the value of that money is derived not from its being made of something that is itself deemed valuable (such as a precious metal), but rather from the fact that a powerful institution – for example a government – decrees that it has value, and guarantees it.

We will turn to the vital role of governments in a moment, because that role has a pivotal geographical component; but it is important to recognize that in a fiat money system the power to create money is not vested uniquely in governments and (at least nominally) independent central banks such as the US Federal Reserve or the European Central Bank. Private commercial banks *also* create money, and can do so in volumes unconstrained by the money creation activity of state and quasi-state institutions – which, given the power embedded in money, is no small privilege. Mainstream economists have long refuted such private power, but no less mainstream an institution than the Bank of England recently admitted it: "banks create money, in the form of bank deposits, by making new loans. When a bank makes a loan, for example to someone taking out a mortgage to buy a house, it does not typically do so by giving them thousands of pounds worth of banknotes. Instead, it credits their bank account with a bank deposit of the size of the mortgage. *At that moment, new money is created*" (McLeay, Radia, and Thomas 2014, p.3; original emphasis).

Finance, needless to say, is no less saturated in relations of power. Consider the lending of money at interest. When we borrow money we create and enter into a particular form of social relation, between creditor and debtor, and that relation is by its very nature a power relation – whether the power of the creditor is overt and socially abjured (as in the case of a loan shark or payday lender) or more subtle and socially accepted (as in the case of a high-street bank). Lending money gives the creditor power both to make demands – for the payment and repayment of interest and principal, respectively – and, if those demands are not met, to take actions, of which the seizure of property is a well-recognized one.

Moreover, geographic space very often explicitly mediates the relations between power, on the one hand, and money and finance, on the other. Those relations are spatial relations. This explains the particular importance of a critical *geographic* perspective. As Harvey (1989) argues, money can be used to command space and place just as readily as it can labor, and command over space "can easily be parlayed back into command over money" (p.186). In short, space, power, and money – and, as we shall see, finance – are inextricably intertwined. Hence the title of one of the seminal publications in this area of economic geography: *Money, Power and Space* (Corbridge, Martin, and Thrift 1994).

Harvey (1989) uses the example of a property speculator to illustrate this co-constitutive relationship, but much more striking is the example alluded to earlier: that of the nation-state. If we take power and money first (before then turning to power and finance), the state's significance is clearly writ large inasmuch as the nation-state is the dominant, though not only, monetary scale. The power of the state enables it to create money and legitimate its circulation. But money is not only a mere expression of state power. It has historically been crucial to the very consolidation of national space and to the territorialization of state power that this consolidation represents. The creation of *national* currencies helps affirm territorial coherence by (re)producing *national* "identities, values and communities" attached to, among other things, the "the very images produced on the currency, which … forge the nation as an 'imagined community'" (Gilbert 2005, p.375). Recognition that the coherence of the nation-state at its extant scale is "tightly bound to the value and stability of domestic money" leads Mann (2010, p.617) to argue that "if the modern state lives on (in many ways robustly) when it was supposed to have withered in the face of international capital, it lives through its money."

If this sounds unduly abstract, consider, in turn, the geographical nexus of the nation-state, its power, and *finance* – specifically, the dominant form of state financing, taxation. As Cameron (2006, pp.240–241) observes, any system of taxation "must of necessity be expressed in spatial terms – both to define internal jurisdiction" (i.e., where, and thus from whom, the sovereign state has the power to collect taxes) "and to distinguish one 'sovereign' space from all others." Such power over financial space, moreover, has historically been the locus of considerable contestation and conflict, Cameron noting the "violence and militarism" widely associated with "the gradual growth of taxation as a means of funding the nascent state." This securing of financial space was arguably of a significance equal to that of the securing of monetary space in the historic consolidation of territorialized state power.

One of the most fascinating, significant, and problematic ongoing developments in relation to this particular nexus concerns its apparent fraying at the edges. Globalization

(Chapter 8) has fundamentally challenged the security and coherence of the nation's tax-based financial space. Within which sovereign space of taxation do multinational corporations "belong," and thus where should they pay corporation tax? This question has precipitated not only a "race to the bottom" among nation-states – vying to attract corporate headquarters by levying competitively lower taxation rates – but also allegations of corporate "tax arbitrage," which entails the use of creative accounting practices such as disproportionately recognizing profits (on which corporate taxes are levied) in low-tax states. Many major multinational corporations have been fingered, and it is not hard to see why.

Consider Figure 9.1, which shows data for Boeing, the world's largest aerospace company. Note two things, which may or may not be connected to one another. First, almost all of Boeing's profits are recognized in the United States, despite less than half its revenues being generated there; in other words, the company claims that its US operations are highly profitable, while its non-US operations, year after year, barely break even. Second, thanks to various forms of tax relief (e.g., research and development credits and a US manufacturing activity tax benefit), the tax rate Boeing actually pays on its reported US profits is considerably lower than the statutory rate.

The integrity of national taxation space is, in short, in danger of unraveling (if it has not yet already done so) as the spatial constitution of sprawling multinational corporations enables them to escape the power of the nation-state to tax them. The revenues of (some) nation-state governments are arguably endangered as never before; Boeing, for example, generates vast revenues in Europe and Asia, totaling more than $30 billion in each of 2013, 2014, and 2015, but the governments of the relevant countries earned hardly a penny in taxation from Boeing in any of those years. And insofar as the vulnerability of state tax income imperils state expenditures (on infrastructure, social services, and the like) that justify the state's very existence, so the viability of the nation-state itself comes under threat. The modern nation-state may, *pace* Mann, live on through its money; but who is to say that we will not yet see it die through its (non-)financing?

9.6   The Seductions of Finance

9.6.1   Uncritical geographies

A "critical" approach to the geographies of money and finance, therefore, is one that recognizes the rooting of those geographies in relations of power, and which attunes its analytical radar accordingly. By the same token, we can posit that at least one of the characteristics that can make geographical enquiry *un*critical is its neglect of power. Either it is unable to see such power, or it is unwilling to delve into the nature and implications of the power that it does see.

In the unstable wake of the global financial crisis, at a time when many economic actors (including several of the aforementioned nation-states, especially in the eurozone) are still palpably *in* financial crisis, now is a particularly resonant moment at which to reflect on the critical attributes of economic geographic scholarship on money and finance (cf. Christophers, Leyshon, and Mann 2017). Is such scholarship successfully endeavoring to speak the truth to power?

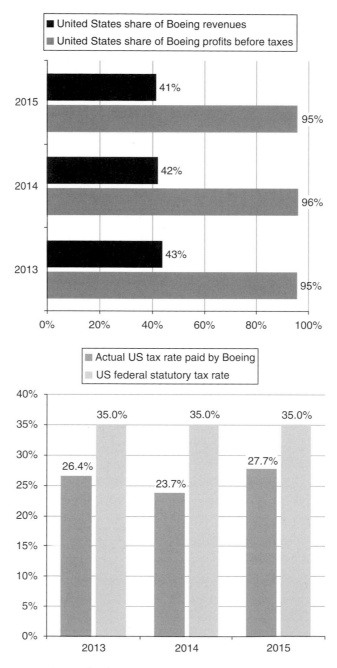

Figure 9.1 Key Boeing financial indicators, 2013–2015. Source: Boeing 2016, pp.68, 69, 106.

This question is given added weight by the fact that commentators other than us have previously expressed doubts. The financial geographer Sarah Hall (2012, p.91), for example, wonders openly about "the extent to which the financial geography in the 2000s … was seduced by the growing power of finance." Instead of questioning this power, did financial geography and geographers come to naturalize it, to regard it as normal and thus unworthy of analytical consideration? Did they perhaps even begin to sidle up to it?

Hall's suspicions relate in part to a shift that economic geographic work on money and finance underwent from the mid-1990s. Having previously been heavily guided by political economic theory, and concerned as such with "the risks, inequalities, and instabilities that characterize money and finance," the sub-field moved toward "a greater concern for the social, discursive, and cultural constitution of money and finance" (2012, p.91). In doing so, Hall conjectures, it "lost sight of" financial power and was perhaps even seduced by it.

Though Hall does not say it, this shift away from political economy was epitomized in the writings of two of the leading exponents of financial geography, Andrew Leyshon and Nigel Thrift. Assembled over the course of a decade, their *Money/Space* (1997) collection moves explicitly from a political economic framing in its early (and earlier) chapters to, in the later ones, "a discursive approach, which emphasizes the role of culture" and in which the "Marxian traces" were "thrown away" (p.xii). But Leyshon and Thrift were notably careful in those later chapters to keep questions of power firmly in view. The specific question that Hall's provocation raises, therefore, is whether other economic geographers, in building out such "discursive" approaches over the following decade (i.e., the decade leading up to the beginning of the financial crisis in 2007), did the same – or whether power got lost by the wayside.

Let's nail our colors to the mast. We think Hall was absolutely right to be concerned. Over the decade in question, economic geographic work on money and finance *was* widely, though certainly not universally, seduced by finance and its power. It failed in large part to analyze such power and hence to *be* critical. The deep interest in power and its geographical dimensions that informed Leyshon and Thrift's work and also economic geographer Linda McDowell's (1997) brilliant study of the gendered cultures of the City of London's varied workspaces essentially disappeared from view. And this was true not only of research into the cultures and discourses of finance but also of research drawing on institutional economics and taking more of a "stocks and flows" approach to financial geography (e.g., Clark 2005).

But blindness to power is not the only marker of an uncritical perspective. Following lines of argument developed to varying degrees in earlier chapters of this book, we want to highlight two others that have also tended to characterize recent economic geographic work on money and finance, especially in the pre-crisis period.

The first of these relates to *what* is studied. As we have stressed throughout, a core attribute of critical scholarship of any sort is reflexivity about one's objects of inquiry: asking not only "Why do we/I deem this to be a worthy research object, who or what makes it so?" but also "How did that object come to assume such apparent worthiness in the first place?" From the late 1990s through the late 2000s, as the power and influence of dominant financial institutions, markets, and practices became increasingly evident, so in turn these were simply presumed to constitute collectively not just *a* but *the* natural locus of geographic enquiry. And this naturalization of research objects was nothing if not geographically exclusive and exclusionary. Since these objects all belonged in "the heartlands of high finance (particularly the financial districts of London and New York)," researchers focused on these heartlands "whilst largely neglecting the experiences of other places" (Hall 2012, pp.95–96).

The second marker we have in mind relates to readings of historical and geographical change. How has the world of money and finance been changing over time and across space? What have been the primary drivers of such change? And what have been its main implications – politically and socially as well as economically? A critical geographical take on money and finance is one that asks all of these questions. An uncritical approach, by contrast, is one that ignores them; instead, it regards change as inevitable and unproblematic, something that simply "is" and should therefore (be let) be.

This second tendency, too, characterized much economic geography of money and finance in the pre-crisis period. It, like the related tendencies to neglect social power and to take a narrow, uncritical perspective on legitimate research objects and spaces, is particularly evident in Gordon Clark's (2005) figuring of "the geography of global finance." Because Clark's intervention is emblematic of the mood and focus of the period, but also because it has subsequently proven influential, we examine it in more detail.

9.6.2   Exhibit A: "Money Flows like Mercury"

"The Geography of Global Finance" is the subtitle of Clark's (2005) article. The main title is "Money Flows like Mercury." Why? Clark felt that a new way of "conceptualising global finance" was needed. This was because he deemed existing theoretical conceptualizations inadequate. Specifically, they "impoverish our geographical imagination" (p.100). Clark's own conceptualization was especially concerned with "the spatial and temporal logic of global capital flows," and to help explicate this logic he invoked a guiding metaphor: "money flows like mercury."

The first point to make about Clark's conceptualization is that his frame of reference is severely constrained. His avowed interest is in "market patterns and processes." This leads him to consider two primary sets of actors (pp.102–108). The first are "pension funds, insurance companies, banks and corporations," or, more generically, "large financial institutions with global reach." The second comprises individuals "implicated in the process of global institutional transformation whether as participants in funded pension plans, as purchasers of mutual fund investment products, or as the consumers of commodities brought to the metropolitan core from the margins of the global economy" – that is, "sophisticated consumers of global financial services."

Clark's global institutions and "sophisticated" consumers are of course entirely legitimate objects of research within global finance. But they surely cannot be the only legitimate such objects. The issue is what, or who, is left out of Clark's "global finance." Occupying a necessarily peripheral position in his research agenda, presumably in the "margins" of a spatial economy inevitably pivoting on the "West," is everyone and everything else: unsophisticated consumers and non-global institutions, not to mention the billions of people who are not financial "consumers" in any meaningful sense but whose livelihoods are nonetheless impacted by "global" finance, for example through food price volatility linked to market speculation (Field 2016).

Indeed, in Clark's figuring, those "others" tend to come into view only as the presumed beneficiaries of an ineluctable process of global financial integration extending out from the Western core and bestowing benefits in kind. In other words, global financial integration driven by Western financial institutions is deemed inevitable and unproblematic. "Not only," insists Clark (2005, p.109), "is global financial integration a necessity for securing the welfare of the baby boom generation, it is also a necessity for improving the welfare of the vast majority of people living on the margins of the global economy." In fact, so self-evident is this reality to Clark that he finds it "surprising to be confronted with arguments to the effect that poor communities of less developed countries have an interest in remaining outside the developing global financial architecture."

Now, we should be very clear here about what we are and are not arguing. We are not saying is that Clark is necessarily wrong about the upsides of financial integration, although he may very well be. After all, there is a large body of social-scientific research demonstrating massively destructive consequences from financial globalization, especially for those in formerly colonized territories (e.g., Bagchi and Dymski 2007); the tragic story of SKS Microfinance (see Box 9.3: SKS Microfinance and Rural Suicides) is but one example. Furthermore, *even the International Monetary Fund* (IMF), for so long the main international cheerleader, architect, and enforcer of globalized finance in the form of openness to cross-border capital flows, has admitted in retrospect that this globalization has incurred heavy social costs (Ostry, Loungani, and Fourceri 2016). (The title of its infamous *mea culpa* – "Neoliberalism: Oversold?" – is a bit misleading: it is specifically financial openness that the IMF admitted to having oversold.)

Our point rather is that Clark's argument is deeply uncritical in the specific sense that it takes the dogma of beneficent global finance for granted. He does not *show* that global financial integration is beneficial (or even, as it happens, that such integration is indeed "global," leaving nobody, so to speak, "behind"). This truism is simply *stated* uncritically. The notion that integration into Westernized financial circuits may be deleterious, or even merely of ambiguous benefit, is simply dismissed out of hand, partly because peoples in those "marginal" spaces are not even part of the formal research agenda. Their exclusion from that agenda means that they cannot be seen to be harmed or otherwise negatively impacted. So complete is this exclusion that if it transpires that in certain "unregulated or ungoverned" scenarios the financial system, like mercury, "may be poisonous," then this is the case "*especially* for those charged with the responsibility of its management" (Clark 2005, p.105; emphasis added).

In sum, Clark's figuring of "the geography of global finance" embodies all the main attributes of what we take to be an uncritical approach. Indeed, it is telling that what he sought to envision was *the* geography of global finance, not *a* geography of global finance, and not the geograph*ies* of global finance either.

It would be a stretch to say that Clark's approach exactly mirrored, and thus serves as a perfect proxy for, the wider body of research and writing on the geographies of money and finance in the decade leading up to the financial crisis. It did and does not. Important critical work was published on, for instance, the geographies of microfinance services provided to the poor in regions such as South Asia (e.g., Rankin 2003). Here there was assuredly no seduction going on. In the sense that it *did* talk about power, about "alternative" research objects and spaces outside of the financial centers of the Global North, and about the causes and consequences of global financial integration, such work was the very antithesis of uncritical. This is why we highlight it here. But such exceptions were rare.

For the most part, Clark's vision *was* broadly representative. Researchers focused on major Western financial firms, their cultures (as well as their economics), and their professional "knowledge networks"; on the fates of and relations between major financial centers; and on the investment horizons and subjectivities of the West's "sophisticated" financial consumers. To the extent that "other" spaces and places figured, for example, they did so mainly insofar as they were discursively "imagined" – typically as "emerging markets" – and

Box 9.3   SKS Microfinance and Rural Suicides

"Microfinance" is the generic term used to refer to the provision of financial services to low-income people, particularly, although not only, in the Global South. It has for several decades been a central plank of the development agenda of non-governmental organizations, regarded as essential to help pull "unbanked" people, namely those excluded from mainstream financial services networks, out of poverty. But since the turn of the millennium, microfinance has undergone important transformations, of which a crucial one has been the growing involvement of for-profit financial institutions headquartered in the Global North. Planning scholar Ananya Roy (2010) invokes the concept of "poverty capital" to refer to this microfinance-based form of Western financial globalization.

One example of this wider historical transformation is provided by the Indian company SKS Microfinance, now renamed Bharat Financial Inclusion Limited. SKS was founded in 1997 as a conventional, non-profit microfinance lender. But over time, with the support of Western venture capital, it transitioned to a for-profit model, leading to an initial public offering (IPO) on the Bombay Stock Exchange in 2010 – attracting investors such as the US banks Goldman Sachs and Morgan Stanley. Not everyone was impressed. Muhammad Yunus, a pioneer of non-profit microfinance, saw this as "pushing microfinance in the loansharking direction," continuing: "By offering an IPO, you are sending a message to the people buying the IPO there is an exciting chance of making money out of poor people" (DealBook 2010).

Yunus's warning would prove prophetic. SKS hit the news again the following year when it was reported that more than 200 people in the south Indian state of Andhra Pradesh had killed themselves and the state government pinned the blame on over-lending by microfinance companies, including SKS. The Associated Press (AP) claimed that "internal [company] documents … as well as interviews with more than a dozen current and former employees, independent researchers and videotaped testimony from the families of the dead, show top SKS officials had information implicating company employees in some of the suicides. An independent investigation commissioned by the company linked SKS employees to at least seven of the deaths" (Kinetz 2012).

The AP report makes for grim reading (Kinetz 2012):

One woman drank pesticide and died a day after an SKS loan agent told her to prostitute her daughters to pay off her debt. She had been given 150,000 rupees ($3,000) in loans but only made 600 rupees ($12) a week.

Another SKS debt collector told a delinquent borrower to drown herself in a pond if she wanted her loan waived. The next day, she did. She left behind four children.

One agent blocked a woman from bringing her young son, weak with diarrhea, to the hospital, demanding payment first. Other borrowers, who could not get any new loans until she paid, told her that if she wanted to die, they would bring her pesticide. An SKS staff member was there when she drank the poison. She survived.

> An 18-year-old girl, pressured until she handed over 150 rupees ($3) – meant for a school examination fee – also drank pesticide. She left a suicide note: "Work hard and earn money. Do not take loans."
>
> In all these cases, the report commissioned by SKS concluded that the company's staff was either directly or indirectly responsible.

hence rendered investable *by* those Western economic actors (Sidaway and Pryke 2000). This, bluntly stated, and recalling the four main "narratives" of economic geography identified in Chapter 2, was very much economic geography in the "Geographies of Business" style. The financial "risks, inequalities, and instabilities" (reusing Hall's words) central to 1990s texts such as *Money, Power and Space*, *Money/Space*, and Stuart Corbridge's seminal *Debt and Development* (1993) had largely receded from view as the economic geography of money and finance became essentially uncritical.

9.7 Geographies of Money and Finance after the Crisis

If one good thing can be said to have come out of the global financial crisis, it is that it has re-enlivened research into the geographies of money and finance. Having been widely seduced by the hollow triumphs of finance in the run-up to the crisis, such research has reassumed a critical, theoretically informed edge.

As much as anything else, the geographical horizons of such research are expanding all the time. Hence, recent analysis of the substantive role of urban financial centers in mediating wider financial circuits includes not only Lai's (2012) examination of the three main Chinese financial centers but also work by Bassens, Derudder, and Witlox (2011) on the role of Gulf cities – Dubai, Kuwait City, and Manama – as financial centers. This work deepens and extends our existing understanding of how place matters for finance both relationally (Bassens et al. focus on how interlocking directorates in the Shari'a advisory boards of Islamic financial services firms serve to connect financial centers) and in more absolute terms (the dominant position of Manama derives from its role as a standard-setting city for Shari'a-compliant investments).

New studies have also substantially enriched our knowledge of the ways in which place figures as an investment opportunity per se. Significantly, such studies point to the increasing centrality of nature, environmental change, and climate-related risk to the uneven landscapes of post-crisis investment potential. Leigh Johnson (2014), for instance, invokes the concept of "returns on place" to capture how place-specific physical *vulnerabilities* associated with so-called peak perils such as US hurricane and earthquake risks are translated through securitization techniques into investable financial *assets*.

Recent research in this vein has begun to tell us much more than we previously knew about the often ambiguous ways in which securitization and cognate techniques, through their role in connecting parties, entangle key stakeholders in complex new socio-spatial relations. Some of the most revealing findings emerge from research into a domain which alongside environmental and climate risk is increasingly seen as one of the key contemporary investment growth poles: urban infrastructures, and particularly public urban infrastructures such as transportation routes and hubs. With public finances under severe and growing pressure in many territories, and with austerity many governments' favorite watchword, innovative mechanisms for financing new public infrastructures and for privatizing and "releasing value" from existing ones are being explored. One such mechanism that Ashton, Doussard, and Weber (2016) analyze in relation to US cities like Chicago is leasing existing infrastructure to global investment consortia. The latter gain access to the income stream (e.g., toll fees) generated by those assets. The authors demonstrate that such deals enmesh city governments in financial relationships with a range of non-local actors, exposing those governments to a range of new and only partially acknowledged liabilities.

While infrastructure and climate risk are therefore diversifying investor portfolios, implicating significant new actors such as the local state, a mainstay of economic geographic research on money and finance remains households and especially their mortgage debts. And this research shows that in the run-up to the financial crisis, redlining was alive and well. The work of Wyly et al. (2009) has been particularly influential in this respect. Framed, like studies of environmental risk and infrastructure, by the spatial connections forged through securitization, Wyly et al.'s analysis shows that exploitative, predatory, place-differentiating lending behavior in subprime working-class and racially marginalized communities across hundreds of US metropolitan areas fed global networks of debt and investment through Wall Street securitization conduits. For borrowers, place really did (and does) matter.

But not always – or at least, place is not always the sole or even primary relevant geographical scale for the pricing of household or consumer credit risk. Other scales are also relevant; and, it transpires, we cannot understand the constitution of financial relations at the local scale without factoring in the production and reproduction of such relations at regional or even national scales. This, for instance, is the central theme of Mark Kear's (2014) discussion of US legislative efforts in the 1990s and early 2000s to create a national space for the purchase and sale of consumer credit risk. Specifically, he charts a series of amendments in the Fair Credit Reporting Act that aimed to fashion a national market in order to enable the pricing of risk explicitly in "place-free" terms – the very opposite of Johnson's "returns on place." The effective place-based redlining identified by Wyly et al. needs to be read against the backdrop of such efforts, which, as Kear also emphasizes, highlight the significance of geographical scale – and its active production and reproduction – to processes of financial subject formation.

Studies such as these have begun to reinvest "financial geography" with a forceful critical sensibility. Questions of power and financial "progress" are back on the agenda, as are reflexive questions about the taken-for-grantedness of research objects.

Yet, in other post-crisis geographical research on finance, we see reasons to sound a note of caution. Clearly, much of the post-crisis political and popular debate on finance and the financial sector has been extremely hostile, castigating perceived irresponsibility, greed, and recklessness. This is to be expected. What is problematic, however, is when such judgementalism permeates and begins to suffocate scholarly analysis – when, in short, the latter becomes a thinly veiled form of moralism.

Our reservations here, therefore, are very different from those identified above in relation to the pre-crisis "geography of global finance" figured and championed by Clark. In fact, in his "money flows like mercury" intervention Clark (2005) offered a criticism of certain existing strains of scholarship on finance that closely mirrors the one we want to make here. "Rereading [Fredric] Jameson and others concerned about the power of Anglo-American financial capitalism," Clark (p.103) reflected, "one is struck by a familiar line of moral outrage albeit clothed in theoretical gesture."

In relation to much of the post-crisis scholarly critique of finance – non-geographical as much as geographical – we agree: there is considerable moral outrage, albeit, yes, sometimes clothed in theoretical gesture. Among those writing explicitly from a geographical perspective, one prominent researcher whose work strongly features such moralism is Ewald Engelen. He openly decries the "outright evilness" (2014, p.254) of finance and financiers.

While such outrage may be understandable and is sometimes justifiable, it does not really get us very far as a critical strategy. Mann (2013, pp.2–3) explains why in counseling against the "smug radicalism ... of self-proclaimed rebels repeating conspiracy stories and sweeping generalizations with which their listeners already agree: 'Banks rule the world!' 'Capitalism = greed'":

> Perhaps the CEOs of Shell Oil or Citibank are indeed cruel profiteers and super-rich megalomaniacs. Perhaps they really are bad guys. That is not, and cannot be, the basis of a critique of capitalism. Capitalism is neither made nor defended by profiteers and super-rich megalomaniacs alone, nor did they produce the system that requires the structural position they fill. In reality, capitalism is produced and reproduced by elaborate, historically embedded, and powerful social and material relationships in which most of us participate.

A critique of financiers' "evilness" is a critique not of capitalism but of *capitalists*, or a perceived immoral fraction thereof. Root out all such individual evilness, the argument implies, and all will be well in the world. To cast critique in these moralistic terms is not only to imply that actions flow straightforwardly from underlying psychological predispositions – a debatable postulate – but to ignore the *political economy* of capitalism within which "evilness" flourishes. And this misreads the problem. "The problem," as Mann (2013, pp.4–5) says, "is not that capitalism is a conspiracy of greedy people. The problem is that capitalism, as a way of organizing our collective life, does its best to force us to be greedy – and if that is true, then finger-pointing at nasty CEOs and investment bankers may be morally satisfying, but fails to address the problem."

While we therefore agree with Clark that moral outrage is no substitute for careful critical analysis of economic geographies, it is nonetheless the case that Clark's own (2011) reading of the causes of the financial crisis succumbs to a strikingly similar failing. For Clark, however, it is not evilness so much as "myopia" – a fundamental human predisposition to short-termism, translating into a hazardous investor emphasis on short-term returns – that primarily explains the crisis. Yet, the problem is still posited as humanity and its psychology, only with immorality being substituted for short-sightedness. Either way, a curbing of dangerous human instincts – Clark laments "the apparent inability or unwillingness of many (but not all) market agents to exercise a modicum of self-control" (p.5) – is what is required. By this way of thinking, the essential problem is found not in historically and geographically contingent aspects of the prevailing political economic system but in the nature of (some) people. As such, the problem is theoretically capable of being contained, if not eradicated, within the context of the macro-level socioeconomic status quo.

9.8   Conclusion

Money and finance and their seemingly pervasive influence over and colonization of different dimensions of contemporary social life are clearly critical issues *de jour*. This chapter has had one overriding objective: to demonstrate the indispensability of a geographical imagination in coming to terms analytically with these issues. Money and finance cannot be systematically unpacked without due consideration being given to their geographical dimensions because quite simply they are always and everywhere explicitly geographic phenomena. Geographies are integral to their very constitution and operation. Change the geographies, and you change the phenomena.

But we have also hoped to show that some approaches to understanding those geographies are more analytically defensible, insightful, and (perhaps a bad word) profitable than others. In accordance with the critical perspective that frames the whole book, we have made a case for the merits of research that is alive to questions of power, reflexive to its objects of inquiry and their conditions of possibility, and concerned to interrogate both the forces underlying change in the world and the consequences of such change. In all these regards, money and finance are matters especially warranting critical awareness.

In illuminating some of the main ways in which the geographical – broadly conceived – shapes the monetary and financial, the chapter has demonstrated the centrality not just of "place" in general but of *urban* place, and places, in particular. Such places figure materially both as the commanding locations from which the financial industry operates – in an unequivocal refutation of O'Brien's "end of geography" thesis with which we began – and as places where finance (drawing on Clark's flow metaphor) repeatedly pools and seeks to exploit through investment and distribution strategies. Moreover, as the subprime fiasco and ensuing recession have exemplified, the nexus of the urban and the financial appears to be especially material in times of economic crisis. Indeed, in an interview conducted during the crisis, Harvey (2009) labeled it a "financial crisis of urbanization."

The next chapter argues that there was and is nothing accidental about this convergence. Money and finance and their geographies are no less intimately linked to the phenomenon at the heart of that chapter – urbanization – than they are to the central phenomenon of the previous chapter – globalization. In fact, not only can none of the individual themes of the various chapters in Part II of this book be understood except in and through its geographies, but none can be fully understood independently of all the other geographies. In the same moment that geographies of finance are often geographies of globalization, so too they are frequently geographies of urbanization.

References

Ashton, P., Doussard, M., and Weber, R. 2016. Reconstituting the State: City Powers and Exposures in Chicago's Infrastructure Leases. *Urban Studies* 53: 1384–1400.

Bagchi, A.K., and Dymski, G.A., eds. 2007. *Capture and Exclude: Developing Economies and the Poor in Global Finance*. New Delhi: Tulika Books.

Bassens, D., Derudder, B., and Witlox, F. 2011. Setting Shari'a Standards: On the Role, Power and Spatialities of Interlocking Shari'a Boards in Islamic Financial Services. *Geoforum* 42: 94–103.

BIS (Bank for International Settlements). 2016. Global OTC Derivatives Market. http://www.bis.org/statistics/d5_1.pdf (accessed June 26, 2017).

Boeing. 2016. 10-K Annual Report for the Fiscal Year Ended December 31, 2015. https://www.sec.gov/Archives/edgar/data/12927/000001292716000099/a2015 12dec3110k.htm (accessed July 3, 2017).

Bryan, D., and Rafferty, M. 2006. *Capitalism with Derivatives: A Political Economy of Financial Derivatives, Capital and Class.* Basingstoke: Palgrave Macmillan.

Cameron, A. 2006. Turning Point? The Volatile Geographies of Taxation. *Antipode* 38: 236–258.

Christophers, B. 2009. Complexity, Finance, and Progress in Human Geography. *Progress in Human Geography* 33: 807–824.

Christophers, B. 2013. *Banking across Boundaries: Placing Finance in Capitalism.* Oxford: Wiley-Blackwell.

Christophers, B. 2015. The Limits to Financialization. *Dialogues in Human Geography* 5: 183–200.

Christophers, B., Leyshon, A., and Mann, G., eds. 2017. *Money and Finance after the Crisis: Critical Thinking for Uncertain Times.* Oxford: Wiley-Blackwell.

Clark, G.L. 2002. London in the European Financial Services Industry: Locational Advantage and Product Complementarities. *Journal of Economic Geography* 2: 433–453.

Clark, G.L. 2005. Money Flows like Mercury: The Geography of Global Finance. *Geografiska Annaler: Series B* 87: 99–112.

Clark, G.L. 2011. Myopia and the Global Financial Crisis: Context-Specific Reasoning, Market Structure, and Institutional Governance. *Dialogues in Human Geography* 1: 4–25.

Corbridge, S. 1993. *Debt and Development.* Oxford: Blackwell.

Corbridge, S., Martin, R., and Thrift, N., eds. 1994. *Money, Power and Space.* Oxford: Blackwell.

DealBook. 2010. SKS I.P.O. Ignites Microfinance Debate. http://dealbook.nytimes.com/2010/07/29/sks-i-p-o-sparks-microfinance-debate/?_r=0 (accessed June 26, 2017).

Engelen, E. 2014. Geography Can Explain Much, If Not All… *Environment and Planning A* 46: 251–255.

Ferguson, N. 2009. *The Ascent of Money: A Financial History of the World.* London: Penguin.

Field, S. 2016. The Financialization of Food and the 2008–2011 Food Price Spikes. *Environment and Planning A*, 48: 2272–2290.

Gilbert, E. 2005. Common Cents: Situating Money in Time and Place. *Economy and Society* 34: 357–388.

Hall, S. 2012. Theory, Practice, and Crisis: Changing Economic Geographies of Money and Finance. In T.J. Barnes, J. Peck, and E. Sheppard (eds), *The Wiley-Blackwell Companion to Economic Geography.* Oxford: Wiley-Blackwell, pp.91–103.

Harvey, D. 1989. *The Urban Experience.* Oxford: Blackwell.

Harvey, D. 2009. Their Crisis, Our Challenge. http://www.redpepper.org.uk/Their-crisis-our-challenge (accessed June 26, 2017).

Henderson, G. 1998. *California and the Fictions of Capital.* Oxford: Oxford University Press.

Ingham, G. 2004. *The Nature of Money.* Cambridge: Polity Press.

Johnson, L. 2014. Geographies of Securitized Catastrophe Risk and the Implications of Climate Change. *Economic Geography* 90: 155–185.

Kear, M. 2014. The Scale Effects of Financialization: The Fair Credit Reporting Act and the Production of Financial Space and Subjects. *Geoforum* 57: 99–109.

Kinetz, E. 2012. Lender's Own Probe Links it to Suicides. http://www.usnews.com/news/world/articles/2012/02/24/ap-impact-lenders-own-probe-links-it-to-suicides (accessed June 26, 2017).

Lai, K. 2012. Differentiated Markets: Shanghai, Beijing and Hong Kong in China's Financial Centre Network. *Urban Studies* 49: 1275–1296.

Leyshon, A., and Thrift, N. 1995. Geographies of Financial Exclusion: Financial Abandonment in Britain and the United States. *Transactions of the Institute of British Geographers* 20: 312–341.

Leyshon, A., and Thrift, N. 1997. *Money/Space: Geographies of Monetary Transformation*. London: Routledge.

Mann, G. 2010. Hobbes' Redoubt? Toward a Geography of Monetary Policy. *Progress in Human Geography* 34: 601–625.

Mann, G. 2013. *Disassembly Required: A Field Guide to Actually Existing Capitalism*. Edinburgh: AK Press.

McDowell, L. 1997. *Capital Culture: Gender at Work in the City*. Oxford: Blackwell.

McLeay, M., Radia, A., and Thomas, R. 2014. Money Creation in the Modern Economy. *Bank of England Quarterly Bulletin* Q1.

O'Brien, R. 1992. *Global Financial Integration: The End of Geography*. London: Chatham House.

Ostry, J.D., Loungani, P., and Fourceri, D. 2016. Neoliberalism: Oversold? *Finance & Development* 53: 38–41.

Rankin, K.N. 2003. Cultures of Economies: Gender and Socio-Spatial Change in Nepal. *Gender, Place and Culture* 10: 111–129.

Roberts, S. 1994. Fictitious Capital, Fictitious Spaces: The Geography of Offshore Financial Flows. In S. Corbridge, R. Martin, and N. Thrift (eds), *Money, Power and Space*. Oxford: Blackwell, pp.91–115.

Roy, A. 2010. *Poverty Capital: Microfinance and the Making of Development*. New York: Routledge.

Sell, F.L. 2001. *Contagion in Financial Markets*. Cheltenham: Edward Elgar.

Sidaway, J.D., and Pryke, M. 2000. The Strange Geographies of 'Emerging Markets.' *Transactions of the Institute of British Geographers* 25: 187–201.

Williams, P. 1978. Building Societies and the Inner City. *Transactions of the Institute of British Geographers* 3: 23–34.

Wyly, E., Moos, M., Hammel, D., and Kabahizi, E. 2009. Cartographies of Race and Class: Mapping the Class-Monopoly Rents of American Subprime Mortgage Capital. *International Journal of Urban and Regional Research* 33: 332–354.

Chapter 10

Cities and Urbanization

10.1 Introduction

It might seem odd to have a chapter on cities and urbanization – the latter being the process by which more and more people in a society come to live in urban rather than rural areas – in a book on economic geography. Isn't that the terrain of urban geography? Not only urban geography. We will demonstrate in this chapter that economic geography cannot be indifferent to cities and urbanization.

Consider a striking and provocative formulation from the geographer Dick Walker. Reflecting on the global financial crisis, Walker (2012, p.54) referred to "the essential unity of urban and economic geography." What did he mean by this? Did he mean that urban and economic geography are one and the same? That all urban geography is necessarily economic (geography) and that all economic geography is necessarily urban (geography)? We don't think so. After all, urban geography examines cities' cultures and politics as well as their economies, and economic geography is about the country as well as the city (Williams 1975). We think, rather, that Walker had two things in mind. It is these two things that the present chapter pursues.

First, human settlement patterns are not incidental to economic processes and outcomes. The geographical configurations into which societies organize themselves fundamentally shape the form and development of those societies' economic activities. This is especially true of cities and the economies they engender. To understand the economy, in other words, it is essential to understand the economy of cities.

Economic Geography: A Critical Introduction, First Edition. Trevor J. Barnes and Brett Christophers.
© 2018 John Wiley & Sons Ltd. Published 2018 by John Wiley & Sons Ltd.

Second, if cities and urbanization matter to the economy, so too the economy matters to cities and urbanization. A society's economic character – capitalist or socialist, mercantile or industrial, manufacturing or service-led, and so on – has profound implications for how its cities develop and what they look like. To understand the geographical phenomena that are cities it is imperative to do so in terms of economy and economics.

In sum, there is a reciprocal, mutually constitutive or *dialectical* relation between the economic and the urban – a relation that economic geography, concerned as it is with economy *and* geography, is ideally positioned to illuminate. This chapter critically explores some of this project's key elements.

10.2   Cities and Economies through History

10.2.1   Early hydraulic city systems

Let's start with the emergence of the world's first cities in the valleys of the Nile, the Indus, and the Tigris–Euphrates some five millennia ago. Historians and archaeologists have long argued that this development must be understood in terms of fundamental transformations in the economies of those regions. In an influential article on the "urban revolution," for example, the archaeologist Gordon Childe (1950, p.3) wrote of a "new economic stage in the evolution of society." Economic change gave rise to, or at least facilitated, the birth of the city. How so?

The key was the transition from a more or less hand-to-mouth system of socioeconomic reproduction lacking a meaningful "division of labor" – the term used by economists to refer to the division of society into groups of people specializing in different tasks (see Chapter 12) – to one generating storable surpluses of food and beginning to disaggregate into different types of specialist laborers. This transition was predicated upon the development of new irrigation techniques affording year-round water supply in regions with rich floodplain soils, which transformed crop cultivation and productivity and for the first time enabled the farmer to produce "more food than was needed to keep him and his family alive" (1950, p.6).

This economic transition had a number of profound geographic consequences. Most importantly, it reduced the need for seasonal migration in search of food, allowing sedentary living. And it yielded a division of social labor because the surpluses were "large enough to support a number of resident specialists who were themselves released from food-production" (1950, p.8). Add to these developments contemporaneous innovations in transportation (wheeled vehicles and pack animals, as well as water-based transport modes) that "made it easy to gather food stuffs at a few centres," and the conditions for city formation were present. "Thus," says Childe, "arose the first cities": Alexandria, Harappa, Ur, and others. Only relatively large-scale settlements could accommodate a sedentary society increasingly organized on the basis of specialist labors and thus characterized by economic interdependence.

The key point to take away from this historical snapshot is that such cities were artefacts and representatives of a set of historically and geographically specific economic relations. *All cities are*: we must always read and understand them – their existence, role, and form – in the light of those economic relations.

But there is also a second important point to take away, one equally strongly conveyed by the literature on early city formation. It concerns cities and social power. Food surpluses could not feed cities if the farmer outside the city could not somehow be persuaded to relinquish them. Exploiting surpluses to seed a division of labor requires the power to collect them. Cities, in short, embody power from the get-go.

10.2.2   Merchant cities

We now leap forward to the emergence at a much later stage of human history, in a different part of the world, of a very different urban phenomenon: the medieval cities of Western Europe studied by, among others, the great Belgian historian Henri Pirenne. Having documented the relative decline of the cities of the Roman Empire caused by the Germanic and Islamic invasions, Pirenne's *Medieval Cities* (2014 [1925]) argued that substantial cities only reemerged in Europe in the eleventh century. And they did so, crucially, not primarily because of agricultural surpluses but in the formative context of a very different economic milieu.

These medieval cities were *merchant* or mercantile cities. They were pivotal hubs of an increasingly *trade-based* economy linking Europe to the Islamic world, which developed over the course of the next five centuries into a fully-fledged configuration of what economic historians call "merchant capitalism." It was the earliest phase in the development of capitalism as a socioeconomic system. These cities served as the centers where merchants came together to trade, making the process vastly more efficient than if it were geographically dispersed.

The cities in question had a distinctive and novel economic function, whether it was the great textile-trading cities of Bruges, Ghent, and Ypres in Flanders or the Italian cities of Amalfi, Genoa, and Venice that came to dominate Mediterranean-based trade. Subsequent research may have modified the timeline of Pirenne's account, suggesting that multiple trade routes linked Western Europe to the Islamic world already by the second half of the ninth century, but the essential coupling of city development to a trade-oriented economic role has not been challenged.

10.2.3   Industrial cities

What, then, of cities closer in time to our own age? Can we similarly illuminate their genesis, development, and social life with reference to their "positioning" in the economies of the day? In a sense, the label applied to the penultimate city type we will look at – "industrial cities" – answers our question. These cities were, and in many parts of the world still are, *industrial*: defined, that is to say, by their economic function (to house "industry," i.e., machine-based manufacturing), or at least by the nature of the economic world (industrial capitalism) they inhabit and characterize (see also Chapter 12).

Indeed, industrialization is closely associated not just with the industrial city, its particular urban embodiment, but with the city – urbanization – per se. The idea that industrialization and urbanization have historically been joined at the hip and have proceeded in tandem with one another is one of social science's "stylized facts" (broadly accepted generalizations). The few exceptions, such as large-scale rural industrialization in China in the 1950s through 1970s prior to market-oriented-reforms (Sigurdson 1977),

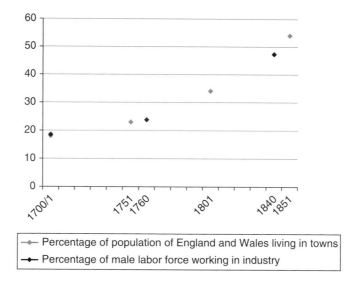

Figure 10.1 Urbanization and industrialization in Britain. Source: Thompson 1990, p.8; Crafts 1985, p.62.

prove the rule. Generally, industrialization meant urbanization and vice versa. Figure 10.1 provides an illustration of this symbiosis for the paradigmatic, original case: Britain, the first industrial nation and the first also to substantially and simultaneously urbanize, in its case between the mid-eighteenth and mid-nineteenth centuries.

How should we understand the dynamics of the relationship between industrialization and urbanization? On the one hand, the development of industrial cities that first emerged in northwestern Europe in the second half of the eighteenth century mirrored the development of the ancient hydraulic urban civilizations studied by Childe. They were originally fed by the creation of surpluses elsewhere and were places where the accumulation of surpluses was concentrated. Only now the surpluses in question were not just of food. First, a combination of incipient capitalist industrial manufacturing activity and centuries of capitalist trading activity generated surpluses – quantities not needed for immediate use – of something that ancient cities never knew: money – specifically, money as a store of value (Chapter 9). Money surpluses could be and were used in the construction of new urban infrastructures. Second, developments in the countryside also created huge surpluses of another commodity: labor power. In eighteenth-century England, most notably, the privatization or "enclosure" of lands with common rights created a landless class who were now "free" to sell their labor power to industrial capitalists (Chapter 12). Like surplus money, this human surplus also moved to new urban centers, in its case, in search of paid work.

Yet if industrialization did not begin in the industrial city, the latter was nonetheless crucial to its full-blown development. The industrial city rapidly became the beating heart of industrial capitalism: an extraordinary geographical phenomenon without which large-scale industrialization would arguably have been unthinkable. David Harvey (1985) coined the concept of "supply-side urbanization" to capture the significant relation between the geographic (the industrial city) and the new economy (industrialization). The industrial city uniquely *supplied* raw materials for industrial-scale capitalist production to take place and thrive. One such raw material was the urban factory system; another, no less critical, was a substantial pool of readily accessible labor. "A manufacturing establishment," Friedrich

Engels (2009 [1845], p.66) famously observed in the heyday of industrial capitalism and from one of the premiere industrial cities of the time, Manchester, "requires many workers employed together in a single building, living near each other and forming a village of themselves in the case of a good-sized factory." The dense new urban geography of the industrial city facilitated the new, mass-production-oriented economy of factory-based industrialism. Industrial cities were "the workshops and forcing houses of industrial capitalism," the very "engine of capital accumulation" (Amin 2000, p.115).

And in some parts of the world they still are. Look at China. Many enormous cities sprang up during what the geographer You-tien Hsing understandably calls its *Great Urban Transformation* (Hsing 2010). The period from 1980 to 2002 saw some 250 million people added to the urban population as the urban share of the total population leapt from 20 to 40% (p.2). They are emphatically industrial cities, even if this is often "light" rather than "heavy" industry. Industry was and remains central to the state-driven project of turning China into the workshop of the world and then maintaining that status. Led by industrial behemoths such as Foxconn (see also Chapter 1), this project has seen the longstanding economic geographic model of the city as crucible of intensive industrialization develop in dizzying new directions (see Box 10.1: City Foxconn).

10.2.4 Postindustrial cities

In other parts of the world, however, particularly in much of Western society, cities have changed enormously since the industrial period. Manufacturing industry, and especially heavy industry, began disappearing from Western cities in the 1960s and in many of them it now represents a marginal activity at best. We will turn in a moment to the question of what has taken its place, but first let's look at why this has happened.

There are a number of reasons for the change, often closely bound up with one another. One explanation relates to geographic transformations within Western societies. Some of the industrial activity departing the city has simply gone elsewhere within the same country, to suburban or ex-urban locations. Improved transportation and communications technologies have clearly contributed to this development, making non-city locations more connected. So too have differential movements in land prices. In recent decades the cost of urban land has typically inflated more rapidly than that of non-urban land. This is partly a result of widespread ongoing processes of urban gentrification, which have seen capital and the middle classes move "back to the city" (Smith 1979). Industry, which is particularly land-hungry, is less able to compete for expensive inner-city urban locations.

Then there are important underlying shifts in the very structure of Western economies (see Chapter 12). Look again at Figure 10.1, which shows levels of industrialization and urbanization in Britain increasing more or less in lockstep through to the mid-nineteenth century. What happened thereafter? Did Britain and other Western societies further industrialize and urbanize? No. While urbanization continued its upward path, with Britain and the United States now close to 80% urbanization, industrialization did not. Whether we analyze economic structure on the basis of different sectors' relative contributions to economic output or of their respective numbers of employees, Britain's economy was arguably *never more industrialized* (focused on manufacturing activity) than it was in 1851. Data from the Office for National Statistics (ONS 2013) show that the proportion of all working people (women and men) employed in manufacturing remained remarkably

Box 10.1   City Foxconn

Foxconn Technology Group is a Taiwanese electronics manufacturer, specifically the largest "contract" electronics manufacturer – making products under contract for third parties such as Apple (iPhones and iPads), Amazon (Kindle), and Sony (PlayStation) – in the world. But it is also a manufacturer of cities, especially in China.

Pun Ngai and Jenny Chan (2012) describe Foxconn as a "mega world workshop" (p.385). It has more than a million employees in China alone. They work in a series of "factory compounds" dotted across the country. With the number of people working at these compounds ranging from approximately 20,000 to upward of 400,000, the largest of them represent industrial cities in and of themselves. Ngai and Chan (2012, p.392) elaborate as follows: "A Foxconn 'campus' – as the company managers like to call it – is a distinctive dormitory factory regime, which organizes the sphere of production and the sphere of reproduction." Engels would have been impressed; in fact it is almost as if the designers of Foxconn's campuses had read his above-cited blueprint for the industrial city.

The biggest compound of all is the famed goliath of Shenzhen Longhua (famed not just for its size but for a spate of reported employee suicides in 2010–2011), which Ngai and Chan estimate as having more than 430,000 workers. Their depiction of the campus merits citing in full (2012, p.394):

> This 2.3-square-kilometer campus includes factories, warehouses, twelve-story dormitories, a psychological counseling clinic, an employee care center, banks, two hospitals, a library, a post office, a fire brigade with two fire engines, an exclusive television network, an educational institute, bookstores, soccer fields, basketball courts, a track and field, swimming pools, cyber theaters, supermarkets, a collection of cafeterias and restaurants, and even a wedding dress shop. This main campus is divided into ten zones, equipped with first-class production facilities and the "best" living environment since it is the model factory for customers, central and local-level governments, and visitors from media organizations and other inspection units.

constant for over a century: it was 38.5% in 1851, and still 36.3% in 1961, when the industrial city remained, just about, the British norm. But manufacturing then began a precipitous decline: to 23.1% of the workforce by 1981 and just 8.9% by 2011. Britain, in other words, like most other Western societies, deindustrialized, and today is largely "postindustrial" (Chapter 1); most people work in the services sector, which in 1961 still employed less than half the working population but by 2011 gave jobs to over 80%. And as the long history we have parsed in this section of the chapter has already shown, with a different type of economy one typically gets a different type of city: in this case, the "postindustrial city."

Of course, the fact that many Western societies today have postindustrial economies and cities is not unconnected to the fact that many other societies, including China, have economies and cities characterized by greater levels of industrial intensity. Commentators often figure national economies' historic progression through different phases of development – from a dominant "primary" sector based on agriculture and mining, through a "secondary" phase focused on manufacturing, to a "tertiary" services-based economy, and in some cases, like the United Kingdom and United States, even onto a "financialized" fourth phase oriented heavily toward specifically *financial* services (Chapter 9) – as if they really are *national*, or bounded, economies. But as Chapter 8 showed, they are not. National economies are interlinked, and in recent decades increasingly so. One of the reasons why industrial activity in Western countries and their cities declined so rapidly from the 1960s is that manufactured goods were increasingly imported from countries and cities in other parts of the world. Sometimes this reflected Western-headquartered multinationals outsourcing or offshoring production to lower-cost locations. Sometimes it was a question of Western manufacturers simply being outcompeted. Either way, the Western industrial city was able to become postindustrial partly because cities elsewhere generally did not.

So if the merchant city was quintessentially where people traded goods, and the industrial city was quintessentially where they made them, what do people do in the postindustrial city? What type of city does the postindustrial economy crystallize and nourish? What type of city does it call forth, to keep it ticking over and growing?

Inevitably, given the contemporary dominance of the services sector, today's Western city is archetypally a city of services. And it is harder to be much more specific than that. Indeed, the fact that the word "postindustrial" remains the preferred epithet – hence why we use it here – is instructive in itself. The designation "postindustrial" is telling you it is *not* industrial, but it doesn't tell you what it *is*: today's city is defined by the economy that it does not belong to or characterize (the industrial economy) rather than one it does. What comes after the industrial economy and city? Nobody is quite sure. Plenty of attempts at a more positive definition have been tendered: today we are said to live in a knowledge or information economy and, therefore, in an "informational city" (Castells 1989); alternatively, this is the age of the creative economy and the "creative city" (Landry and Bianchini 1995). But none has stuck. Why? Because, as the geographer Ash Amin (2000, p.116), in a more accurate if less exotic assessment, says, "the urban economy appears to have become a mixed bag of activities":

> Today's urban hustle and bustle seems to be associated with servicing the needs of consumers and residents through such activities as shopping, leisure, and tourism, supporting the productive economy with business services such as banking, insurance, and accountancy, facilitating social reproduction through welfare services such as education, health care, sanitation, and transport and providing activities related to local and national public administration and governance. The urban economy, increasingly, is associated with the production, exchange, and consumption of services of one sort or another.

10.3 The Economy of the City

10.3.1 Foundational propositions

When he conjectured about the type of social geography most conducive to industrial productivity, Engels was onto something theoretically. His argument was essentially that a productive industrial economy required the agglomeration of its main raw material: labor. Cities provided this. In doing so, they became the center of gravity of industrial capitalism's dynamism.

Subsequent generations of scholars have largely concurred with Engels. They have explored in depth the question of why productivity – the efficiency of production, or the level of bang (what comes out of production) for one's buck (what goes in) – varies from place to place. And they have concluded that there *is* something in the nature of the city, postindustrial as well as industrial, that augments economic productivity. But what exactly is it? How does the geographic constellation of people, institutions, and built environments that we associate with cities shape economic dynamics in such a way as to ordinarily, but not always, boost productivity and output? It is in the range and depth of answers they have given to this question that later scholars have progressed far beyond Engels's basic proposition.

That cities *are* typically the main motors of the economy, the places where output is disproportionately concentrated, is beyond question. One recent report found that in 2014 the world's 300 largest cities accounted for 20% of the world's population but nearly half (47%) of global economic output (Parilla, Trujillo, and Berube 2015, p.1). Even accounting for a possible "selection effect" – perhaps cities attract inherently more productive people and firms rather than making them more productive – such data are hard to ignore.

The extent to which cities outperform non-urban areas varies greatly from place to place. Figure 10.2 demonstrates this, showing, for each of the world's major regions, cities' share of regional economic output (gross domestic product: GDP) and of regional population in 2007, along with projections for how these respective proportions will have changed by 2025. Our main interest is in the 2007 figures. Urban outperformance on productivity – signaled by an urban GDP share greater than the urban population share – is generally most pronounced in regions that are relatively less industrialized and, inevitably, in regions that are relatively less urbanized (as regions tend toward 100% urbanization, the urban contribution to output also, of course, tends toward 100%, thus narrowing any differential between cities' population and output contributions). This includes all of the "Developing regions" shown on the chart. And where there is a marked exception to this pattern, there is usually an obvious explanation. Australasia is a good example. With only 65% of the population living in cities, one might expect cities' outperformance to be more pronounced than in North America, where 77% live in cities. But it is not. Why? Because the Australian economy is more oriented to mining and resource-based activities than any other Westernized country, and these activities are rural.

Many of the most important core insights into the productivity occasioned by urbanization were developed a long time ago – indeed, not so long after the period when Engels was writing. Two figures were especially important to these intellectual developments.

One was the English economist Alfred Marshall discussed already in Chapters 3 and 4. He argued that cities benefited from agglomeration economies, which explained their economic domination (Chapters 4 and 12). Marshall (1890) posited that three factors were crucial. The first of this "trinity" was a refinement of Engels's point about the urban congregation of workers (though Marshall would doubtless have blanched at such a comparison): firms benefit from sharing a local labor pool. The second key advantage of agglomeration involved the cost and ease of sourcing "intermediate" goods – those used as inputs in the production of other products. Cities support a denser network of (shared) suppliers of these inputs, reducing costs to buyers. And the final benefit of agglomeration identified by Marshall concerned information diffusion. In cities, the secrets to improved business performance are generally hard to conceal and one does not have to travel far to

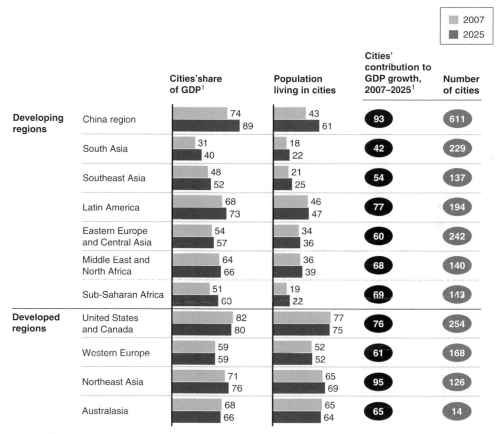

| | | 2007 |
| | | 2025 |

		Cities'share of GDP[1]	Population living in cities	Cities' contribution to GDP growth, 2007–2025[1]	Number of cities
Developing regions	China region	74 / 89	43 / 61	93	611
	South Asia	31 / 40	18 / 22	42	229
	Southeast Asia	48 / 52	21 / 25	54	137
	Latin America	68 / 73	46 / 47	77	194
	Eastern Europe and Central Asia	54 / 57	34 / 36	60	242
	Middle East and North Africa	64 / 66	36 / 39	68	140
	Sub-Saharan Africa	51 / 60	19 / 22	69	143
Developed regions	United States and Canada	82 / 80	77 / 75	76	254
	Western Europe	59 / 59	52 / 52	61	168
	Northeast Asia	71 / 76	65 / 69	95	126
	Australasia	68 / 66	65 / 64	65	14

1 Predicted real exchange rate.

Figure 10.2 The role of cities varies from region to region, % (except number of cities).
Source: Exhibit from "Urban World: Mapping the Economic Power of Cities," March 2011, McKinsey Global Institute, www.mckinsey.com. Copyright © 2017 McKinsey & Company. All rights reserved. Reprinted by permission.

uncover them; they are, as Marshall himself famously said, "in the air." Scholars today refer to these positive diffusion effects as "knowledge spillovers."

The second key figure was another Alfred: Weber (see Box 4.2: The German Location School). First published in 1909, Weber's *Theory of the Location of Industries* (1929) made several claims about urban agglomeration comparable to those of Marshall, albeit without once mentioning the work of his predecessor. But it differed, *inter alia*, by virtue of trying to offer a generalized *theory of location* of manufacturing industry; indeed, Weber argued, perhaps in a veiled critique, that any attempt to understand what he referred to as "the rush to the city" (i.e., agglomeration) remained premature "when we do not possess as yet any real knowledge of the general rules determining the location of economic processes" (p.3). He tried to establish such rules. And, in doing so, he identified four main sets of what he called "agglomerative factors" (pp.127–131). The first consisted of the advantages, cost-related and otherwise, of firms essentially timesharing access to specialized technical equipment; the second consisted of the pooling of labor power (cf. Marshall), and the division of the large labor pool into specialized fractions; the third consisted of cost savings realized both in sourcing raw materials (cf. Marshall's intermediate goods) and in selling finished products, the latter since "the concentrated industry produces a sort of large unified market for its products" (p.130); and the fourth consisted of savings on general overheads relating to infrastructural items such as transportation networks, buildings, energy utilities, and so forth.

Together, Marshall and Weber laid out a series of core principles concerning the economic gains from urban agglomeration. And as Walker (2015, p.141) says: "Such agglomeration economies are still the foundation stones of any theory of urban concentration." Indeed they are. Read any modern book on cities' outsized contribution to economic output, not least the influential *Triumph of the City* (2011) by Harvard economist Edward Glaeser, and the ghosts of Marshall and Weber haunt its pages.

10.3.2   Postindustrial refinements

Still, it would be wrong to suggest that understandings of the economy of the city have not progressed since the two Alfreds' days. They have. One reason is that the city itself has changed. The sources of cities' economic vitality, what the economic geographer Michael Storper calls the *Keys to the City* (2013), are not exactly the same in the postindustrial as in the industrial age, even if agglomeration economies remain fundamental.

In the more contemporary literature on the economy of the city, two significant themes stand out. The historic shift from an industrial to a postindustrial economy in which knowledge is deemed the key source of competitive advantage is one, which highlights the economic gains conferred by modern cities' *social* characteristics. The fact that these cities throw people and firms together in close proximity is thought to create fertile social conditions for experimentation, learning, and innovation (Chapter 12). The kind of dense, loosely organized, and interactive social networks that AnnaLee Saxenian (1996), studying Silicon Valley, found to be conducive to flexible adjustment and collaborative learning by firms and employees materialize most readily in urban environments. Geographic concentration facilitates the ready circulation and exploitation of both codified and "tacit" knowledge (Chapter 12). The latter, which is knowledge that is hard to articulate (codify) in speech or writing, is by definition harder to transfer *except* through being there, witnessing. Cities are thus ideal "learning regions" and the hubs, today, of the much-vaunted "informational economy." It is easy to recoil from the commonly boosterish nature of such discourse and its litany of buzzwords. This is especially true when it becomes the mulch of often highly profitable policy prescriptions offered to cities somehow missing out on the knowledge-led growth bonanza, as it does in the hands of figures such as the "creative cities" guru Richard Florida. Nevertheless, there are important insights in the literature.

There are similarly important insights in the literature on the second significant theme we want to highlight: the concept of *increasing returns* or *internal economies of scale* (Chapter 12). When economists talk about "returns to scale" they are referring to a technical property of production that predicts what happens to output if the quantity of all input factors is increased by a given amount. If output increases by the same amount then the returns to scale are constant, but if output increases by more than the uplift in inputs then the returns are increasing. What has this got to do with cities? In the 1980s, Paul Krugman and other economists turned their attention to urban economies, traditionally a peripheral concern for the discipline, and concluded that the concentration of economic activity in urban centers – which Krugman (1991, p.5) deemed no less than "the most striking feature of the geography of economic activity" – represented "clear evidence of the pervasive influence of some kind of increasing returns" (Chapter 4). The elaboration of formal models demonstrating such returns, and the interweaving of these with the "traditional" agglomerative factors identified by the likes of Marshall and Weber, became a staple of the "Geographical Economics" or "new economic geography" that we identified in Chapter 2 as one of today's four main narratives of economic geography.

10.3.3   Urban discontents and diseconomies

If all this sounds like an altogether too rosy picture of the economy of the city – a happy harmony of firms, workers, and cities jointly reaping the productivity benefits of urbanism – then that is because it is. First of all, let's go back to Engels for one second. He believed that dense urban settlement made industrial workers more productive. But even leaving aside for a moment the often appalling living conditions for workers in mid-nineteenth-century industrial Manchester, did Engels consider workers' productiveness positive *for them*? Hardly. He shared Marx's view of the matter. "It is," Marx (1963, p.225) maintained, "a misfortune to be a productive labourer." Why? Simple. Because "a productive labourer is a labourer who produces wealth *for another*" – the capitalist. Living and working in cities may help make workers more productive, in other words, but since the laborer produces wealth for the capitalist it is the latter who realizes all and any productivity gains. That, at least, is Marx's theory.

Whether we take the Marx and Engels view of things or the more sanguine perspective on working life under capitalism of most non-Marxian economics, one *cannot* ignore the question of living conditions that we previously parked – the existence of "poverty, hardship, informal economic activity, and hand-to-mouth survival" (Amin 2000, p.120). The reason is not just ethical or social, but also economic. As Amin (2000) explains, this "other" city is not some unwanted, unproductive, and unnecessary appendage to the productive city of Marshall, Weber, and Storper, but is rather an essential *and* productive part of it: "the dynamism of the central business districts themselves derives in part from their proximity to the 'other' city." We see this symbiosis in many of the cities of the Global South. For example, in India the productivity of the "formal" urban economy is predicated on work performed in a thriving if unmeasured "informal" economy that cannot be meaningfully distinguished from it (Breman 2016). We also see it in many cities of the Global North, for example, in "the dependence of the superstressed professionals and business elites upon the labor [of] low-income, displaced, and ethnic minority groups" (Amin 2000, p.120).

And there is a perhaps more fundamental reason still for questioning the conventional association of urbanity with economic productivity. To be sure, as Figure 10.2 showed, urban populations collectively demonstrate above-average productivity in every major world region. But that is not the same thing as saying that all cities are productive hothouses. Figure 10.2 provides data for regional groups of cities collectively, and therefore masks differences in levels of productivity between cities within the same region, as well as between regions. More granular data show not only that such differences are often substantial – some city economies, unsurprisingly, are much more productive than others – but that urbanity does *not* always enhance productivity. Sometimes, it appears to detract from it.

The United Kingdom again represents a striking case. Office for National Statistics (ONS 2016) data show that in some of the country's best-known cities, productivity levels are not just low but considerably lower than the national average that includes non-urban as well as urban areas. For 2014, the ONS's preferred measure of productivity – gross value added per hour worked – was 13% lower than the national average in both Birmingham and Manchester, 14% lower in Liverpool, and fully 26% lower in Nottingham. So much, it would seem, for generic urban agglomeration benefits. Indeed, if UK cities collectively are characterized by higher economic productivity than the country's rural areas, then we can largely explain this feature with one word: London. London accounts for over 20% of UK economic output, and over 25% of England's, whilst the equivalent figures for population are approximately 13 and 16. By ONS calculations, productivity in London (in 2014) is 30% higher than the UK average.

How then might we explain the fact that cities clearly are not always cradles of superior economic performance? After all, this appears to run directly counter to conventional economic wisdom. One possible explanation is that while the agglomeration we see in cities certainly has the capacity to generate positive economic effects, it can also generate negative ones associated with crowding, congestion, crime, pollution, and (as mentioned earlier) inflated land costs. Where these less helpful urban attributes – referred to by economists as "negative externalities" – outweigh the positive aspects of agglomeration, economies of scale can easily become diseconomies thereof (Richardson 1995). Economists therefore regard one of the most important roles of the state and the planning system vis-à-vis the city as being to cultivate the inherent benefits of agglomeration *and*, through devices such as congestion charges, to curb the associated disbenefits.

But agglomeration diseconomies are not the only possible explanation for marked real-world variance in the economy of the city. Let's step back for a second to see where we are in our discussion. Our analysis of the dialectical relation between the economic and the urban has so far focused primarily on what "cityhood" does to economic dynamics. But this represents a narrow, because bounded, perspective on things. Cities, after all, do not stand alone, and neither do their economies. Individual urban economies are not independent organisms evolving in isolation from economic developments occurring outside the city's borders. What goes on in one, politically, culturally, and economically, is related to what goes on elsewhere, both in other cities and in non-urban areas. Cities shape the economy not only individually in terms of their "internal" characteristics, but *relationally* in terms of their "external" interactions. These interactions are the subject of the chapter's next section, which begins with their potential significance to the issue at hand: cities' differential economic performance.

10.4   The Economy of Cities

10.4.1   Parasitic cities

What if the much-hyped economic success of cities is not – or is not primarily – based on the geographical attributes (density, proximity, and so forth) of cities themselves? What if instead the ability of firms and people in "successful" cities to outperform firms and people elsewhere is based primarily on their relations *with* those others?

Given what we have already learned earlier in this book, such a possibility should not appear outlandish. Think back in particular to Chapter 8. There, we encountered a rich tradition of theories of uneven development. These theories attempt to explain why the capitalist economy develops in geographically uneven ways. They have generally been applied to uneven international development, but in most cases, they are no less applicable to uneven development at other geographical scales, for example, in understanding varying economic fortunes between urban and rural areas, or between different urban centers. Moreover, these theories mostly posit an inherent reciprocity: healthy development in one place is not just linked to but *dependent upon* retarded or "under" development elsewhere. Might we therefore be able to understand Glaeser's "triumph of the city" (or at least of *some* cities) in these terms?

The development economist Bert Hoselitz thought so. In trying to conceptualize the relation between cities and the regions or countries in which they were located, he drew a distinction between what he called "generative" and "parasitic" cities (Hoselitz 1955). A city is generative "if its formation and continued existence and growth is one of the factors accountable for the economic development of the region or country" (p.279). It is parasitic if it exerts an opposite impact. The productivity and growth displayed by a parasitic city are based in substantial part on leeching economic potential from elsewhere and therefore hindering the development of those connected spaces. Hoselitz believed that many of the urban settlements of the British East India Company in India had a parasitic quality. So too did the Spanish colonial capitals in Mexico and elsewhere in Latin America. His arguments fit within a longer tradition of economic thought that distinguishes between economic "parasites and producers" (Boss 1990). Parasites dispossess value created by others rather than creating it themselves. Parasitic cities are concentrated embodiments of this tendency.

Notably, Hoselitz claimed that the most enduring case of a parasitic city had probably been the imperial city-state of Rome in the third to fifth centuries. Financial tribute was exacted from far and wide. And as the historical geographer Gary Brechin (2006, p.xxxi) writes, Italy's long experience with city-states eventually saw the notion of urban parasitism, or something like it, become part of the language. There was the city and its *contado*, "the territory that the city could militarily dominate and thus draw upon," and which "provided the city with its food, resources, labor, conscripts, and much of its taxes." Brechin deploys this concept in his own study of *Imperial San Francisco* (2006), highlighting the role of powerful urban elites in coordinating that city's history of extraction of "food, minerals, timber, and, above all, water and energy from the contado" (the city is often parasitic on nature, too). This calls to mind the ancient hydraulic cities from the beginning of this chapter, and the power of the urban elite to arrogate freshly produced food surpluses.

Is it far-fetched to conceive the strong productivity of "successful" postindustrial cities in a similar way? Not according to Doreen Massey. Though she does not use the word "parasitic," the London she depicts in her *World City* (2007) displays just the attributes identified by Hoselitz. Its 30% productivity premium is not self-generated. Rather, London's wealth is garnered through the exploitation inherent in uneven geographic development: through creating, sustaining, and benefiting from North–South divides both national *and* international. Massey describes London's elites, particularly the City's financial elites, "draining … other regions of professional workers" and "picking up the threads of old Empire to build a new one through which financial tribute could once again be collected" (p.214). No wonder researchers increasingly describe London as nothing less than a modern city-state (Ertürk et al. 2011), one whose elites enjoy comparable power to the city-states of old. And, of course, if parasitism helps explain the extravagant "success" of some modern cities like London, it also helps explain differential urban productivity: other UK cities not only bleed economic potential to London but lack its power to exploit national and international hinterlands.

10.4.2   Networked cities

Other scholars, however, cleave to a more benign view of why some cities benefit more than others from the webs of external economic relations within which all cities are to one degree or another enmeshed. Take the historical geographer William Cronon's *Nature's Metropolis* (1991), one of the seminal texts of urban economic geography. On the face of things, this book is similar to Brechin's book about San Francisco. It, too, focuses upon the historic relations between an important growing US city – in this case, Chicago in the second half of the nineteenth century – and the areas, primarily rural, connected to it. Cronon maps the vast flows of meat, grain, and lumber that fed Chicago and its growth, much as Brechin charts the various commodity flows that resourced a growing San Francisco.

But Cronon's is not a tale of parasitic urban elites and nor in his telling is the Great West part of a Chicago-centered "empire." Instead, city and country enjoyed a rather more harmonious or "symbiotic relationship" (1991, p.34) within a unified spatial economy. Urban-rural commercial interaction was, for Cronon, the very motor of westward US economic development. And Chicago "succeeded," becoming the preeminent urban center of the region, not because it exploited the countryside's resources better than other cities did but rather because the geography of the United States' burgeoning transportation system made it uniquely competitive. The national railroad system consisted of two largely autonomous, eastern and western, networks, and Chicago – located at the "breaking point" (p.83) between them – was able to exploit the growing need for the one to be coordinated and connected with the other. Its rise to regional preeminence was a function more of positioning than of power.

The late Allan Pred was another brilliant chronicler of US cities' "external" economic relations and of the significance of these to economic development more generally. In a series of remarkably original studies, Pred (1973; 1977; 1980) found that for all the undoubted importance of a city's interdependence with its rural hinterland (as later demonstrated by the likes of Brechin and Cronon), cities' primary economic interactions, from the very outset of industrial urbanization, have actually been with one another. They are networked phenomena. If this was a surprising conclusion, so was Pred's (1973) discovery, decades before it became fashionable to talk about "informational" capitalism and cities, that information has for a long time been as much a part of inter-urban circulations in the United States as physical commodities have. These heterogeneous circulations, Pred argued, are integral to the dynamism of wider national economies. The latter, one might say, are stitched together by cities' interactions with one another.

And so too, scholars now suggest, is the *global* economy. In other words, where Pred saw the vitality of flows between New York, Chicago, Los Angeles, and so on as pivotal to the vitality of the United States' economy at large, other scholars regard the efficient networking of *international* cities as pivotal to the health of the economy at a global scale. For the proliferating connections between countries that go by the name of globalization (Chapter 8) do not materialize in spatially random or evenly distributed ways in the countries in question. Whether the connections are of money, policies, people, goods, or cultural forms, these connections tend to be spatially concentrated, crystallizing primarily in cities; globalization represents in considerable measure the process whereby international cities, not (just) nation-states, come to be linked to one another.

Not unlike Chicago as discussed by Cronon, those cities that "win" today tend to be those that are best positioned or "networked" internationally. They emerge as the global "command and control" centers for multinational corporations that the sociologist Saskia Sassen famously discussed in her book *The Global City* (1991). Such "global" or "world cities" (Taylor 2004) are pivotal nodes for the strategic coordination and direction of international networks of production and exchange. The collective nerve center of the global economic system, these cities are also the pivotal hubs for international flows of knowledge and capital, the places where these flows pool to create the world's leading financial and information/data markets. To recognize the significance of these cities – the Londons, Hong Kongs, New Yorks, and so forth – and the links between them is to add another important dimension to the complex geographies of globalization delineated in Chapter 8.

10.4.3   Competitive cities

We need to be very careful, of course, when we use adjectives such as "parasitic" or "networked" to refer to cities. Cities are not things that act in the world. Cities are people, social institutions of various kinds, and the relations between them. To speak or write about cities as if they are spatial containers or aggregates that themselves effect social outcomes is to "fetishize," or assign causal power to, space. It is "spatial fetishism." A city, for example, does not exploit its hinterland; if exploitation is taking place then it must be people and their social institutions doing the exploiting. An inquiry into how relations between cities shape the economy is always an inquiry into the economic effects of relations between thoroughly social urban actors.

To this point, our focus in this section of the chapter has been mainly on one particular set of urban actors: capitalist firms. But there are other important urban actors, too. In the remainder of this section we will therefore consider some of the main ways in which scholars have approached economic relations between cities from the perspective of two other key sets of actors: workers and, first, states or governments.

How might we understand the economy collectively constituted by cities-as-their-governments? David Harvey (1989) has influentially argued that the answer, at least in Western societies, is in terms of *competition*. Cities, or their governments, compete with one another, with critical implications for contemporary Western economies.

Before considering how city governments compete and the effects of this competition, the first crucial question is: why? Examining the experiences of the United Kingdom and the United States, Harvey (1989) argued that prior to the 1970s, local governments were generally well funded and free to focus on their "managerial" role of improving social conditions and resolving collective needs through the delivery of public services within the city region. But for political and economic reasons, funds eventually dried up and the local

Targeted economic activity	Local government objective	Available local government tools	Characteristic spaces / artifacts
Production	■ Seek to stimulate growth in local production of goods and services	■ Infrastructure investment ■ Financial incentives (e.g., tax breaks) ■ Worker training	■ New industrial districts ■ Urban enterprise zones
Consumption	■ Seek to stimulate growth in local retail and consumption expenditures	■ Upgrading of the urban environment and amenities ■ Subsidized construction of consumption spaces ■ Hosting of cultural events	■ Shopping malls ■ Sports and events stadia
"Command and control" functions	■ Seek to attract government departments and professional service companies (banks, law firms, consultants, etc.)	■ Investment in transportation and communication networks ■ High-level educational provision (e.g., business and law schools)	■ Business parks ■ Investment agencies ■ Professional development programs
Public expenditure-based activities	■ Seek to attract institutions funded from public sources (e.g., health and education institutions, military and defense contractors)	■ Infrastructure investment ■ Public-private partnerships	■ Affordable housing programs ■ Anchor institutions

Figure 10.3 Components of inter-urban competition. Source: Adapted from Harvey 1989.

state was compelled to become "entrepreneurial": to focus more and more on attracting private-sector investment to stimulate urban economic development. This put it in direct competition with other city governments, targeting the same pool of investment opportunities.

City governments compete to attract different types of economic activities to their regions. Figure 10.3, adapted from Harvey's article, shows what these activities are, the kinds of tools that the state can use to entice them, and their characteristic urban manifestations. Such inter-urban competition is by no means restricted to the Western societies studied by Harvey. Ngai and Chan (2012) provide a telling example in their discussion of Foxconn in China where local governments have used a range of mechanisms to try to induce the company "to set up new factory compounds in their territories so as to boost GDP growth under their jurisdiction" (p.385). These mechanisms have included: the provision of land and transportation infrastructure; loans; free labor recruitment services; waiving of rent and tax; and the "dispatch" of students from vocational schools to work for Foxconn as interns unprotected by labor laws.

Where does this government-driven inter-urban competition lead economically? Inevitably, to another "race to the bottom," just like the competitive lowering of corporation tax rates by nation-states discussed in the previous chapter; and, with it, to a deepening of existing patterns of uneven geographic development. Competition to lure firms encourages and subsidizes, and therefore also accelerates, the very dynamic that renders cities vulnerable to disinvestment and retarded development: namely, the geographical mobility of capital. As yet another city government jumps on the rolling bandwagon by offering larger tax incentives or building a shinier new business enterprise zone, capitalist firms are tempted to flee from cities no longer offering best-in-class inducements or where the main profits to be made from such inducements have already been extracted and exhausted.

10.4.4   Cities in equilibrium?

Last but not least in our discussion of relations between cities, what of the position of workers? In recent times, one of the most forceful economic geographic arguments on this subject has come from urban economics, which we can think of as a version of "Geographical Economics" (Chapter 2). The argument runs broadly as follows.

As workers, when we choose where to live and work we aim to maximize our own individual "utility," which is economists' measurement of our particular preferences among a given set of goods or services. In the context of workers' locational decisions, utility is not just about wages. Other "goods" we factor into our decision-making are things like the cost of housing and the availability of various "amenities," which can be anything from the quality of the weather to the quality of local services. So, if I were weighing up which Spanish city I wanted to live and work in, I would think about the "goods" offered by Seville, Madrid, Barcelona, Valencia, and so forth strictly in relational terms. Which one offers the best combination of "goods" to match my personal preferences? To what extent am I willing to trade the sea air but lower wages of Valencia for the congestion and higher wages of Madrid? Where is my utility maximized?

Urban economists argue not only that this is the best way to think about the economy of cities from the worker's perspective, but that in seeking to maximize their individual preference profiles workers collectively succeed in bringing urban systems – systems of cities between which workers are assumed to enjoy freedom of movement – into a state of "spatial equilibrium." By this they mean a situation in which workers of similar profiles enjoy equivalent levels of utility. Sure, those living in some cities have lower wages than others, but superior local provision of other "goods" compensates. Nobody can be made better off by moving; if they could have, they would have. Everyone is a winner; or at least, nobody is a net loser.

In theory, this is an elegant model of the spatial economy. Workers clearly do make decisions about where to live and work; their respective preferences for the different bundles of "goods" offered by different cities clearly factor into these decisions; and, most importantly from our perspective, these decisions and the residential moves they precipitate clearly shape the evolution of the economy. Again, we are back to the central point of this section: that we need to understand economic dynamics in terms of how individual cities relate to other places – here as alternative and competing utility offerings – as well as in terms of their "internal" geographic characteristics.

The problem, however, is that the model is flawed, even if it provides insight into urban economic geographic developments. Those who object on principle to this kind of formal, model-based approach to economic geography would likely call out the unrealistic assumptions written into the model – not least the presumption of perfect worker mobility – and go no further. But Thomas Kemeny and Michael Storper (2012) have undertaken the painstaking analytical task of figuring out whether utility *is* indeed equivalent between cities in the country to which the arguments about spatial equilibrium have typically been applied: the United States.

Their answer: Utility levels are *not* comparable, even for similar workers. As Kemeny and Storper (2012, p.87) write, "households in larger American cities enjoy greater access to amenities, higher nominal wages, and most importantly, higher incomes after accounting for differences in housing costs." They are, in plain language, better off in every sense of the term.

Furthermore, utility does not appear to be getting more equal either. In other words, the mythical mobile, utility-maximizing worker is not moving to places offering superior utility in the way that the theory suggests she should and would; "population growth rates over the 20-year period between 1980 and 2000 are not significantly different in cities with low and high real wages and amenities" (2012, p.87). So not only is the system not in equilibrium, but it is not even trending in that direction.

What then should we conclude from this? The simple lesson we take is that just as utility levels are not equal, nor, as we have emphasized throughout this book, are all approaches to economic geography. As with the question of levels of economic development in different countries addressed in Chapter 8, and as with the question of levels of economic productivity in different cities addressed earlier in the present chapter, a satisfactory approach to levels of utility in different cities must be able to account for marked and enduring unevenness. Some cities win while others lose; some *people* win while others lose. Approaches in the "Geographies of Capitalism" mold are generally better equipped to explain this fact than those within "Geographical Economics." And the former emphasizes not equilibrium but *dis*equilibrium.

10.5   Urban Transformation as Economy

10.5.1   Building cities

Thus far we have been concerned with "the urban" primarily as the geographic context of "the economy," by which we mean the place or places where economic processes play out. Hence, we have looked at how different economic systems at different stages of economic development give rise to different urban formations, and, relatedly, at how cities – both individually and collectively – configure economic dynamics. Yet economic processes involve not just doing things *in* cities, or *between* cities (linking them together); they also involve doing things concretely *with* cities. The word "concretely" is pivotal. Manipulation of the physical, spatial fabric of the urban built environment is itself a profoundly significant economic activity, whether this involves building, rebuilding, or even destroying cities.

Consider the basic activity of city-building (Fainstein 1994), construction. Construction, which is mostly urban, constitutes an important economic activity in its own right in all countries in all periods bar the very deepest of depressions. It accounts for a significant proportion of total economic output or "value added." Figure 10.4 charts this proportion for the period since 1990 for each of the world's 10 largest national economies, showing that while it has ranged from a maximum of nearly 11% (Japan in 1990) to a minimum of 3.5% (the United States in 2013), it has typically hovered in the 5–7% range.

How should we understand the economy of city-building and its historical fluctuations? Why, for example, is the history of city-building punctuated by periodic building booms and busts (Weber 2015)?

Economists tend to favor demand-side explanations. Levels of construction, they argue, ultimately respond to levels of demand in the wider economy, in the long run if not always in the short run. There is an obvious appeal to this argument. The built environment is built because it is needed, whether to house manufacturing industry (factories), retail operations

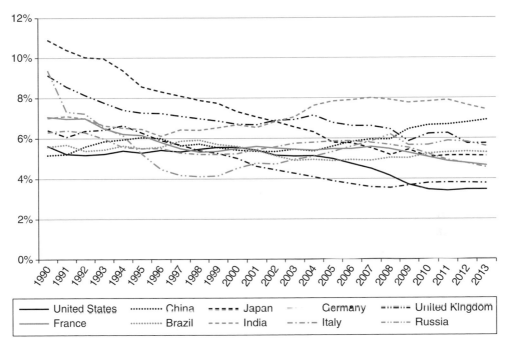

Figure 10.4 Share of construction in total economic value added. Source: United Nations Statistics Division (National Accounts Main Aggregates Database). Note: Share shown in constant 2005 prices.

(shops), professional service firms (offices), workers/consumers (homes), or for other reasons. This is not to deny that speculative building – building that *speculates on* the possibility of future demand – sometimes occurs. But for advocates of demand-side explanations such speculation is aberrational, and soon rooted out by the discipline of the market. If you build speculatively, watch out, because you will rapidly get found out, and burned.

Nor do demand-side theorists deny that governments sometimes interfere in demand-driven cycles of city-building. This clearly happens. Governments can inhibit building for which demand does exist, for instance through planning regulations. They can also stimulate building where demand seemingly does not exist. One reason governments have sometimes done the latter is that they have seen the construction sector as a fruitful source of so-called economic multiplier effects, the idea being that an increase in state spending grows total spending by a multiple of that increase. This was part of the logic behind New Deal housing policy in the United States in the 1930s in the wake of the Great Depression. Government intervention was designed to bolster the housing construction sector *and* generate growth in employment and economic output more broadly (Florida and Feldman 1988).

But not everyone finds demand-side theories persuasive, especially when it comes to accounting for prolonged and largely speculative city-building booms. As the geographer Rachel Weber (2015) shows in her study of the decade-long boom in commercial real estate construction in Chicago that preceded the global financial crisis, the empirical evidence not only does not support but flatly contradicts the notion that the boom was demand-led.

Instead, Weber, like many other economic geographers, sees a supply-side explanation as more credible, at least in terms of explaining construction booms. The most influential such explanation is David Harvey's (1978) concept of "capital switching" (tied to his larger theory of accumulation; Chapter 3.5) Harvey says that capital switching occurs in the context of an actual or imminent crisis of capital "overaccumulation" (see Box 10.2: Overaccumulation).

Box 10.2   Overaccumulation

Overaccumulation is an important concept developed within certain versions of Marxian political economy. To understand what is meant by the term, two other terms must first be introduced.

One is *surplus value*. According to Marx, workers produce value when laboring for capitalists. Of course, they generally get paid wages for doing so. But crucially, the amount they are paid in wages is less than the value they produce. The difference is surplus value, which accrues to the capitalist.

The second key term is *accumulation* (one can't have overaccumulation without accumulation). Accumulation is one of the uses to which capitalists put surplus value that they realize. They have alternatives – they can hoard the surplus, or spend it on consumption – but accumulation, Marx argues, is hardwired into capitalism (Chapter 3.5). It involves the *reinvestment* of at least some of the surplus value created by workers into a new round of production. Further, it is a necessity for what Marx calls "expanded reproduction," or what most of us simply call "economic growth."

What, then, is *over*accumulation? Essentially it means that too much surplus value is produced. Of course, this sounds like a thoroughly wrongheaded idea at first blush: how can there be too much? Capitalism and capitalists are all about maximizing growth and profits, are they not? Shouldn't they invest as much as they can? Overaccumulation? Don't be ridiculous.

The amount of accumulation that can be carried out profitably is limited, though. For accumulation to be profitable there must be consumer demand for the goods that that accumulation creates. Here's the problem. By increasing the amount of surplus value taken from workers, and then spending it in the form of accumulation, capitalists take away money from workers who would otherwise have spent it as consumers buying goods. Overaccumulation occurs when capitalists take too much surplus value and cannot immediately find opportunities to invest it because there is insufficient consumer demand to buy the goods that would then be produced. The extra surplus value takes the form of an aggregate pool of capital waiting to be used, overaccumulation. As Harvey (1985, p.4) says, overaccumulation occurs when "too much capital is produced in aggregate relative to the opportunities to employ that capital."

How might we recognize such a scenario in practice? There is no simple answer to this question. There is no "test" that demonstrates that capital is or is not overaccumulated. But there are telltale *signs*: commodity gluts, falling rates of profit, idle productive capacity, unemployment, and cash surpluses. If all or several of these are present, political economists would say that there is likely a *crisis of overaccumulation*.

Investment of overaccumulated capital is "switched" from the production of everyday goods and services for which demand is absent into production of the urban built environment. In other words, this construction occurs not because new buildings are needed, *pace* the demand-side explanation. It occurs because a suitable investment outlet is required for over-accumulated capital. In what sense is this a "supply-side" explanation? In the sense that the *supply of capital* explains the boom. Developers overbuilt in Weber's Chicago "because financial markets were making more capital available" (2015, p.2).

A number of observations concerning this thesis are in order. The first is that Harvey provides his own particular twist on the demand-side/supply-side distinction, and if one is not careful it is easy to be confused. Specifically, while capital switching is a supply-side *explanation* for urbanization (urbanization in the form of construction of the urban built environment), it nonetheless generates what Harvey (1985) calls demand-side *urbanization*. How so? Let's go back to the industrial city. Harvey argues that the industrial city evidences supply-side urbanization insofar as it *supplies* the necessary geographic conditions (agglomeration and so forth) for industrial capitalism to prosper. Now (re)read our text box on overaccumulation. Overaccumulation is about demand, or more exactly a lack thereof. City-building resulting from capital switching provides the demand for the use of overaccumulated capital that is otherwise absent.

In other words, Harvey's ultimate claim is that we are dealing with nothing less than two fundamentally different conjunctures of capitalism and urbanization, or of economy and space. The industrial city historically enabled modern industrial capitalism to take off. It provided the space for unprecedented rates of capital accumulation to occur; it put the kettle on the boil. The city of capital switching enters the picture when capitalism comes to boiling point. When too much capital has accumulated, new ways must be found to absorb it lest the system overheats. Switching capital into production of the urban built environment provides this release valve. Rather than generating the steam to power capitalism, it enables capitalism to *let off* steam.

Second, inasmuch as it is precipitated by crisis conditions (overaccumulation) and takes explicitly geographical form (city-building), capital switching represents a quintessential form of "spatial fix" (Chapter 5): a "spatial manifestation of the needs of capital accumulation", in Weber's (2015, p.25) words. One manifestation of this fix came in the immediate postwar period in the United States. The massive building of new urban and especially suburban infrastructures helped alleviate pressing conditions of overaccumulation. Their building was, in fact, *the* primary solution to the threat of imminent economic crisis – the "suburban solution," as Walker (1981) memorably labeled and theorized it.

But like all spatial fixes, the fix occasioned by capital switching is only ever a temporary one. Rather than eliminating crisis tendencies, investing overaccumulated capital in city-building delays the day of reckoning and changes both the form that crisis ultimately takes and the locus of its materialization. Writing in the 1980s, Harvey (1985, p.20) maintained that both of the twentieth century's "global crises of capitalism," in the 1930s and 1970s, had been preceded by spatial fixes in the form of capital switching: namely, "the massive movement of capital into long-term investment in the built environment as a kind of last-ditch hope for finding productive uses for rapidly accumulating capital." Weber (2015) and others (Christophers 2011) have shown that the global financial crisis of 2007–2009 was too. And this particular crisis initially materialized in precisely the way Harvey (1985, p.12) argued that crises delayed by capital switching are wont to do: as "a crisis in the valuation of assets."

Our third and final observation is simply that capital switching is not a generalizable explanation for city-building in all times and places, and nor is it intended to be. Frequently, construction occurs simply because new buildings *are* needed; equally, Harvey's supply-side urbanization, as we saw when discussing the industrial city, remains vibrant in some parts of the world. But capital switching does offer a convincing theory for when other explanations tend to fail, especially to account for city-building booms. So does it potentially provide purchase on the biggest boom of them all – the post-millennial construction of vast spaces of new urban built environment in China, many of which are in use but many of which famously stand idle as "ghost cities"? We do not know. But Harvey (2012, p.65) for one thinks it might:

> The absorption of surplus liquidity and overaccumulated capital in urbanization at a time when profitable opportunities are otherwise hard to come by has certainly sustained capital accumulation not only in China but around much of the rest of the globe over the last few crisis years.

10.5.2   Destroying cities

Something else of signal importance to the economy of urban transformation has also been visible in recent decades in China. This is not the construction of urban space but its destruction.

In Chapter 12, where we will discuss industrial change in more depth than we have in the present chapter, we draw on the economist Joseph Schumpeter's idea of "creative destruction" to capture a key aspect of capitalist development: the fact that for the new to come into being the old must be jettisoned. This frequently applies to cities too. Where an existing urban built environment is in place, the destruction of some of this space may occur before a new round of production proceeds. And this destruction is also a significant economic activity.

The destruction of urban space can be brutal in its intensity and scale. Here China provides striking recent examples. The sociologist Xuefei Ren (2014) relates the devastating experience of Shanghai where "the conservative estimates of the city government" – "the actual numbers are probably much higher" – indicate that between 1990 and 2008 the city demolished 70 million square meters of housing and relocated 1.2 million households from inner-city neighborhoods (p.1082). So economically pivotal has this destruction been that Ren styles Shanghai's a "demolition economy." Through the various economic functions it performs, ranging from enhancing government revenues (by enabling the city government to acquire and then lease out land to private investors) to creating new demand for housing to crystallizing a lucrative debris-recycling industry, destruction has come to play no less than a "central role … in contemporary Chinese urbanism" (p.1081).

But destruction of the physical urban fabric does not have to be intensive and interventionist to be significant; it can be slow, and can result from *in*attention as much as from attention. After all, investment is required not just to grow the value of the built environment but, given inevitable deterioration in the absence of such investment, merely to maintain value. Hence, disinvestment, or the withdrawal of investment, ultimately destroys too. And research shows that such disinvestment has been as economically integral and geographically uneven in many Western cities as demolition has been in Shanghai.

Indeed, disinvestment and its economy represent a powerful explanation for that most emblematic phenomenon of post-1970s Western urban physical transformation: inner-city gentrification. This is the process of neighborhood succession whereby individuals and households of lower socioeconomic status, typically living in run-down rented accommodation, are displaced by higher-income residents whose arrival coincides with wholesale reinvestment in the housing stock. Gentrification is a complex phenomenon and has been driven in part by the desire of suburbanites to move "back" to the city where jobs in knowledge-based industries have substantially replaced manufacturing. But it is also about *capital* moving back to a city where historical disinvestment has created the necessary economic conditions for reinvestment and redevelopment to yield profits. This is the crux of Neil Smith's (1979) influential "rent gap" thesis. In the absence of investment in the built environment – destruction through neglect, as it were – a gap opens up between the rents that can be extracted in the existing (depressed) conditions and the rents that could potentially be extracted if redevelopment were to occur. When the gap is big enough, the economic logic becomes unassailable and investment flows in.

Gentrification, then, is another example of cities being (re)made geographically for structural economic reasons; it is urban transformation *as* an economic process, the means by which value is created. Capital had earlier built the suburbs, at least in part, to fend off the threat of an overaccumulation crisis; decades later, through gentrification, it left those suburbs behind. The "seesaw" of uneven geographic development (Chapter 8) is as much an intra-urban phenomenon as it is an inter-urban and international one. It occurs at all geographic scales.

10.6   Conclusion

We began this chapter with a formulation that the rest of the chapter has tried to exemplify: Walker's "essential unity" of economic and urban geography. Along the way, though, something else periodically threatened to crash this party, to get in on the mix: "nature," or the environment. Consider in particular Brechin and Cronon's books on San Francisco and Chicago, respectively. Assuredly, these testify to the unity evoked by Walker. But they also undermine other dominant dualisms that tend to keep nature out of the picture. The "economy," they show, is not separate from "nature"; it rests rather on nature's commoditization. And neither is the city; Chicago, as Cronon's title says, is no less than *nature's* metropolis, a product of the urbanization *of* nature. Economic and urban processes are alike in that they are both *socionatural* ones.

Instead of a two-part "essential unity" of urban and economic geography, then, perhaps we need a tripartite unity that folds in environmental geography. In any event, the economic and the environmental, the social and the natural, bleed into one another with a particular intensity in the urban milieu.

Where the present chapter has examined the mutually constitutive relation between two elements of this trinity – the economic and the urban – the following chapter reorients our lens toward the relation between the economic and the environmental. The significance of the urban to this particular relation is not neglected, and indeed surfaces at numerous junctures; but it is not our emphasis. Rather, we aim to show that, like the urban, what we typically call "nature" is also produced within, by, and for particular economies and economic formations; and that different economies produce, enroll, and are shaped by *different* natures. Economic geography must at once be environmental geography, for the economy is always and everywhere environmentally embedded and the environment is always and everywhere economically configured. For a long time the notion would have seemed outlandish, but nature – substantively implicated as it is in economic processes – is today just as much the "natural" subject matter of economic geography as industrial location, trade, or uneven geographic development.

References

Amin, A. 2000. The Economic Base of Contemporary Cities. In G. Bridge and S. Watson (eds), *A Companion to the City*. Oxford: Blackwell, pp.115–129.

Boss, H. 1990. *Theories of Transfer and Surplus: Parasites and Producers in Economic Thought*. London: Unwin Hyman.

Brechin, G. 2006. *Imperial San Francisco: Urban Power, Earthly Ruin*. Berkeley: University of California Press.

Breman, J. 2016. *At Work in the Informal Economy of India: A Perspective from the Bottom Up*. Oxford: Oxford University Press.

Castells, M. 1989. *The Informational City: Information Technology, Economic Restructuring, and the Urban-Regional Process*. Oxford: Blackwell.

Childe, V.G. 1950. The Urban Revolution. *Town Planning Review* 21: 3–17.

Christophers, B. 2011. Revisiting the Urbanization of Capital. *Annals of the Association of American Geographers* 101: 1347–1364.

Crafts, N.F.R. 1985. *British Economic Growth during the Industrial Revolution*. Oxford: Clarendon Press.

Cronon, W. 1991. *Nature's Metropolis: Chicago and the Great West*. New York: W.W. Norton & Company.

Engels, F. 2009 [1845]. *The Condition of the Working Class in England*. Oxford: Oxford University Press.

Ertürk, I., Froud, J., Sukhdev, J. et al. 2011. City State against National Settlement. CRESC Working Paper Series No. 101. http://hummedia.manchester.ac.uk/institutes/cresc/workingpapers/wp101.pdf (accessed June 27, 2017).

Fainstein, S. 1994. *The City Builders: Property Development in New York and London, 1980–2000*. Oxford: Blackwell.

Florida, R.L., and Feldman, M. 1988. Housing in US Fordism. *International Journal of Urban and Regional Research* 12: 187–210.

Glaeser, E. 2011. *Triumph of the City*. London: Macmillan.

Harvey, D. 1978. The Urban Process under Capitalism: A Framework for Analysis. *International Journal of Urban and Regional Research* 2: 101–131.

Harvey, D. 1985. *The Urbanization of Capital*. Oxford: Blackwell.

Harvey, D. 1989. From Managerialism to Entrepreneurialism: The Transformation in Urban Governance in Late Capitalism. *Geografiska Annaler Series B* 71: 3–17.

Harvey, D. 2012. *Rebel Cities: From the Right to the City to the Urban Revolution*. London: Verso.

Hoselitz, B.F. 1955. Generative and Parasitic Cities. *Economic Development and Cultural Change* 3: 278–294.

Hsing, Y.T. 2010. *The Great Urban Transformation: Politics of Land and Property in China.* Oxford: Oxford University Press.

Kemeny, T., and Storper, M. 2012. The Sources of Urban Development: Wages, Housing, and Amenity Gaps across American Cities. *Journal of Regional Science* 52: 85–108.

Krugman, P. 1991. *Geography and Trade.* Cambridge, MA: MIT Press.

Landry, C., and Bianchini, F. 1995. *The Creative City.* London: Demos.

Marshall, A. 1890. *Principles of Political Economy.* New York: Macmillan.

Marx, K. 1963. *Theories of Surplus Value, Part 1.* Moscow: Progress Publishers.

Massey, D. 2007. *World City.* Cambridge: Polity.

McKinsey Global Institute. 2011. Urban World: Mapping the Economic Power of Cities. http://www.mckinsey.com/global-themes/urbanization/urban-world-mapping-the-economic-power-of-cities (accessed July 6, 2017).

Ngai, P., and Chan, J. 2012. Global Capital, the State, and Chinese Workers: The Foxconn Experience. *Modern China* 38: 383–410.

ONS (Office of National Statistics). 2013. Percentage of Working People Employed in Each Industry Group, 1841 to 2011. http://webarchive.nationalarchives.gov.uk/20160105160709/http://www.ons.gov.uk/ons/rel/census/2011-census-analysis/170-years-of-industry/rft-tables.xls (accessed June 27, 2017).

ONS (Office of National Statistics). 2016. Labour Productivity (GVA per hour worked and GVA per filled job) indices by UK NUTS2 and NUTS3 subregions. https://www.ons.gov.uk/file?uri=/employmentandlabourmarket/peopleinwork/labourproductivity/datasets/subregionalproductivitylabourproductivitygvaperhourworkedandgvaperfilledjobindicesbyuknuts2andnuts3subregions/current/nutsindeceslevels2016unlinked20160307.xls (accessed June 27, 2017).

Parilla, J., Trujillo, J.L., and Berube, A. 2015. Global Metro Monitor 2014. http://washcouncil.org/presentations-2015/2014-Global-Metro-Monitor-Map.pdf (accessed June 27, 2017).

Pirenne, H. 2014 [1925]. *Medieval Cities: Their Origins and the Revival of Trade.* Princeton, NJ: Princeton University Press.

Pred, A. 1973. *Urban Growth and the Circulation of Information: The United States System of Cities, 1790–1840.* Cambridge, MA: Harvard University Press.

Pred, A. 1977. *City Systems in Advanced Economies: Past Growth, Present Processes, and Future Development Options.* New York: Wiley.

Pred, A. 1980. *Urban Growth and City Systems in the United States, 1840–1860.* Cambridge, MA: Harvard University Press.

Ren, X. 2014. The Political Economy of Urban Ruins: Redeveloping Shanghai. *International Journal of Urban and Regional Research* 38: 1081–1091.

Richardson, H.W. 1995. Economies and Diseconomies of Agglomeration. In H. Giersch (ed.), *Urban Agglomeration and Economic Growth.* Berlin: Springer, pp.123–155.

Sassen, S. 1991. *The Global City: New York, London, Tokyo.* Princeton, NJ: Princeton University Press.

Saxenian, A. 1996. *Regional Advantage: Culture and Competition in Silicon Valley and Route 128.* Cambridge, MA: Harvard University Press.

Sigurdson, J. 1977. *Rural Industrialization in China.* Cambridge, MA: Harvard University Press.

Smith, N. 1979. Toward a Theory of Gentrification: A Back to the City Movement by Capital, Not People. *Journal of the American Planning Association* 45: 538–548.

Storper, M. 2013. *Keys to the City: How Economics, Institutions, Social Interaction, and Politics Shape Development.* Princeton, NJ: Princeton University Press.

Taylor, P.J. 2004. *World City Network: A Global Urban Analysis.* London: Routledge.

Thompson, F.M.L. 1990. Town and City. In F.M.L. Thompson (ed.), *The Cambridge Social History of Britain, 1750–1950.* Cambridge: Cambridge University Press, pp.1–86.

Walker, R. 1981. A Theory of Suburbanization: Capitalism and the Construction of Urban Space in the United States. In M. Dear and A. Scott (eds), *Urbanization and Urban Planning in Capitalist Society.* New York: Methuen, pp.383–429.

Walker, R. 2012. Geography in Economy: Reflections on a Field. In T.J. Barnes, J. Peck, and E. Sheppard (eds), *The Wiley-Blackwell Companion to Economic Geography.* Oxford: Wiley-Blackwell, pp.47–60.

Walker, R. 2015. The City and Economic Geography: Then and Now. In H. Merrill and L.M. Hoffman (eds), *Spaces of Danger: Culture and Power in the Everyday.* Athens, GA: University of Georgia Press, pp.135–151.

Weber, A. 1929. *Theory of the Location of Industries.* Chicago: University of Chicago Press.

Weber, R. 2015. *From Boom to Bubble: How Finance Built the New Chicago.* Chicago: University of Chicago Press.

Williams, R. 1975. *The Country and the City.* New York: Oxford University Press.

Chapter 11

Nature and the Environment

11.1   Introduction

We have not been shy in this book about the scope of economic geography's compass. One of the many reasons economic geography is good for you, we have argued, is that it helps to explain a great deal about the contemporary world, shedding unique light on such "big" important issues as cities, money and finance, and globalization. This expansive analytical remit is in fact a feature of geography more broadly. "To study geography," submits Alastair Bonnett (2008, p.1) in his excellent book *What is Geography?*, "is to study the world." Economic geography cleaves faithfully, albeit sometimes fitfully, to this capacious agenda.

If geography's interest – no doubt an immodest one – is in "the world," it seems important from a critical perspective to ask what, if anything, is special about the *contemporary* world. A vital characteristic of a critical standpoint as we have depicted it is its questioning specifically of the world as it is now: Why is it this way? Does it have to be this way? Could it and should it be another way? One answer to the question of the specificity of the "modern" world is that it is *capitalist*. Capitalism is what distinguishes the present from the (feudal, communist, other) past. Hence the fact that, as we saw in Chapter 2, the answer to the question "What is economic geography?" – especially among those of a critical disposition – has very often been "Geographies of Capitalism."

Yet, an alternative and arguably equally valid perspective on the specialness of the contemporary world has gained currency in the past decade. What is truly special about the modern world is not so much the relation of humans to humans as the relation of humans to the biophysical environment, to "nature." Indeed, it is today frequently claimed that human

Economic Geography: A Critical Introduction, First Edition. Trevor J. Barnes and Brett Christophers.
© 2018 John Wiley & Sons Ltd. Published 2018 by John Wiley & Sons Ltd.

society's relation to the environment in the modern era is sufficiently distinctive for that era to be understood as constituting no less than a new geological epoch, one in which human activities for the first time substantively *change* global ecosystems: the much-touted "Anthropocene." This distinctiveness in the realm of human–environment relations is certainly not unconnected to distinctiveness in human–human relations (although scholars are far from agreed that the Anthropocene is precisely coeval with what some call the Capitalocene – the age of capital). Nevertheless, the implication is that if economic geography is to live up to the billing we have given it, then it must contribute to discussion of the Anthropocene and of the world-defining issues it exhibits: environmental exploitation, transformation, and crisis.

The present chapter demonstrates and assesses exactly that contribution, which it argues is substantial. The chapter's central premise and theme, recalling the general definition of economic geography we provided in Chapter 2, is easily summarized: what we call "nature" is no less substantively implicated in economic processes and outcomes than space, place, scale, and the other fundamental "stuff" of geography. Economic geography's role is to tease out and explain that implication. Take "nature" away or, indeed, change its constitution and complexion and the economy would necessarily look very different. And, in an era of deep-seated environmental change, that is no minor claim.

11.2   The Use of Nature

The economy "uses" nature in all sorts of obvious ways. When economists talk about the "primary sector" of the economy they are referring to economic activities that make more or less direct use of "natural" environmental resources. Standard examples are agriculture, fishing, forestry, and mining. Such activities are "closest" to nature in the sense that they work directly in and on it. Different fates then await the things extracted from nature through such activities. Sometimes they are consumed directly, as is typically the case with fish, for example, or with much agricultural produce. But often they become inputs to other economic activities, in which cases the product ultimately consumed is likely to be something made *from* the natural resource as opposed to the resource per se. The mining sector is a good example of an industry that generates both types of outputs. One type are consumed as they are, such as precious metals like gold, the other as inputs in the "secondary" (product manufacturing) sector of the economy, such as iron ore.

A large part of the contribution of economic geography has been to help dispel or at least refine some of the many stereotypes and simplifications that have come to surround the economic uses of nature under capitalism. A consideration of that "demythologizing" work is therefore a productive method for beginning to appreciate critically economic geographic understandings of nature and environment. Three such popular mischaracterizations, each related to the others, are especially prominent and have elicited forceful critical economic geographic readings.

11.2.1   Misconception #1: distancing nature

The first misconception is that countries necessarily follow a quasi-teleological develop-mental trajectory that sees the primary sector, and thus direct economic engagement with the environment, become less and less important over time. Popular versions of this thesis include the so-called three-sector and stages of economic growth theories associated with economists such as Jean Fourastié and Walt Whitman Rostow, respectively. As an economy develops or "modernizes," its principal focus allegedly shifts from the primary to the secondary to the "tertiary" (service) sector. To be economically modern is to have effectively distanced nature as a source of socioeconomic sustenance and as a sphere of socioeconomic reproduction. "For economists," observes Dick Walker (2001, p.167), "modernization is a long march out of primary extraction through manufacturing into high-tech futures."

Although such models may help capture the broad outlines of certain important historical trends in certain places, they are problematic. For one thing, there are important and obvious anomalies. The region studied by Walker – California – is one such: now "the world's high-tech capital and one of the richest spots on earth," California, Walker reminds us, has historically followed "a resource-intensive road," suggesting that "perhaps the contribution of nature to economic growth is a topic worthy of serious consideration" (2001, p.167). Other apparent anomalies include Australia and Canada: both are extremely wealthy; both are resource-rich; and, in both, the primary sector still contributes a substantial share of national economic output, at least 10% in most years in each place. It was a Canadian economist, Harold Innis, who advanced perhaps the best-known thesis against three-sector-type teleologies. Informed by the Canadian experience, his "staples thesis" claimed that an export-led model built around primary resource staples *can* be economically sustainable, although there is likely to be greater economic volatility than for countries that specialize in manufacturing or services.

Furthermore, the conventional teleological narrative is geographically anemic. It treats national economies as bounded economic spaces. One of the reasons why various "developed" nations have historically been able seemingly to distance themselves from nature and the primary sector – and then also from the manufacturing of the secondary sector – is that they have offshored their requirements to countries in the Global South, increasingly sourcing primary and secondary products from them, and in the process helping to *keep* them "underdeveloped" (see also Chapters 8 and 10). "Modernization," seen this way, is less about escaping the economic attachment to nature than it is about spatially transplanting it.

In any event, "modern" economic activities are rarely as distant and detached from nature as they can appear. The links to nature are there, even if we cannot immediately see them. The geographer Scott Prudham makes this case emphatically in his book *Knock on Wood* (2005), which we will return to shortly. Capitalism, he concludes (p.8), "cannot sustain itself alone." In all its myriad shapes and forms it "will always rely to some extent on nonproduced inputs." In fact this makes for one of capitalism's most notable tensions, one of its most striking curiosities: "It is a system so impressive in its capacity for growth, productivity increase, and innovation, yet in this respect it is curiously dependent and vulnerable, often without our acknowledgment. Capitalism needs nature."

11.2.2   Misconception #2: deterministic nature

Capitalism does need nature. And this need of nature has periodically exercised some of the foremost minds in the history of economic thought, leading, we suggest, to the second key area of mischaracterization concerning nature's economic use. This is the assertion that "natural" resource endowments and their use deterministically shape economic outcomes. The specific *nature* of this determination, however, and especially the question of whether natural resource wealth is deemed a positive or negative attribute, has been the subject of considerable dispute.

At one extreme were the early political economic thinkers subsequently labeled the Physiocrats. Led by François Quesnay and Anne-Robert-Jacques Turgot, these influential eighteenth-century writers identified nature and its cultivation as the source of all wealth creation. "Quesnay," notes the economist Raimund Bleischwitz (2001, p.25), "remarked on agriculturalists as the 'productive class' of society and other groups as 'sterile classes'. Turgot called farmers the 'unique source of all wealth.'" Other sectors of the economy, according to the Physiocrats, merely moved wealth around or parasitically consumed it. An abundance of natural resources and a willingness to work them were the keys to economic success.

At the other extreme, and working much closer to the present day, we find those saying more or less the exact opposite: that natural resource wealth is less a blessing than a curse; it is a "resource curse." It hinders rather than fuels economic growth and development. The rock-star economist Jeffrey Sachs is the most famous proponent of this view, one ostensibly supported by data showing that resource-rich countries have in recent decades tended to grow more slowly than resource-poor countries (e.g., Sachs and Warner 2001). So much for Physiocracy.

At one level, of course, these divergent claims likely reflect in part the nature of the economies of the respective eras in which they were formulated. In the eighteenth century, the primary sector *did* dominate the economy; it would have been absurd to suggest that nature's bounty was anything but indispensable. Today, by contrast, this sector is much less economically significant, especially in the Global North. Today's economy thus represents "natural" soil for the germination of very different perspectives on the keys to economic prosperity.

But as economic geographers such as Walker (2001) in his work on California have shown, it is not just unhelpful but incorrect – a crude form of environmental determinism (Chapter 3) – to suggest that nature's attributes mechanically dictate economic outcomes either positively *or* negatively.

California's natural resource wealth, for example, did not make economic prosperity inevitable, and neither did it doom the region to failure. California's economic "success" (by no means all inhabitants benefited) required allying its resource base with a textbook capitalist sociopolitical order incorporating, *inter alia*, strong property rights and an amenable state. Similarly, Norway, with a contemporary economy built on oil resources, hardly seems cursed, enjoying as it does among the highest levels of per capita economic output in the world – a metric on which resource-rich Australia and Canada also figure strongly.

As for those (often African) countries nominally cursed by resource wealth, another economic geographer, Michael Watts (2004), provides a devastating rebuke to notions of determinism. Nigeria's "curse" has not been oil but the political economic environment into which, from the early 1960s, it began to flow, one that was worlds away from the Californian archetype, shaped, not least, by a colonial past that formally ended in only 1960. The flow of oil into "weak institutions and a charged, volatile federal system," says Watts (p.61), "produced an undisciplined, corrupt and flabby oil-led development." The "curse" was indelibly historical, geopolitical, and institutional – in a word, social – rather than environmental.

11.2.3   Misconception #3: deferential nature

The final important mischaracterization concerning the economic use of nature is that such use is largely unproblematic. Of course, it has long been recognized that capitalism's relation to nature is *ecologically* problematic. That is not our point. Our point is that the economic use of nature has been deemed problematic *only* in an ecological sense. Deal with the ecological repercussions, the conventional view suggests, and everything will be alright.

The environmental historian and philosopher Carolyn Merchant provided a stunning exegesis of this view in its various historical guises in *The Death of Nature* (1980). She showed that as a concept, nature has been thoroughly feminized and, in the process, rendered deferential to a powerful, dominating (and male) "society" and "economy." Nature can be controlled; she can be managed. Sure, she has to be treated tenderly, like a lady, and not physically sullied. But that is all. As long as she is not physically (ecologically) damaged, she will happily let you go about your ordinary business.

Economic geographers and others, however, have shown that this is simply not true. When capitalism uses nature, it encounters economic and social as well as ecological problems. Nature does not simply defer; it reacts. And economy and society have to accommodate, sometimes uncomfortably, to nature's demands. The economic geographer Gavin Bridge provides an instructive example in his study of the US copper industry (see Box 11.1: Contradiction and Copper).

In his study, Bridge's conceptual inspiration comes from the radical economist Elmar Altvater. But scholars making comparable arguments about the socioeconomic problems arising from the economic use of nature have more often turned to a different critic of Western capitalism: the Hungarian Karl Polanyi. Why?

Read his *The Great Transformation* (1944) and it is not hard to see (see also Chapter 4). The book contains a lacerating critique of free-market liberal capitalism and of its relentless imperative to *commodify*, to turn things into objects of trade. Capitalism's commodification of land/nature – Polanyi used the terms interchangeably – alongside labor and money receives particular scorn. We will have more to say about nature's commodification in the next section. The crucial point here is that nature, according to Polanyi, does not meekly submit to being used this way. It kicks back, causing problems that, left unchecked, society could not ultimately bear. Allowing markets free rein over nature (and labor and money), Polanyi says, eventuates not only in the destruction of nature but also in "the demolition of society." For: "Nature would be reduced to its elements, neighborhoods and landscapes defiled, rivers polluted, military safety jeopardized, the power to produce food and raw materials destroyed" (p.76). The very society that commodifies nature comes to realize that limits must be placed on such commodification, *for society's as well as nature's benefit*. We need land use regulations. We need environmental laws. Nature, in short, refuses to defer. Nature says, "enough." And society, in the end, must listen.

One of the several economic geographic studies that lean productively on Polanyi to understand socioeconomic problems arising from nature's economic use is Prudham's *Knock on Wood*. In his examination of industrial logging and the production of timber-based commodities in the Douglas-fir region of Oregon, Prudham demonstrates how nature (the forest) has historically required capitalism to adapt, lest it fails. Nature is a reluctant commodity and society has to deal with this reluctance. For instance, when capitalist production relations are introduced to the forest, both the forest *and* those relations are transformed. Nature's vagaries, such as unpredictable variations in production conditions, necessitate departures from the regularities of factory production. Nature, Prudham claims, "ecoregulates" capitalist labor processes. It does not lie down in deference.

Box 11.1   Contradiction and Copper

Bridge (2000) examines the crisis conditions that came to beset the copper mining and processing industry in the late 1980s in Arizona and New Mexico, US states that traditionally account for some three-quarters of the nation's copper output.

He argues that copper mining is characterized by three key "ecological contradictions" (pp.244–249):

Contradiction 1. *Mining actively consumes existing deposits and, in the process, generates socio-political opposition to the acquisition of new deposits. Yet, access to mineral deposits is an essential precondition for growth.*

Contradiction 2. *Mining firms preferentially select the highest grade ("richest") ore bodies that can be processed profitably using current technology. Yet depletion of high-grade ores over time means that, without innovation to reduce unit costs, production costs will rise.*

Contradiction 3. *Mining and mineral processing are segregating processes that necessarily generate large and complex waste streams: the continued availability and legitimacy of low-cost waste disposal options are therefore essential for growth. Yet, the conventional disposal of mining wastes depletes existing disposal options and generates resistance and opposition from non-mining interests.*

The significance of these contradictions to Bridge's analysis is that collectively they enflamed social conflict, which in turn led to declining profitability and ultimately industry crisis.

The conflict surfaced around two connected issues: access to land and with it raw materials; and waste disposal practices. In the 1980s, the copper industry's disposal practices increasingly created tension with other economic sectors (including real estate, high technology, and tourism) as well as with environmental pressure groups. This tension had damaging repercussions because it undermined a traditional buttress of the industry's health and wealth: support from the state in the form of ready access to federal land and limited penalization of environmental costs. As tension turned into conflict, it directly influenced state decisions over matters such as proposed mine expansions or water quality requirements. The effects for the copper industry were severe.

Nature, in Bridge's reading, had refused to defer. Capitalism's use of nature is contradictory, and eventually hurts not just nature but capitalist institutions.

11.3   The Production of Nature

11.3.1   Beyond "use"

In seeking to understand nature's involvement in economic processes, is "use" the only or best framing available? Is it the case simply that capitalism "uses" nature as a resource? Or is something perhaps more complex, more involved, more specific going on? If not "use," what *is* the most appropriate and illuminating way of figuring capitalism and nature's relations with one another? These questions have been at the center of critical thinking in economic geography for more than three decades, and are the focus of this section.

One of the main reasons that these questions have been asked is that scholars, especially those working in the "Geographies of Capitalism" tradition (Chapter 2), have increasingly recognized the *significance* of economy–nature relations. They have recognized that part of the answer to questions such as "What is capitalism?" and "What is it that makes capitalism different from other systems of socioeconomic organization?" is "a particular relation to nature." How economy and society are organized in relation to nature is not incidental. In capitalism's case, it is integral, making capitalism capitalism no less definitively than other of its notable characteristics – such as workers' requirement to sell their labor power to capitalists who own the means of production – do. Capitalism would not be capitalism, or at least not capitalism as we recognize it today, if its relations with nature were configured substantially differently.

Most geographers would say that "use" is indeed a rather blunt and uninformed verb to articulate these relations. Two other processes, closely connected with one another, have instead generally been highlighted. They amount to "using," but in very particular ways.

The first process is commodification. This is about turning the environment into a commodity – something that can be exchanged in markets – or a set of commodities. timber, minerals, and so on. It stands to reason that capitalism's relation to nature entails commodification. Capitalism's relation with more or less everything does. That is why Karl Marx began with the commodity in *Capital*. Its first line: "The wealth of societies in which the capitalist mode of production prevails appears as an 'immense collection of commodities'; the individual commodity appears as its elementary form" (Marx 1976, p.125).

The second process woven into capitalism's relation with nature is privatization. Privatization concerns ownership. It entails the transfer of ownership from public (government) or communal to private hands. Under capitalism, nature in the form of natural resources like fish stocks has increasingly been privatized.

The relation *between* nature's privatization and commodification is interesting. Privatization is not always accompanied by commodification. Natural resources can be privately owned yet not traded. There is a vivid illustration in Roald Dahl's all-too-real fiction *Danny, the Champion of the World*. Victor Hazell, the villain of the piece, uses his private English woodland to host pheasant-shooting parties for the local aristocracy. He does not sell the birds. Nor does commodification require privatization. Land reform in China, for example, has involved the introduction of leasehold markets and market-like competition for public land, a case of "commodification without privatization" (Hsing 2006, p.169).

Yet capitalist commodification and privatization of the environment do very often go hand in hand. In particular, privatization clearly abets commodification. Once privately owned, nature, accordingly parceled up, can be readily bought and sold in markets; and, crucially, firms can begin to profit from this trade. The *combination* of privatization and commodification is therefore a hallmark of capitalism's relation with nature. It is, we suggest, a striking example of what David Harvey (2003) terms "accumulation by dispossession" (Box 11.2: Accumulation by Dispossession).

But for all the insights provided by work on capitalism's commodification and privatization of nature, there remains an inescapable problem with this way of conceptualizing matters. A particular phrasing we have relied on thus far – "capitalism's relations with

Box 11.2   Accumulation by Dispossession

We know what capital accumulation is (Chapter 10): the reinvestment of at least some of the surplus value created by workers into a new round of capitalist production. So, what, specifically, is *accumulation by dispossession*? And what does it have to do with nature and its commodification and privatization?

Harvey's (2003) "accumulation by dispossession" (ABD) is a reworking of Marx's concept of "primitive accumulation." By "primitive," Marx meant "original," not "ancient." Primitive accumulation is essentially the *first* capital accumulation or, more accurately, the process that originally made accumulation possible. What was this first process? Since accumulation requires surplus value, and surplus value requires laborers who are forced to sell their labor power to capitalists, then accumulation was made possible by the process (or processes) that historically created the divide between laborers and capitalists and that compelled the former to work for the latter. The key in the United Kingdom was the "enclosure" of common lands and with it the expropriation of self-sufficient peasants (see Chapters 10 and 12). This process, which was fundamentally one of privatization (of land rights), was Marx's "primitive accumulation."

Harvey's ABD is not so different. The main reason he prefers a different term is the historical connotation of the word "primitive," which suggests that the kinds of processes that Marx was referring to are merely a thing of the past. Harvey says they are not. They have remained essential to capitalism throughout its history. Whereas primitive accumulation was what made capital accumulation possible from the outset, in capitalism's formative historic contexts such as late eighteenth-century England, ABD represents the processes that have *continued* to make accumulation possible since then.

What are these processes? Privatization and commodification. The latter is no less important than the former, and it was important to primitive accumulation, too – after all, it was no good divorcing peasant producers from their means of production if their newly available labor power could not be bought and sold by capitalists as a commodity. No commodified labor power, no surplus value.

And one of *the* key domains of ABD (privatization and commodification), Harvey insists, is the environment. In recent decades in particular, we have witnessed the privatization and "wholesale commodification of nature in all its forms" (2003, p.148). This dispossession of environmental resources makes capital accumulation possible by making available new sets of inputs to economic processes.

A good example is provided by the sociologist Jack Kloppenburg in his *First the Seed* (1988), which charts in fascinating and painstaking detail the historical transformation of seed from being a public good to a privatized commodity anchoring the accumulation of capital. The book's particular value lies in detailing the range and extent of work involved in these "dispossession" activities. The scientific community played an important role. And so, needless to say, did the state: like all former noncommodities, plant germplasm could not be subjected to a capitalist calculus of exchange until legible and enforceable property rights had been extended to it.

nature" – is indicative. Because, if capitalism or "the economy" can have a relation *with* nature, the implication is that capitalism and the environment are fundamentally separate entities. The latter comprises a "natural" world detached from the social, albeit exploitable by it; the former, the economy, exists in a social realm discrete from, but again not unaffected by, the biosphere. The problem with such a standpoint is that nature is *externalized*. And it is by no means clear that such a dualistic framing is adequate to understanding and explaining the contemporary world and particularly the status of what we call "nature."

11.3.2   Beyond dualism

Numerous writers, many coming from an explicitly economic geographic perspective, have attempted to find a language and a conceptual framing that can transcend the dualistic economy–nature binary and, just as importantly, that can be operationalized in empirical research. An enormous amount could be (and has been) written about such efforts. We do not want to belabor them. They are complex, frequently very abstract, and sometimes frighteningly esoteric. But we do think it is important to review them briefly. They have been, and remain, highly influential.

The main source of theoretical inspiration is often the same: Marx. With good reason. As geographers Morgan Robertson and Joel Wainwright (2013, p.895) remark, not only was "the human-environment relation … fundamental to Marx's concept of capitalism," but his critique of political economy signaled the possibility of thinking beyond dualism. That is, not thinking of "human-environment" *as* a relation at all. Marx theorized commodity production, the very engine of capitalist growth, as "at once social and natural." It is "a process that defies the taxonomy that segregates the social from the natural" (2013, p.895).

Probably the most significant rethinking of nature-and-capitalism along non-dualistic, Marxian lines is found in the work of Neil Smith. In his magnificent *Uneven Development* (1984) and then in a series of subsequent works, Smith formulated the idea of "the production of nature."

As Smith acknowledged, nature "seems to be the epitome of that which neither is nor can be socially produced" (Smith 2007, p.25). But he made a pretty compelling case that this – the production of nature – is in fact precisely what happens under and within late capitalism. The word "within" is key. Nature is not something external to and ravaged by capitalism (as generations of environmentalists have asserted), but is increasingly produced internally to it, as *socionature*. Hence, no dualism.

Smith conceded the argument that when capitalism was in its early stages there *was* still an external "first nature," and that the "second nature" (a Hegelian concept) of socioeconomic institutions was "produced out of and in opposition to" it. The economy originally "used" (first) nature. With its inexorable spread and intensification, however, capitalism "incrementally infiltrates any remnant of a recognizable external nature" to the extent that "first nature comes to be produced from within and as a part of … second nature itself" (2007, pp.30–31). Is anything sacred? Yes and no. Yes, we still have "wilderness" areas, the epitome of "nature." But these are nonetheless internal to capitalism in the sense that they remain "wild" *only because capital has decided to keep them that way.* Their wildness is produced, and therefore *un*natural. Under contemporary capitalism, Smith concluded, nature's production is "a systemic condition of social existence" (p.25).

The practical benefits of thinking about the world in this way are real enough. It allows for a more dynamic and integrated understanding. Instead of a "thing," nature is conceived as an (economic) activity, something capitalism "does," as an "accumulation strategy" in Smith's (2007) terms. Capitalism, in turn, can be conceived as a specific environmental project or "regime" (Moore 2015). And events or phenomena that we are accustomed to reading as social *or* natural become more easily grasped as both, as capitalism's socionatural *products*. Here are two examples:

i.   The *food crises in the West African Sahel* studied by Watts. *Silent Violence* (Watts 1983) is a powerful refutation of environmentally determinist accounts of famine – the idea that they are "natural" disasters. Watts shows that environmental conditions certainly play a part. But so too do political economic conditions. Famine results from particular conjunctures of the environmental with the political economic. They form a conducive fusion of conditions. In Smith's terms, they produce specific capitalist natures.

ii.   The *Chicago heat wave of 1995* studied by the sociologist Eric Klinenberg (2002). Klinenberg's is also an account of the production of a tragic capitalist nature. Over 700 people died in conditions that were physically inhospitable but which nonetheless required overlaying of social inhospitability to produce the devastating results they did. Klinenberg explains the effects of the heat wave with reference to Chicago's dysfunctional political economy and the social disintegration it had historically enshrined.

The idea of the production of nature is particularly salient today because the production of nature – or of *new capitalist natures* – has seemingly become more and more integral to capitalism. The processes that geographer Jason Moore (2015) associates with the social production of nature, namely those "through which new life activity is continually brought into the orbit of capital and capitalist power," are now ubiquitous. Examples include, but range well beyond, "intellectual patents in genes and life-forms, land-grabs and privatization of parks or beaches" (Kallis, Gómez-Baggethun, and Zografos 2013, p.98); wildlife captured for the commodified pet trade (Collard 2014); and microbes and medicinal plants enlisted for corporate drug development (Hayden 2003).

How can we understand this contemporary ubiquity? Harvey (2003) provides a plausible answer, which rests on the link between two concepts we examined earlier: overaccumulation (Chapter 10) and accumulation by dispossession (this chapter). Harvey believes that capitalism has been plagued by underlying conditions of overaccumulation for several decades. A lack of sufficient demand for the products and services that capitalism has the capacity to produce has become more or less congenital. Creating new sources of demand for the idle capital that accumulates under conditions of insufficient demand is one possible "fix" (e.g., capital switching; see Chapter 10). But, Harvey says, it is not the only one: "it is also possible to accumulate in the face of stagnant effective demand if the costs of inputs … decline significantly" (p.139).

This is where accumulation by dispossession enters the picture. It makes renewed capital accumulation and economic growth possible by providing through dispossession new cheap inputs, including, increasingly, new cheap, privatized, commodified *environmental-resource* inputs – such as cheap or even free genes; cheap or free wildlife; and cheap or free medicinal plants. Smith's contribution is to help us think about this

The environmental "resource"	Contemporary capitalist production of this nature	Capital's fallibilities
Water in the United Kingdom	Privatization and commodification of supply in the 1980s and 1990s	Karen Bakker (2003) describes water as an "uncooperative" commodity: • It is dense and expensive to transport • It is a "flow resource" (not easily bounded, either above or below ground), which makes establishing property rights difficult • It is often required to perform several different and not necessarily compatible functions during its circuit through the hydrological cycle It has therefore resisted being (re)produced as a privatized commodity; post-privatization, difficulties in commercialization have been evident in: • Increased political and public intervention • Stricter environmental and economic re-regulation • Significant, unanticipated corporate restructuring
Alaska pollock in the North Pacific	Privatization of fisheries in the 1990s	Pollock, Becky Mansfield (2004) shows, is also a recalcitrant commodity, which resists being internalized within capitalist "second nature": • It is highly mobile, raising property rights issues similar to those of water • Stocks fluctuate, including seasonally, and are difficult to predict • Stocks are difficult to monitor – fish are a "fugitive resource" These and other issues have made privatization a highly imperfect and conflicting exercise, evident in two key contradictions vis-à-vis "free market" ideals: • The imposition of limits on commercial expansion (e.g., catch allocations) • Strict rules regarding the formation and constitution of industry actors
Arctic seas exposed by ice molt	Exploitation of new hydrocarbon sources and shipping lanes from the 2000s onwards	Leigh Johnson (2010) is concerned less with capital's fallibilities in the past and present and more with its likely future fallibilities. In order for the new commercial opportunities opened up by Arctic ice melt to remain possible and profitable, nature must continue to cooperate (p.844): • "increased iceberg production must be minimal" • "newly ice-free Arctic coastlines must remain stable against increased wave action" • "permafrost must not melt" • "sea level must not rise too dramatically"

Figure 11.1   Fallible capitalism.

accumulation-by-dispossession of nature as a process of production: the production, in Moore's (2015) terms, of "cheap nature." If capital switching (Chapter 10) represents a "spatial fix" to overaccumulation, then the production of cheap nature in myriad new forms represents an "ecological fix."

Is this production process an unproblematic one? No. Coupling Smith with Polanyi, we might say that producing nature has always been a socioeconomically as well as an ecologically risky business for capitalism. And if "traditional" economic uses of "traditional" natures like Prudham's forests and Bridge's copper mines engender socioeconomic difficulties, then it is unsurprising that today's new-fangled capitalist socionatures are generally at least as disobliging. Thus although capitalism may *feel* "unstoppably hegemonic" when it "reaches into cellular metabolism, gene frequencies, and the upper stratosphere in its search for new spaces of accumulation" (Robertson, 2012, p.373), in reality things never proceed entirely smoothly. Capitalism at the frontiers of contemporary profit-oriented nature production is "fallible capitalism" (p.373) maybe in its most vivid forms (Figure 11.1).

11.3.3   Beyond production?

Economic geographers continue to grapple with the thorny question of nature and capitalism, or capitalist natures. Smith's theory of the production of nature remains influential but is treated in the way that we suggested, in Chapter 5, all good theories should be: with respect, but with an invitation to critique, extension, supplementation, even rejection. We will finish this section by considering recent developments.

It is helpful to begin with reference to "conventional" approaches to relations between capitalism and nature of the type we critiqued in the first section. These share three core characteristics. First, they conceptualize matters in terms of "relations": capitalism and nature, economy and environment, *are* separate, albeit related. Second, this separateness is a historical constant: using terms introduced earlier, there is still today a "first nature," even if a history of exploitation by "second nature" means it is less extensive than it once was. And third, the relation between economy and environment, second and first nature, is one-way: capitalism uses, even abuses, nature. It is the (only) active agent.

The reason this is helpful is that we can see where Smith's and the other critical approaches we have examined have challenged the conventional common sense and where they have not. Smith only unsettled one aspect of this common sense, the second. Specifically, he said there is now no separate "first nature"; it is produced within "second nature." But there *was* a first nature when capitalism was still immature. And while capitalism today produces nature, nature does not and never has produced capitalism in turn. Smith's nature, as geographer Bruce Braun (2008, p.668) says, is "a passive or inert realm."

What about Bridge, Prudham, and the geographers whose ideas are presented in Figure 11.1? They have further unsettled the conventional common sense by overturning its third aspect, the idea that while capitalism shapes nature, the obverse does not apply. They all show nature kicking up a fuss, refusing to toe capital's line, making capital conform somewhat to *its* demands. Hence, unlike in Smith, "it is no longer the case in these accounts that economic forces are seen to 'produce' nature in any straightforward way" (2008, p.668).

But as Braun says, this still leaves one conventional shibboleth untouched. In these latter accounts nature only begins to influence capitalism once capitalism internalizes it, producing "first" within "second" nature. The problem is "digestive": nature objects to being swallowed. Nature's agency is purely reactive; nature was – once – "outside" capitalism. But was it? What if Smith's history is wrong, and in fact "the liveliness of non-human nature [was] there, in the economy, from the outset" (2008, p.669)? How might this require us to reconceive nature's agency?

It is these kinds of questions that geographers like Moore (2015) have recently been posing, pushing beyond "production"-based theories. Moore, for example, answers Braun's historical question in the affirmative. Consequently, he writes not about capitalism producing nature but about nature performing, and having always performed, work for capitalism. The energy of the biosphere is turned *into* capital. How such ideas can be mobilized in empirical research remains to be seen. At the very least, however, by placing nature at the very heart of the economy both theoretically and historically, they reaffirm economic geography as not only a powerful but an indispensable frame for apprehending the world.

11.4   The Value of Nature

Whether we think in terms of capitalism using or producing nature or of nature doing work for capitalism, we have established that nature is heavily involved in contemporary economic processes. Still, there is an extremely important set of environment-focused economic processes that we have not touched on. These are processes designed not to exploit environmental resources but explicitly to protect or conserve them. Such processes have become an increasingly notable feature of capitalist economies in recent decades. With nature located at their nub, they are an economic geographic concern no less than mining or forestry or wildlife trade. They are our focus in this final section of the chapter, which has three aims.

The first aim is simply to introduce this "protection economy," for want of a better label. By describing one of its principal forms, we give an example of how it functions. The second aim is to identify the intellectual logic that buttresses this economy. And the third is to offer critical reflections.

11.4.1   The protection economy

If a society believes that "nature" needs protecting or conserving, what mechanisms are available to effect such protection? The most obvious, perhaps, is simple prohibition: you must not trespass on this wetland; you shall not burn that coal; we must prevent poachers from killing elephants. Another alternative is to impose limits: you can only hunt this number of wolves per season; you can only emit that much carbon dioxide. This type of mechanism, generally referred to as "environmental regulation," is usually operated by the state or by regulatory agencies to which the necessary powers are delegated.

There is a very different approach, though, that we call the economy of environmental protection or conservation. It reduces or obviates altogether the need for "intrusive" regulatory oversight. In essence it comprises a series of *markets*. Markets are sites of exchange, "places" – they need not have physical locations we can visit and touch, and indeed these days seldom do – where commodities can be bought and sold. To understand the protection economy we need to come to grips with the commodities that circulate in it and the market processes via which they are traded. Let's take an example and then generalize from there.

Our example concerns the paradigmatic institution of the economy of environmental protection, the carbon market. It is somewhat misleading to refer to "the" carbon market because numerous such markets exist around the world, at national, sub-national (e.g., in California), and supra-national (e.g., European Union) scales. But in every case the exchangeable commodity and basic market principles are broadly consistent. The commodity bought and sold on the carbon market is a carbon credit, a permit to emit a given quantity of carbon. This commodity can be created in two different ways. The state creates credits and allocates them to carbon emitters. Additional credits ("offsets") can be formed by physically taking carbon out of the atmosphere, for example through sequestration projects. Two main processes then happen on the market. Carbon emitters trade the credits among themselves according to their relative appetites to emit. And creators of offset credits sell these credits, for example to emitters wanting more of them.

Generalizing from the carbon case, how can we grasp the essence of such markets? The locus of commodification, or the "thing" being commodified, is not nature but what Smith (2007, p.20) refers to as "allowable natural destruction." In traditional markets for physical environmental resources, such as iron ore, or timber, or oil, a price attaches to nature. It costs to buy nature; one gains from selling it. In markets such as carbon markets, a price attaches to the allowable destruction of nature. It costs to buy such allowance (the right to emit carbon); one gains from selling it. Perverse as it may sound, this is the reality. Arguably, then, we should have labeled this economic sector "the destruction economy"; we chose "protection" because protection *is* the objective.

Even if the reality sounds perverse, it is readily explicable. In capitalism, goods typically need to have two related attributes to command a price. They have to be in demand, and they have to be scarce. The right to engage in certain activities of environmental destruction, such as the burning of fossil fuels, has been in demand ever since that right became beneficial to the exercise of a whole suite of profitable economic activities. But until governments introduced mechanisms such as those used for rationing carbon emissions allowances, this right had never been scarce. It now is.

11.4.2 Neoliberalism and nature

The degree to which economic developments are driven by economic ideas has long been a contentious subject in social science. At one extreme are "idealists," those who believe that changes in prevailing ideas about the economy are the primary motor behind changes in how the economy is organized. At the other extreme are "materialists." They maintain that the direction of causality is the opposite. The economic world changes as a result of non-ideational influences. It is ideas that then (sometimes) follow suit.

What has this to do with the economy of environmental protection, or of, after Smith, "allowable natural destruction"? This economy provides unusually powerful substantiation of idealist claims. Economic ideas have been fundamentally important to the development of markets such as carbon markets, not only validating their existence but informing their operation. This does not mean that ideas alone explain the growth of the market-based approach to environmental protection. They do not. As the late political scientist Judith Layzer (2012) showed in her analysis of the ascendance of market approaches in the United States, for example, a mix of ideas and institutions, principles and power, helped propel the market paradigm. But ideas were and are significant. The environmental protection economy is one that cannot be properly understood *except* with reference to the ideas that cultivated and constitute it. This is one of two reasons why we examine these ideas now.

The other reason is that, on the face of things, the existence of markets designed to conserve nature represents a remarkable paradox that demands unraveling. After all, one need only wind the clock back to the 1960s, 1970s, or even the early 1980s to retrieve an ideational milieu in which the mobilization of capitalist markets to *protect* the environment would have been essentially unthinkable. Capitalism and markets and the growth they fostered were the *problem*. Not only "radicals" thought that way. Seminal environmentalist texts such as Rachel Carson's *Silent Spring* (1962) and the Club of Rome-commissioned *The Limits to Growth* (1972) were read widely and approvingly across the political spectrum. If you had told someone that in due course markets would be seen as the answer to conservation rather than the precise opposite, they would likely have furrowed their brow. This paradox requires explaining. How did markets come to be seen as part of the solution?

There happens to be a very short and actually largely accurate answer to this question: neoliberalism (see Chapter 1). Although there are many different definitions of neoliberalism in circulation, most emphasize the growing use of markets as organizational and allocative mechanisms in all manner of spheres since the early 1980s. One particular "primer" on neoliberalism by the economic historian Philip Mirowski (2009) provides an especially useful diagnosis. For neoliberals, Mirowski says, markets provide the solution not only to problems caused by, for instance, incompetent states, but also to those *seemingly caused by markets in the first place* (pp.439–440). This act of apparent intellectual contortion rests on the thesis that where markets fail to deliver optimal solutions it is because they have been arranged or operated inadequately. In other words, the fault lies not with markets but with their human creators and managers (just as with the "halfhearted" adopters of globalization we encountered in Chapter 8). Or in terms of "marketization," the problem is deemed not "too much" marketization but "not enough." Rather than abandoning markets, we must simply do them *better*.

This, we suggest, is exactly how we should understand the turn to markets in the sphere of environmental governance. It is a singularly neoliberal transformation. If market-based capitalism has historically been ecologically destructive, then that is because we have not gotten markets right, not because markets themselves are problematic

As a now vast literature on neoliberalism shows, neoliberal ideas take different forms and develop in different ways in different contexts. So, while the short answer to our question – How did this happen? – is indeed "neoliberalism," it is important to put flesh on that explanatory skeleton.

11.4.3   Valuing nature

While the rise of neoliberal thinking about environmental governance began in the 1980s, it had important intellectual antecedents. Arguably the most important of these was the ecologist Garrett Hardin's (1968) notion of "the tragedy of the commons." Hardin said that environmental "tragedy" was rooted in communal ownership of and access to natural resources. As the economic geographer James McCarthy (2012, p.619) glosses: "uncontrolled access to a resource would lead inevitably to the ruin of both the resource and its users, as each individual user followed the inescapable economic logic of receiving all of the gain from each additional unit they took from the resource while bearing a fraction of any associated costs, and assuming that any unit they left behind, another would take."

What was the answer to the "tragedy" of communal access to resources such as fisheries, beaches, forests, and so forth? Hardin proposed two different alternatives: privatizing nature or subjecting it to state control. Writing as he was on the cusp of the neoliberal revolution, it was perhaps fated that the former would become the default ownership option for mainstream environmental conservation programs, notwithstanding isolated instances of states opting for the latter (e.g., Sayre 2002). Where the markets in environmental protection that emerged from the late 1980s depended on privatization as well as commodification of nature, at any rate, Hardin supplied the rationale for the first of these two elements.

The rationale for commodification in aid of protection, meanwhile, came from the new field of environmental economics. This field and its core ideas shaped the new market-based approach to environmental protection from the outset. In the United Kingdom, for example, the first book in the formative Blueprint for a Green Economy series was originally prepared as a report for the Department of the Environment. Its lead author, David Pearce – the book (Pearce, Markandya, and Barbier 1989), a touchstone of environmental economics, became known simply as The Pearce Report – was for several years one of the government's chief environmental advisors. The explanation that environmental economics gave for capitalism's legacy of environmental destruction was quintessentially neoliberal: the fault was not with markets but with their "imperfect" existing forms. Already by the early 1990s, the field's central thesis was set in stone: "the problem, however framed, reduces to a matter of making markets work better. Societies overexploit natural resources because markets for environmental services are imperfect" (Howarth and Norgaard 1992, p.473).

The exact nature of this existing imperfection was critically important, as was the notion of "environmental services." Often also referred to as "ecosystem services," the latter refer to the various benefits that society realizes from the existence of different "natural" ecosystems. Clean air, in other words, is deemed a "service" that nature provides to society, as is a tolerable level of atmospheric carbon dioxide concentration. Environmental economists argued that capitalism had failed historically to protect the environment because markets had not recognized these services and their value; this was their (markets') signal imperfection. "The water quality provided by a forest, the habitat provided by a wetland, the climate mitigation provided by a grassland sequestering carbon" – such things, notes Robertson (2012, p.376), were clearly valuable, yet markets had comprehensively neglected them.

They had also therefore neglected the damages – so-called negative externalities – inflicted when businesses operated in such a way as to impair the environment's "service provision" – for example, by emitting pollutants that increased atmospheric CO^2 levels or that despoiled the water quality provided by the forest and the habitat provided by the wetland. The history of environmental destruction under capitalism was thus reconceived as a history of market failure: the reason we have not better protected nature is that under capitalism we have neither been rewarded for doing so nor penalized for not doing so. By incorporating environmental services into an economic calculus, businesses would be incentivized *not* to pollute because polluting would now cost.

Still, the key question was *why* markets had not factored in environmental services. The environmental economists' answer was that these services' monetary value – the universal currency of capitalist markets – was unclear. Indeed, it was not clear whether they even *had* a monetary value. The central project of environmental economics, therefore, was to be twofold. First, to make the intellectual case for expressing nature's worth *in* monetary value terms. And second, to develop methods for imputing monetary values *to* the various "services" provided by nature. Only if these twin tasks were achieved could markets be mobilized in the service of effecting environmental protection. "If we knew the value of environmental services," Howarth and Norgaard (1992, p.473) wrote of this logic, "we could determine how to allocate their use efficiently." Conservation would follow axiomatically from recognizing, calculating, and trading nature's previously hidden, but now legible, economic worth.

11.4.4   False premises and promises

But the recent historical record suggests that conservation has not in fact followed from creating markets in environmental services. There is no evidence that in those parts of the world where approaches to environmental preservation have moved toward market-based mechanisms there has been a net reduction in environmental degradation. Rather there is considerable evidence to suggest that degradation has continued apace, and sometimes worsened. This has led critics to conclude that the vaunted ecological benefits of marketization of environmental management were in large measure "false promises," as the subtitle of an influential collection on the subject puts it (Heynen et al. 2007). Furthermore, the negative consequences of the turn to markets have, as we will see, been social as well as ecological.

How then can we explain this apparent failure? Four main sets of explanations have been put forward:

i.   Hardin's critique of communal forms of natural resource ownership and access, which market-based regimes have in many instances replaced, is itself based on false premises. Communal systems are not doomed to fail socially and ecologically. Hardin's argument was based on inappropriate assumptions, including that of individual utility-maximizing behavior (on which see Chapter 10) (Harvey 1996, p.154). Thus, in the real world, the experiences of societies using communal resource pools simply do not mirror Hardin's dystopia. People do not necessarily act in self-interested ways; they can create effective rules of access that are neither centralized nor marketized. While the work of the late Nobel prizewinner Elinor Ostrom (e.g., 1990) contains the best-known critique of the "tragedy" thesis, hers was far from a lone voice in the wind. Commons-based approaches to environmental management are not irredeemably defective, and thus need not necessarily be replaced by what may turn out to be inferior approaches.

ii. Whatever the merits of market mechanisms as instruments of environmental policy, they are rarely if ever sufficient in and of themselves. Prudham (2004) provides a case study that thoroughly repudiates the neoliberal conceit that market approaches are adequate substitutes for state-led environmental regulation. In 2000, several people in Ontario, Canada, died from poisoning caused by contamination of drinking water. Prudham shows that neoliberal reforms of provincial environmental governance, leading to failures of oversight and accountability, were deeply implicated in this tragedy.

iii. Money is not a "neutral" instrument of valuation and market coordination. As we argued in Chapter 9, it is a form of social power and as such it has, in Harvey's (1995, p.155) words, "a certain asymmetry to it – those who have it can use it to force those who do not to do their bidding." This clearly has implications for market-based environmental policy: "Excessive environmental degradation and costs, for example, can be visited upon the least powerful individuals or even nation states" (p.155). The sociologist Michael Goldman (2005) provides a compelling analysis of such visitation in his study of World Bank-led "green neoliberalism" in the Global South. This neoliberal project, which Goldman treats as a form of imperialism, trades explicitly on the argument that sustainable development requires "proper" valuation of the environment in place of existing "distorted" valuations and thus poor utilization of natural resources. In Laos, for example, "what counts as biodiversity" is now defined "by actors other than the people who live there," and the results are stark:

> When human populations are scientifically isolated from their environments and categorized as slash-and-burn cultivators, poachers, illegal loggers, and failed rice farmers, and when new rules and regulations prohibit hunting, fishing, semi-nomadism, swidden cultivation, and forest use in large swathes of inhabited forest, these changes affect not only epistemic politics but also ontological and material realities. The new authoritative logic of eco-zone management that is carving up the Mekong region is designed to ensure that there will be ample high-value hardwood supplies for export, depopulated watersheds for hydroelectric dams, and biodiversity preservation for pharmaceutical firms and eco-tourists. (Goldman 2005, pp.177–178)

iv.   Putting a price on nature's "services" is incredibly messy and arbitrary and therefore markets in these services will always generate arbitrary and unforeseeable outcomes. Whereas natural ecosystems are complex, diffuse, and holistic, environmental economics must contrive clean and discrete monetary values – the value of *this* particular wetland or *that* particular species – to enable market trade. Sometimes the very definition of the resource to be protected is itself a subject of scientific dispute and the commodity can only be "stabilized" in readiness for trade by actively silencing this underlying disagreement (Robertson 2006, pp.368–369). Estimating a price – even more so the right price – is always fraught with practical difficulties. The work of environmental monitoring, measurement, and certification required to translate physical nature into monetary value is performed not by infallible robots but by all-too-human individuals. Prices are ultimately subjective, creative, ad hoc, contingent, uncertain, unstable and, therefore, nonreplicable. Capitalist markets demand information about the natural world that "scientists *cannot provide* in an uncontroversial way" (p.382).

Even if the neoliberal holy grail of making the currently flawed markets in environmental protection work better than they presently do *were* attainable (it is always held out as a tantalizing possibility), two arguably larger and deeper concerns continue to haunt this approach to environmental policy.

The first is that by its very nature, the discipline that provides the logic and method for this approach – mainstream economics – is ill-equipped to deal with the most important environmental issues. The ecological economists Richard Howarth and Richard Norgaard argue that these issues demand social and political commitments that market valuations can never guarantee: "incorporating environmental values per se in decision-making will not bring about sustainability unless each generation is committed to transferring to the next sufficient natural resources and capital assets to make development sustainable" (Howarth and Norgaard 1992, p.473). It is extremely hard for economics to escape core assumptions like product substitutability and the reversibility of choices. This matters. A "wrong" valuation cannot be changed once a species is extinct; future generations simply have to live with it. And just as "there are species or resources without which ecosystems cannot be sustained," so there are ones "for which there are no adequate substitutes or equivalents" (Kallis, Gómez-Baggethun, and Zografos 2013, p.98).

Second, and relatedly, there are abiding philosophical and ethical concerns that should not simply be waved away. Philosophers such as John O'Neill (2001) and Mark Sagoff (2004) have led critiques of market approaches. They argue that environmental economics unilaterally applies the individualist, instrumental, and utilitarian calculus of money, benefits, consumer demand, and so forth to a set of issues literally unthinkable in isolation from moral and aesthetic values and collective, rights-based principles. "It is hard in the light of these problems not to conclude," writes Harvey (1995, p.155), "that there is something about money valuations that makes them inherently *anti-ecological*, confining the field of thinking and of action to instrumental environmental management."

11.5 Conclusion

In this chapter we have demonstrated that the "economy" and "nature" are, and always have been, deeply entwined with one another. Indeed, so intimate is their relation that many scholars have argued against any type of conceptualization that arbitrarily draws boundaries between them and keeps them apart. To understand today's economy, at any rate, it is essential to understand how nature variously nourishes and shapes it. Similarly, to understand today's environment, one cannot ignore the centrality of capitalism in producing and reproducing it.

All of this argues for the central relevance and value of economic geography. And this relevance and value will only heighten in the years ahead. Consider climate change. Here is the dialectic of economy and nature writ large. To understand climate change it is plainly essential to engage with the history of capitalism and its dependence on fossil fuels (Malm 2016), while it is inconceivable that we can conceptualize capitalism's present and future except in the context of climate change and the ways it is impacting, and will increasingly impact, global political economy. In fact, if ever there was an unassailable argument for why economic geography is good for you and for your understanding of the world, then climate change – a phenomenon located precisely at the nexus of the economic and the geographic – may well turn out to be it.

References

Bakker, K.J. 2003. *An Uncooperative Commodity: Privatizing Water in England and Wales.* Oxford: Oxford University Press.

Bleischwitz, R. 2001. Rethinking Productivity: Why Has Productivity Focussed on Labour Instead of Natural Resources? *Environmental and Resource Economics* 19: 23–36.

Bonnett, A. 2008. *What Is Geography?* London: Sage.

Braun, B. 2008. Environmental Issues: Inventive Life. *Progress in Human Geography* 32: 667–679.

Bridge, G. 2000. The Social Regulation of Resource Access and Environmental Impact: Production, Nature and Contradiction in the US Copper Industry. *Geoforum* 31: 237–256.

Carson, R. 1962. *Silent Spring.* Boston: Houghton Mifflin.

Collard, R.C. 2014. Putting Animals Back Together, Taking Commodities Apart. *Annals of the Association of American Geographers* 104: 151–165.

Goldman, M. 2005. *Imperial Nature: The World Bank and Struggles for Social Justice in the Age of Globalization.* New Haven, CT: Yale University Press.

Hardin, Garrett. 1968. The Tragedy of the Commons. *Science* 162: 1243–1248.

Harvey, D. 1996. *Justice, Nature and the Geography of Difference.* Oxford: Blackwell.

Harvey, D. 2003. *The New Imperialism.* Oxford: Oxford University Press.

Hayden, C. 2003. *When Nature Goes Public: The Making and Unmaking of Bioprospecting in Mexico.* Princeton, NJ: Princeton University Press.

Heynen, N., McCarthy, J., Prudham, S., and Robbins, P., eds. 2007. *Neoliberal Environments: False Promises and Unnatural Consequences.* London: Routledge.

Howarth, R.B., and Norgaard, R.B. 1992. Environmental Valuation under Sustainable Development. *The American Economic Review* 82: 473–477.

Hsing, Y.T. 2006. Global Capital and Local Land in China's Urban Real Estate Development. In F. Wu (ed.), *Globalization and the Chinese City.* London: Routledge, pp.167–189.

Johnson, L. 2010. The Fearful Symmetry of Arctic Climate Change: Accumulation by Degradation. *Environment and Planning D* 28: 828–847.

Kallis, G., Gómez-Baggethun, E., and Zografos, C. 2013. To Value or Not to Value? That Is Not the Question. *Ecological Economics* 94: 97–105.

Klinenberg, E. 2002. *Heat Wave: A Social Autopsy of Disaster in Chicago.* Chicago: University of Chicago Press.

Kloppenburg, J.R. 1988. *First the Seed: The Political Economy of Plant Biotechnology, 1492–2000.* Cambridge: Cambridge University Press.

Layzer, J.A. 2012. *Open for Business: Conservatives' Opposition to Environmental Regulation.* Cambridge, MA: MIT Press.

Malm, A. 2016. *Fossil Capital: The Rise of Steam Power and the Roots of Global Warming.* London: Verso.

Mansfield, B. 2004. Rules of Privatization: Contradictions in Neoliberal Regulation of North Pacific Fisheries. *Annals of the Association of American Geographers* 94: 565–584.

Marx, K. 1976. *Capital, Volume 1.* Harmondsworth: Pelican.

McCarthy, J. 2012. Political Ecology/Economy. In T.J. Barnes, J. Peck, and E. Sheppard (eds), *The Wiley-Blackwell Companion to Economic Geography.* Oxford: Wiley-Blackwell, pp.612–625.

Meadows, D.H., Meadows, D.L., Randers, J., and Behrens, W.W. 1972. *The Limits to Growth.* Washington, DC: Potomac Associates.

Merchant, C. 1980. *The Death of Nature: Women, Ecology, and the Scientific Revolution.* New York: Harper and Row.

Mirowski, P. 2009. Postface: Defining Neoliberalism. In P. Mirowski and D. Plehwe (eds), *The Road from Mont Pèlerin: The Making of the Neoliberal Thought Collective.* Cambridge, MA: Harvard University Press, pp.417–456.

Moore, J.W. 2015. *Capitalism in the Web of Life: Ecology and the Accumulation of Capital.* London: Verso.

O'Neill, J. 2001. Markets and the Environment: The Solution is the Problem. *Economic and Political Weekly* 36: 1865–1873.

Ostrom, E. 1990. *Governing the Commons: The Evolution of Institutions for Collective Action*. Cambridge: Cambridge University Press.

Pearce, D.W., Markandya, A., and Barbier, E. 1989. *Blueprint for a Green Economy,* volume 1. London: Earthscan.

Polanyi, K. 1944. *The Great Transformation: The Political and Economic Origins of Our Time*. Boston, MA: Beacon Press.

Prudham, S. 2004. Poisoning the Well: Neoliberalism and the Contamination of Municipal Water in Walkerton, Ontario. *Geoforum* 35: 343–359.

Prudham, S. 2005. *Knock on Wood: Nature as Commodity in Douglas-Fir Country*. New York: Routledge.

Robertson, M. 2006. The Nature that Capital Can See: Science, State, and Market in the Commodification of Ecosystem Services. *Environment and Planning D* 24: 367–387.

Robertson, M. 2012. Measurement and Alienation: Making a World of Ecosystem Services. *Transactions of the Institute of British Geographers* 37: 386–401.

Robertson, M., and Wainwright, J. 2013. The Value of Nature to the State. *Annals of the Association of American Geographers* 103: 890–905.

Sachs, J.D., and Warner, A.M. 2001. The Curse of Natural Resources. *European Economic Review* 45: 827–838.

Sagoff, M. 2004. *Price, Principle, and the Environment*. Cambridge: Cambridge University Press.

Sayre, N.F. 2002. *Ranching, Endangered Species, and Urbanization in the Southwest: Species of Capital*. Tucson: University of Arizona Press.

Smith, N. 1984. *Uneven Development: Nature, Capital, and the Production of Space*. Oxford: Blackwell.

Smith, N. 2007. Nature as Accumulation Strategy. *Socialist Register* 43: 19–41.

Walker, R. 2001. California's Golden Road to Riches: Natural Resources and Regional Capitalism, 1848–1940. *Annals of the Association of American Geographers* 91: 167–199.

Watts, M. 1983. *Silent Violence: Food, Famine, and Peasantry in Northern Nigeria*. Berkeley: University of California Press.

Watts, M. 2004. Resource Curse? Governmentality, Oil and Power in the Niger Delta, Nigeria. *Geopolitics* 9: 50–80.

Chapter 12

Industrial and Technological Change

12.1   Introduction

Joseph Schumpeter (1883–1950), the Austrian economist, who taught at Harvard, used the phrase "a gale of creative destruction" to describe the dynamic character and driving force of capitalism (Schumpeter 1942, pp.82–83). For Schumpeter capitalism proceeded in fits and starts, propelled by technological change, new innovations. For him, this outcome was built into the very logic of a market economy. Once you made profit the be all and end all, entrepreneurs would ceaselessly search for "the next big idea," a technological innovation that when implemented would transform the world, and potentially make them enormously rich. Most of the time entrepreneurs would fail. But very, very occasionally, they hit the jackpot: like Thomas Edison (with the lightbulb among many other inventions); Henry Ford (with the Model T); Walt Disney (with full length color animated movies); Jack Kilby and Robert Noyce (the integrated circuit); and more recently, Steve Jobs (Apple), Bill Gates (Microsoft), Mark Zuckerberg (Facebook), Larry Page (Google), and Jeff Bezos (Amazon).

Once the next big idea is put into practice it can transform the existing economic landscape, both literally and figuratively. Jeff Bezos's idea potentially does away with shops, Larry Page's with encyclopedias, telephone directories, atlases, and much, much more, Mark Zuckerberg's with mail and telephone calls. The old is destroyed by the cyclonic forces of the creatively new. Hence, Schumpeter's oxymoron, "creative destruction." For the new to arise, the old must be simultaneously demolished. Marx and Engels (1969 [1848], p.16) used the phrase, "all that is solid melts into air." Nothing lasts forever. Sooner rather than later technological change occurs, annihilating the past, bringing forth a brand-new shiny future.

Economic Geography: A Critical Introduction, First Edition. Trevor J. Barnes and Brett Christophers.
© 2018 John Wiley & Sons Ltd. Published 2018 by John Wiley & Sons Ltd.

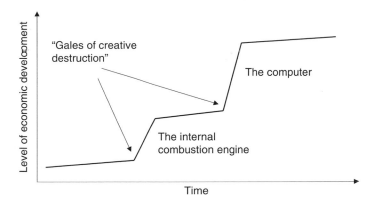

Figure 12.1   Joseph Schumpeter's conception of the "gales of creative destruction."

If Schumpeter's vision were to be put into a diagram it would look something like Figure 12.1. There are periods of stability, the relatively flat lines, punctuated by intense tumult and change, represented by the sharp kinks upward in the diagram. In Figure 12.1 the first kink might be the invention of the combustion engine, the second the invention of the computer. These inflection points are where the gales of creative destruction rage.

Schumpeter thought that behind the technological innovations that create disjunctural economic moments are daring even foolhardy entrepreneurs, single-minded capitalists who undertake extraordinary risks to enrich themselves. If they were not capitalists, they would have a hard time fitting into normal society. But in the world of capitalism, said Schumpeter, their extreme risk-taking is celebrated. They are treated as superstars, as famous as anyone on Hollywood's A list, advisors of presidents and prime ministers. Schumpeter thought those entrepreneurs would remain the driving force of the economy provided the market remained free and there was the prospect of limitless financial rewards. However, once the state becomes involved, Schumpeter believed, especially if it imposes some form of socialism that curbs the market or limits rewards, the bloom is off the rose. The entrepreneurial impulse is blunted, innovation sputters, and capitalism ends not with a bang, as Marx predicted, but with a whimper. Ironically, as we will see, since World War II the state has been intimately involved in the innovation process, but rather than sputtering, the innovation economy has moved at warp speed.

Consequently, this chapter is less interested in Schumpeter's predictions than in his analysis of the role of innovations within the economy. The chapter takes up his depiction of capitalism as a system incessantly interrupted by bursts of intense novel technological change. Think for example of the invention of the steam engine (developed during the eighteenth century) that in many ways was a catalyst for the transformative changes of the industrial revolution that occurred in the United Kingdom (the world's first industrial nation) during the same period. As an innovation, the steam engine epitomized creative destruction. An older rural agricultural feudalism was dismantled as a new industrial urban capitalism was erected. Or, think again of the role of the internal combustion engine that emerged a hundred years or so later that also reorganized the world. A novel landscape came into being of highways, suburbs, strip malls, and international airports, as well as giant factories such as Henry Ford's Rouge Complex in Dearborn, Michigan, set up to produce the internal combustion engine in the commodified form of the car (Chapter 1). Out went the horse and cart, in came the Model T. Or, yet again, think of what became the central innovation of the second half of the twentieth century, and now the new millennium, the computer. It is everywhere, suffusing even our most mundane activities. Likely we spend directly or indirectly half our waking lives interacting with this machine, maybe even more. The ultimate crisis will be when the machine stops. It will be not only the end of capitalism as we know it, but the end of life as we know it.

Technological innovation has been crucial to the industrial economy, changing it profoundly. Schumpeter's interest in creative destruction was primarily temporal. In his 1942 book *Capitalism, Socialism and Democracy* he developed the idea of creative destruction to warn against what might happen if future governments meddled in the market. Focusing on the temporal dimensions of creative destruction allowed Schumpeter to justify his pro-market position and politics. But creative destruction can also be read along a different dimension, our concern in this chapter, giving a different modulation – laterally along the plane of geography. Doing so directs us instead to recognize the uneven geographical consequences that market-produced creative destruction generates. We will contend that creative destruction on the ground is often a geography of profound disparity, of uneven development, with rusted, emptied factories in one place, and shiny, new campuses of high-tech production with gyms, play rooms, and massage suites in another (Chapters 8 and 10).

The chapter is divided into two main sections. The first provides a schematic historical review of the last 200 years or so of the main technology-driven, large-scale forms of change within industrial capitalism. We focus especially on the relationship between labor and the technology of the industrial process, critically associated with issues of power, control, and discipline. We begin with early forms of industrial technology using Adam Smith's (1776) late eighteenth-century example of the pin factory, move to nineteenth-century machinofacture, then twentieth-century Fordism, and finally, to twenty-first-century post-Fordism or flexible production. The second section focuses on the leading edge of technological change in the contemporary economy, producing variously the new economy, the creative economy, the innovation economy, or, our preferred term, the hi-tech economy. Economic geographic relationships, we will see, are integral to the process of technological innovation, to coming up with the next big idea. Here there has been a lot of work by economic geographers, both theoretical and empirical, to explain the geographically clustered form that hi-tech production invariably takes. It is a connection that has attracted interest from different levels of government, particularly local ones keen to know how the economic geographic trick of developing hi-tech clusters is done so that it can be repeated in their place. From a critical perspective, though, this has presented problems, stemming from the destructive part of fomenting creative destruction. While capitalism is enormously energetic and vibrant, creating technological winners, it also leaves in its wake, as Walter Benjamin (1940) once put it, "a pile of debris … [which] grows skyward." The hi-tech economy necessarily takes the form of uneven geographical development (cf. Chapter 8). Asheim and Gertler (2006, p.291) write, "innovation activity is not uniformly distributed across the geographical landscape." Much like globalization, it produces spatial inequality, a few prosperous peaks of knowledge-intensive activity connected by vast lowland plains in between of mediocre economic performance, or worse.

12.2   A History of Industrial Change

12.2.1   Industrial beginnings

The term "industry" meaning the manufacturing of goods arose only in the eighteenth century. Before that it meant simply to work hard (Williams 1976, p.165). One of the earliest users of the new meaning was the world's first recognized economist, the Scotsman Adam Smith. In his 1776 *Wealth of Nations*, Smith discussed the conditions "for the maintenance of industry" (Smith 1776, book II, p.iii). Book I, Chapter 1, of Smith's volume opens with a detailed description of early industrial production, the internal operations of a pin factory. There he is mainly concerned with detailing the separate tasks performed by workers within the factory (there are 18), which he then develops into the important principle of the division of labor (discussed below). He is keen to recognize also the part played by the "machinery" of production, that is, technology, tools, and implements (and illustrated in Figure 12.2 from a contemporary engraving of an eighteenth-century French pin factory, and possibly the model for Smith's account).

The "industrial revolution" that Smith witnessed, beginning in the mid-to-late eighteenth century, marked a cliff-edge break from the earlier agricultural and rural economy. From the start, the revolution was crucially bound up with technological change, associated with a long list of inventions, primarily in the form of new machines that transformed how goods were produced, and more fundamentally how the larger society and its economic geography were organized. Some of the more famous of those machines included: Kay's flying shuttle (1733), Hargreaves' spinning Jenny (1770), and Crompton's mule (1779), all of which transformed the British textile industry; Cort's rolling and pudding iron mill (1783–1784), which made available large quantities of high-quality bar-iron used for everything from bridges to industrial machines and even the Eiffel Tower; and Newcomen's (1712) and later Watt's (1783) steam engines that not only transformed transportation (the railway, the steamboat), but liberated manufacturing location – factory steam engines could be built now at almost any geographical site.

Industry was undertaken at the factory, a specialized building in which workers labored in conjunction with those newly invented bulky machines of production (putting paid to the possibility of home-based manufacture – Watt's steam engine could never fit into a worker's cottage!). Industry also meant town and city (Chapter 10). These were sites that already did,

Figure 12.2   Labor and machines in an eighteenth-century pin factory, from the *Encyclopédie* edited by Denis Diderot and Jean le Rond d'Alembert, 1751–1772. https://thecharnelhouse.org/2013/11/08/adam-smith-revolutionary/. Source: http://www.alembert.fr/.

or could, attract the other important factor of industrial production: labor. It was a chicken-and-egg relation: factories located in towns and cities because they were sites of labor, and labor came to towns and cities because there were factories. In England, there was an additional force directing people to towns and cities: the Enclosure Acts. Beginning from the early seventeenth century, Enclosure Acts passed in the British Parliament changed the legal status of formerly common land accessible and usable by everyone, for instance a furrow or two in a tilled field to make some kind of living (see the earlier discussion on accumulation by dispossession, Chapter 11). Once Enclosure Acts were passed, rights to use the land were severely restricted, forcing those living in the country to go to towns and cities to find work in the factory. The consequence was the rapid growth in population within the industrial cities of England. Manchester increased its population tenfold during a 40-year period, from 35,000 (1801) to 353,000 (1841); Birmingham multiplied its population almost eightfold (23,000 in 1801, 183,000 in 1841); and London, already the largest city in the world (1,096,784 in 1801), more than doubled its population (2,207,653 in 1841). New World cities also burgeoned, albeit later, in this case from mass international immigration. By the turn of the twentieth century, urban immigrants formed the core of America's industrial working class employed in factories in such cities as New York (a population of over 3 m by 1900), Chicago (1.7 m in 1900), and Philadelphia (1.3 m in 1900).

12.2.2   Machinofacture

The work carried out in increasingly teeming industrial English and American cities was highly regulated. Each worker had their specific task. This was Adam Smith's division of labor. As Smith argued, by each worker specializing in a single task, productivity massively increased. In the pin factory, Smith estimated that on average each worker made 4,800 pins a day. Without the division of labor, though, with one person undertaking all 18 separate tasks themselves, Smith reckoned the worker would produce "perhaps not one pin in a day" (Smith 1776, p.9).

In turn, the division of labor was fundamentally connected to the use of machines, the technology of production. Marx called the early form of industrial production "machinofacture." In this system, workers tended the machines, feeding inputs, taking outputs from them. That is, workers functioned as "Hands," the common name for industrial workers during the nineteenth century. Memorably satirized in his novel *Hard Times* (1854), set in that ultimate (fictional) gritty industrial city, Coketown, Charles Dickens used the label "Hands" as a shorthand to express the cruelty of industrial capitalism, which reduced the wholeness of a worker to only two bodily appendages. Marx, who lived at the same time in the same city as Dickens – London – never felt the need to resort to fiction. He used contemporary accounts provided by English factory and public health inspectors to describe in his first volume of *Capital* (2010 [1867], chapter 10) the full horrors of urban working life under a regime of machinofacture: child labor (beginning at six and seven years old); 15-hour work days (6 a.m.–9 p.m.); horrendous worker accidents (death and maiming – the youngest children, called "scavengers," were employed to crawl under machines as they operated to clean, to clear jams, to remove fallen inputs, and to repair broken machinery parts); industrial disease ("pneumonia, phthisis, bronchitis, and asthma"; chapter 10, section 3); and appalling pay (a seven-year-old was paid "three shillings and sixpence" – UK 17.5 new pence or roughly under current exchange rates a US quarter – for a 90 hour week[1]). The form of labor control was called the "drive system." There was very little differentiation in worker skill – laborers were all grunt workers, employed for their "Hands" and for nothing else. Foremen and forewomen then drove the Hands to their limits.

12.2.3   Fordism and mass production

Early twentieth-century Fordism and the regime of mass production that followed the drive system was an extension of machinofacture (see also Chapter 1). The oppression of workers was not so ruthlessly brutal, but the technology of production enforced a level of control and discipline at least equal to that of machinofacture, if not greater. Fordist machines were also even larger than those under machinofacture (part of a broader historical process under capitalism called "capital intensification," that is, the increased investment in fixed capital, machines, and buildings compared to labor). Moreover, the machines were also physically connected by a conveyor belt, a moving assembly line. Work was continuously in process, moving from one machine to another within an integrated production system. Raw materials entered at one end and a new Model T popped out at the other 28 hours later. In between was the assembly line, linking in the right order the various necessary component processes of manufacture, many of them under the same roof (called "vertical integration," meaning that all parts of the production process were undertaken at the same site).

In this system, workers were no longer simply tenders of the machine but became part of the machinery itself. Workers' swirling bodies moved in synch with the rhythms of the technology of production. That's seen in Figure 12.3, "Detroit: Man and Machine," a mural to be found in the Detroit Institute of Art. It is of Ford's Rouge plant (Chapter 1). It was painted by the Mexican muralist Diego Rivera between 1932 and 1933. Rivera was a communist, but he was entranced by the power and might of industrial production found in factories such as the Rouge. He thought both human and natural worlds were propelled by common waves of energy. That's what is depicted in Figure 12.3. Humans and machines are similarly animated by the same irrepressible surging life force, and are moved to ever-greater heights of productivity.

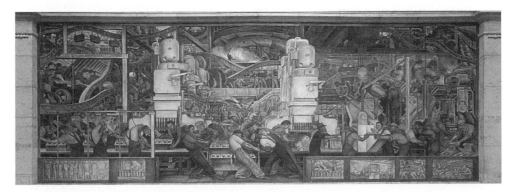

Figure 12.3   Detroit industry mural: Man and Machine by Diego Rivera, North Wall, Detroit Institute of Art.

Box 12.1   Taylorism

Frederick W. Taylor.

Trained as a mechanical engineer, Frederick W. Taylor (1856–1915) recognized early on that workers were not laboring as much as they potentially could. He called this down time "soldiering" (now we would say "slacking off" or "skiving"). This wasted time arose because workers either were inefficient, or they deliberately pursued strategies of not working (e.g., idling between jobs, or taking long breaks going to the toilet, or pretending to work when they were not). Taylor's purpose was to design a set of general workplace strategies that would exact from workers the greatest amount of work in the least amount of work time, that is, to minimize soldiering. To formulate those strategies, Taylor carried out experiments on workers, varying their work conditions and pay, timing and recording their performances (which became known as "time and motion studies"). From those results he developed three principles of "scientific management" that could be applied to all laborers, maximizing the time they worked. The first was disassociating the labor process from the skills of the laborer. This was later called "deskilling." Taylor contended that to maximize productivity workers should be employed to carry out only a minimal set of repetitive simple manual tasks that required no skill. You employed workers only for their brawn, for their ability to perform single physical functions over and over again. If you employed them for their skill, you could lose control over the work process. That was why deskilling was a necessary Taylorist strategy. The second was separating conception from execution. Workers were paid not to think (conception) but only to do (execution). Taylor found from his experiments that thinking took time, producing inefficiencies, and requiring elimination. Workers were most efficient when they

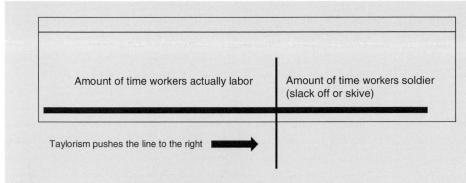

Figure 12.4 The purpose of the scientific management of labor (Taylorism).

became part of the machinery, undertaking without thought rote mechanical acts. The third was to use "task cards" to specify exactly what was to be done, how it was be done, when it was to be done, and how long it should take. Applying all three strategies, thought Taylor, would minimize "soldiering." From Figure 12.4, applying all three principles shifts to the right the line between the amount of time workers actually work within a day, and the amount of time spent soldiering, thereby increasing their output and raising profitability for the employer.

Those heights were Himalayan. Using the moving assembly line, the Rouge factory record for assembling the chassis, that is, the guts of the car including the wheels, frame, and transmission, was 93 minutes (Schoenberger 2015, p.131). More than anything, Ford's production technology was about delivering speed. Ford memorably said, "any customer can have a car painted any color that he [sic] wants so long as it is black." He said that not because he was a killjoy, or was morbid, or was color blind. Black paint dried more quickly than any other color.

Fordist production technology not only delivered speed but also volume. By 1921–1922, Ford produced one and a quarter million Model T vehicles, representing 60% of all cars sold in the United States that year (Schoenberger 2015, p.133). Partly this was because Ford's cars were so cheap: $355–440 per vehicle (still cheap even in today's prices, $4,365–5,365). There was also another chicken-and-egg relation. Ford produced so many cars because there was a great demand for the cheap vehicles he manufactured, but he could produce cheap vehicles because there was such a great demand for them. The secret was increasing internal economies of scale, that is, cost efficiencies stemming from increasing production volumes (cf. Chapter 10). It is internal economies of scale that create the so-called virtuous cycle in which the more that is produced, the lower is its cost; the lower is its cost, the higher is its demand; the higher is demand, the more that is produced, which then starts the cycle all over again. The sources of increasing internal economies of scale from mass production are various, but the most important are: securing cheaper input prices because of bulk purchase; mechanizing parts of the production process that would not be possible at lower levels of output; and, going back to Adam Smith book 1, chapter 1, maximizing the extent of the division of labor, making it as fine-grained as it can possibly be.

The division of labor goes to a final point about the nature of industrial technology under Fordism, and its use and control of workers. Fordist labor became part of the machinery of production. Laborers did one task, and one task only, before work moved on to the next person down the line. Ford, for example, divided the assembly of the flywheel magneto, the device that generated electric current to a car's spark plugs, into 29 separate tasks. Each task was undertaken by a different worker. It was as granular division of labor as was possible, yielding great economies. While it took one person working alone 20 minutes to assemble one flywheel magneto, during the same time 29 people could collectively assemble 126 of them. With division of labor, per worker productivity improved by 400%. In the process, however, workers became robots, every facet of their labor scientifically managed: where they stood, how they stood, how they moved, when work arrived, when work left.

The scientific management of workers was called Taylorism, after the late nineteenth-early twentieth-century time and motion expert, Frederick W. Taylor (Box 12.1: Taylorism). Taylor was concerned with designing workplace strategies that would squeeze from workers the greatest amount of work in the least amount of work time. To formulate those strategies, Taylor carried out experiments on workers, varying their work conditions, pay, bodily comportment, and work layout, recording their performances under different conditions. The principles Taylor derived from those experiments were later taken up and deployed in factories of mass manufacturing such as Ford's Rouge plant. Workers stood in place along the assembly line and robotically carried out the same single minimal task over, and over, and over again. In Charlie Chaplin's spoof of Taylorist work in his film *Modern Times* (1936), Charlie got to pull a gigantic lever that finally brought the conveyor belt to a halt.[2] The machine stopped. But that was a Hollywood ending. In reality the machine stopped only when Fordism began to stop, that is, when its factories started to close at an alarming rate from the 1970s when a new technology of production replaced it.

Just as the geographical was key to the reproduction of machinofacture – urban locations guaranteed necessary plentiful labor (Chapter 10) – so it was also key to Fordism, and at all scales. It began with the self-consciously Taylorist organized micro-spaces of the factory floor permitting assembly-line connections; it broadened to the metropolitan region, which acted as a magnet for labor, capital, and resources; it related to the broader Fordist industrial region in which manufacturers formed a larger interlinked network, the manufacturing belt in the United States, the industrial Midlands and the North in Great Britain, allowing agglomeration economies of scale; it connected to the United States and the United Kingdom as premiere industrial nations; and finally, it linked to a larger international division of labor in which Global Northern countries like the United States and the United Kingdom specialized in Fordist manufacture, and the Global South in primary resources (Chapter 8).

12.2.4   Flexible production

What came after the collapse of Fordism was given various names. We call it flexible production. It is also known as lean production or post-Fordism, where the word "post" signals both coming after Fordism and a difference compared to Fordism. While Fordism delivered the goods like no other industrial technology before it, it was characterized by a series of unbending rigidities. It could produce only standardized goods, with only standardized machines, and with only standardized work practices. All of that changed after flexible production.

Key to that change were advances in computer microelectronics, originally sparked by World War II. The first electronic programmable computer was designed to facilitate the breaking of Nazi military secret code, which was carried out by a branch of British military intelligence based at Bletchley Park, just outside London. Formative were contributions by the English mathematician Alan Turing (1912–1954). Turing was also important as a theorist. He later founded the field of artificial intelligence that compared machine-based and human reasoning. His test – the Turing test – said that if you could not tell whether it was a human or a machine that answered your questions posed from a computer terminal, the machine's intelligence was the same as a human's. There are still no computers that consistently pass the Turing test. But they are getting close. It is the growing intelligence of computers over the last 50 years that has driven flexible production, including associated labor practices.

In the 1960s, there was another critical technological development that joined with computerization to consolidate further flexible production technology. "Packet switching" enabled digital units of information ("traffic") to be transferred from one computer terminal to any other or even many terminals at the same time. Like the development of the computer, the origins of packet switching were with the military. In the late 1960s, the US Defense Advance Research Project (DARPA) developed the Advanced Research Project Agency Network (ARPANET) to allow communication both among the military's computers and with computers at research institutes and universities with which the military had contracted work. ARPANET was the prototype for the internet. The critical technique of packet switching is almost unchanged 50 years later. The circumstances of its development, along with those of the computer, clearly belie Schumpeter's assertion of the dulling effect of the state on innovation. The state, in the form of military funding, was the very foundation for what became the most important gale of creative destruction of the second half of the twentieth century (Markusen et al. 1991).

The destructive force of flexible production took the form of aiding and abetting the demolition of old-style Fordism (already on the run because of the new international division of labor; Chapters 1 and 8). By scientific management Fordism had extracted from its workers as much work time as was humanly possible (in terms of Figure 12.4, the vertical line was located as far to the right as it could go). There was no more slack to squeeze from within the Fordist structure. The only way to increase productivity was to transform the structure, to go back to the drawing board. That is what flexible production did, not competing with Fordism on its own ground, but cutting the ground from under its feet.

The automobile industry provides a clear example of before and after the arrival of flexible production. An exemplary case was the General Motors (GM) Fordist plant in Fremont, California (Box 12.2: NUMMI). In 1984, the plant was reconfigured, getting the flexible production treatment, becoming a NUMMI[3] jointly operated by GM and Toyota. The old Fordist practices were entirely remade to conform with flexible production, taking four forms.

- The first was flexibility in product. Fordism was brilliant in its ability to produce mass quantities of the same good – *but exactly the same good*. Standardization was the name of the game. It went to the other reason Henry Ford said you could have any color of car

Box 12.2   NUMMI

In 1984, a failed Fordist GM plant in Fremont, California, located in the East Bay of San Francisco, underwent a radical experiment. In a new joint venture between GM and Toyota, the plant, now named NUMMI, got the flexible production treatment. GM had closed the old Fordist factory, laying off its workers two years before. Those workers had a terrible reputation. Even Bruce Lee, the President of Local 1364 of the United Auto Workers (UAW) union that represented those workers, considered them "the worst workforce in the automobile industry in the United States." There was absenteeism (1 in 5 workers each day failed to turn up), pranks (inserting coke bottles so they would rattle in door frames), strikes, labor grievances, even pervasive criminal activity. Nonetheless, two years after the closure, the plant reopened with many of the original workers rehired. In the interim, Toyota combined with GM to become a new joint owner. Toyota did that partly to gain a production foothold in America (it had no prior operations in the United States), and partly to test whether American workers could be managed within its own Toyota, post-Fordist flexible production system (if Toyota could manage Fremont workers they could manage any American workers). And GM was a willing partner because it meant that it could learn Toyota's secrets. GM had been losing market share since 1960 when it controlled more than half of the American automobile market. Could GM build a better car using Toyota's flexible production methods? The Fremont factory was gutted of its old Fordist machines, which were replaced by flexible ones, and its workers retrained. In groups of 30, Fremont laborers were sent on two-week paid working holidays to Toyota City in Japan. They went out as American Taylorist workers, and returned as Japanese flexible workers.

Flexibility was the name of the game, based on Toyota's production technology, also called lean production or the Japanese or Toyota Production System. It was flexible in multiple senses, in contrast to the rigidities of Fordism. It meant a capacity: to change machines and their operational specifications; to use the latest computerized technology, which with a few key strokes could be reprogramed to produce varied products and runs of different lengths; to assign employees to any work station in the plant with the expectation they would carry out the assigned task while also thinking about the work (workers both executed and conceived); and to require that workers attend so-called quality circles to discuss work procedures and ideas for production enhancement.

This goes to another aspect of flexible production, the concern with quality (the Japanese "*kaizen*," that is, the quest for continuous improvement). Under the Fordist system, the only cardinal rule was to keep the assembly line moving. "The line could never stop. Never stop the line …. Someone have a heart attack, kick him out of the way, keep that line running."[5] At the old Fremont plant, stopping the assembly line was the only action for which you were fired. Even the most monstrous looking, mangled cars could continue rolling down the production line. There were "cars with engines put in backwards, cars without steering wheels or brakes. Some were so messed up they wouldn't start, and had to be towed off the line."[6] In contrast, in the new flexible production Fremont plant, it was quality rather than quantity.

The "andon cord" running the entire length of the factory, when pulled twice, stopped the conveyor belt. It was Charlie Chaplin's giant lever. If workers saw a problem, they were expected to pull the cord, to stop production and to fix the problem. No more automobile monsters, only quality, flexibly produced motor cars.

you wanted provided it was black. By repeatedly producing, in this case, the same color vehicle, internal economies of scale were maximized. More generally, Fordist production relied on designs and machines that rarely changed. Machines were "dedicated," capable of doing one task, in one way, with one set of specifications. New computer-based production technology, CAD/CAM (computer-aided design/computer-aided manufacturing), changed all that, facilitating flexibility. With a few well-aimed keystrokes at a computer terminal, designs for goods and the machines that manufactured them were completely reconfigured. The same machine could produce multiple goods, guided by multiple designs, using multiple techniques, set to multiple specifications, and capable of producing multiple quantities. You can now have any color of car you want. You can have any kind of car.

- The second was flexibility of machines. CAD/CAM techniques first introduced in the automobile and aeronautical industries during the 1960s were just the beginning. Increasingly, by the new millennium, robotics, even mobile robots on wheels, began to take over jobs formerly done by humans. These new machines were now very close to passing the Turing test. The limit of this trajectory is called "lights-out manufacturing" in which there is no need for humans at all. Everything is automated, carried out by robots. FANUC, a Japanese lights-out-manufacturing company, even uses robots to produce robots, about 50 of them are created every 24 hours. A FANUC factory runs with no lights, and neither heat nor air-conditioning, and goes for 30 days unsupervised.

- The third was flexibility with respect to suppliers of inputs. Under Fordism you knew what inputs were required for manufacture because it was known how many goods would be made during a given period. Consequently, arm's-length suppliers were contracted by a standing order to bring a set amount of inputs at fixed dates to the plant. It was arm's-length because supplier firms had no special relationship with the anchor Fordist firm. They simply fulfilled the order, which was then often repeated unaltered for years to follow. In contrast, the parallel system of suppliers under flexible production was constructed to allow for rapidly shifting demands, involving numerous subcontracting firms each of which was in a close relationship both to other subcontractors and to the anchor firm. This larger scheme of procuring inputs within flexible production is known as just-in-time, or the *kanban* system. Rather than the anchor plant holding an inventory of inputs, suppliers bring them exactly when they are needed, to the minute, just-in-time. For this system to operate successfully, not only must suppliers and the anchor firm trust one another, but in order to coordinate delivery firms need to be privy to the internal workings of the others' operations (forging relations that are tight-knit rather than arm's-length). There is also a strong geographical imperative: firms must be close to one another to ensure deliveries reach their destinations just-in-time, and this creates economic geographic clustering.

- The final form was flexible labor. To realize flexible production, workers need to carry out more than only one pre-programmed task. They must be allowed to think and not only to do. Flexible production meant the end of Fordist workers, and more generally, the end of Taylorism and hierarchical scientific management. In the brave new world of flexible production, workers had to be "functionally flexible." This meant being able to turn their hand to many tasks. Thinking and being creative were crucial, and not the sharp brake on worker productivity feared by Taylor. Thus, quality circles and work teams became critical elements of the flexible production workplace. The shop floor worker job description now included stopping the line, sitting around in a ring talking, brainstorming, resourcefully innovating, and openly discussing factory problems, coming up with imaginative remedies. For this to happen required a different hierarchy of decision-making, much flatter than under Fordism, where voices of those on the production line were encouraged to speak rather than stay silent.

Work, of course, is still work. Workers under flexible production continue to be managed and disciplined. Their employer persists in extracting as much value as possible from them as it can, although the means are now different. Is it better to be nicely exploited or harshly exploited? The American autoworkers at Fremont (Box 12.2: NUMMI) preferred to be exploited as flexible production employees rather than as Fordist employees.[4] The early twentieth-century American wit Dorothy Parker once said, "I've been rich and unhappy, and poor and unhappy, and I know which I prefer." The Fremont functionally flexible workers preferred to be happy and exploited, rather than unhappy and exploited. The larger problem, though, was that there were not enough happy-and-exploited jobs available. While flexible production partly replaced Fordism in the Global North, the replacement jobs it generated were woefully insufficient to make up for the much greater losses in Fordist manufacturing employment brought about by the wider process of deindustrialization (Chapter 1). The Fremont workers were incredibly fortunate in 1984 to have their jobs saved, to change from being Fordist to post-Fordist workers. Tens of millions of Fordist workers in the Global North were not so lucky. When they left their factory jobs they ended up flipping burgers, or becoming grocery store shelf-fillers, or having to do the Full Monty (Chapter 1). In 2009, though, NUMMI was closed. GM went into bankruptcy protection (part of the fall-out of the 2008 financial crisis), and Toyota sought lower wage costs and state subsidies by moving its operations from California to Mississippi. In 2010, however, Fremont workers had another amazing stroke of luck. Tesla Motors, operated by another one of those "the-next-big-idea" entrepreneurs, Elon Musk, took over the Fremont plant. It saved some jobs, at least for the time being.

Those Fremont workers were fortunate in another sense, living in the Bay Area. As we will discuss in the next section, the Bay Area became *the* place in the world for industrial experiments involving new computer technology and its application. Much of where Fordism declined in the United States, however, was in the manufacturing belt, primarily the Upper Mid-West and the North East, far from San Francisco. Moreover, there was little compensating post-Fordist, flexible production at those sites. The resulting geography in cities like Detroit, Cleveland, and Buffalo was primarily destruction, not creation (since 1950, Detroit has lost over 60% of its population, Cleveland 58%, and Buffalo 56%). Creative destruction ignited by new technologies of production has produced severe economic geographic inequities. It is the inconstant and uneven geography of capitalism (Storper and Walker 1989) (see also Chapter 8).

12.3   The Hi-Tech Economy

At the eye of the current storm that is the gale of creative destruction is hi-tech. Coined first in the late 1950s, but not in general circulation until the late 1960s, hi-tech refers to economic activities driven by cutting-edge technology, whatever that might be. Often that technology has turned on computer electronics. But it need not. It could be a DNA gene-slicing technique, or a new solar power cell, or the fabrication of nanomaterial (a material whose physical structure can be manipulated at the molecular scale).

By its very nature, hi-tech is risky business. There are no guarantees of success. The hi-tech landscape is littered with carcases of firms that fall by the wayside. Those firms think they have the next big idea, but sooner rather than later, it is clear they don't. Even when it looks like a hi-tech firm might succeed, chances are that it will be bought by a larger firm. The consequent rewards can be enormous, though. For example, in 2012, Instagram, an online photo-sharing service, was sold to Facebook for US$1 billion. Or again, in 2014, WhatsApp, a firm specializing in developing software for social media messaging, was sold to Facebook for an astonishing $22 billion. Some even said that firm was sold too early. If they had waited, the owner, Kevin Systrom, could have received even more money. Clearly, the former owners of Instagram and WhatsApp will never have to work again. But lower down the hi-tech feeding chain, the entire careers of some hi-tech entrepreneurs are defined by their continuous starting, and then selling, of hi-tech firms. They do this partly to prove to themselves that the first time was no fluke, partly to make yet more money, and partly to scratch some fundamental creative entrepreneurial itch.

The satisfying of that creative entrepreneurial itch takes on a distinctive economic geographic pattern. The high-tech economy is highly spatially concentrated, clumped in a relatively few sites across the globe. The most famous in Silicon Valley, just outside of San Francisco (Box 12.3: Silicon Valley). But there are others, like Route 128 around Boston, or the Durham-Raleigh-Chapel Hill research triangle in North Carolina, or the Kitchener-Waterloo region of Ontario, or around Cambridge in the United Kingdom, or in Samsung City in Korea, or in Shenzhen in China, or at HITEC City in Hyderabad, India. It is in this sense that, as Asheim and Gertler (2006, p.291) put it, "geography is fundamental, not incidental." The interesting economic geographic theoretical question is what accounts for this spatial bunching of hi-tech?

12.3.1   Theorizing high-tech

We already discussed one form of the concentrated geographical pattern found among high-tech firms. We referred to it in terms of industrial agglomeration or industrial districts (Asheim 2000; see also Chapters 4 and 10). It was recognized first by Alfred Marshall in the early twentieth century, and his explanation was couched in terms of agglomeration economies, the cost advantages brought about by firms locating close to one another and produced by external economies of scale. An external economy is a cost advantage

Box 12.3   Silicon Valley

Coined by a journalist in 1971, Silicon Valley is the name of the region south of San Francisco in Santa Clara County centered on San Jose that is home to the world's largest and richest high-tech companies. *The Economist*, which rarely engages in hyperbole, says "that the 50-mile stretch [that is Silicon Valley] … is the most productive and innovative land mass in the world." In 2015, *The Economist* calculated that there were 99 hi-tech corporations located there that were each worth a billion dollars or more (any start-up company worth more than a billion dollars, like Uber, Airbnb, or Dropbox, is called a "unicorn") (Figure 12.5). The total worth of those 99 companies

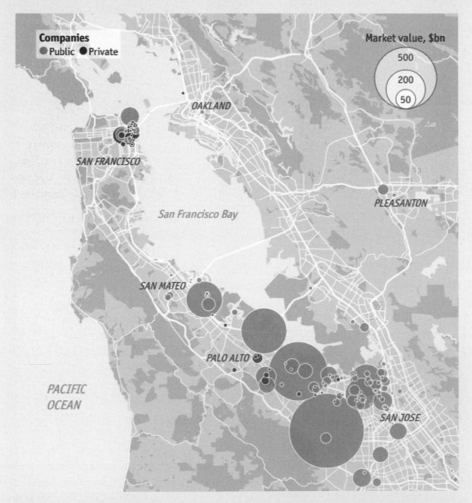

Figure 12.5   Tech-tonic shifts: Valley of the Kings. Tech companies in Silicon Valley worth $1 billion or more. From *The Economist* July 23, 2015. http://www.economist.com/blogs/graphicdetail/2015/07/daily-chart-mapping-fortunes-silicon-valley.

was US$2.8 trillion (the entire US GDP in 2015 was just shy of US$18 trillion). Those companies are all the familiar everyday global hi-tech corporate giants. That explains why *The Economist* also dubs Silicon Valley the "Valley of the Kings" after the one in Ancient Egypt where large numbers of Pharos were laid to rest along with their fabulous wealth. The billionaire and often multibillionaire club in Silicon Valley includes: Apple, Adobe, Airbnb, Cisco, Dropbox, eBay, Electronic Arts, Facebook, Intel, Google, HP, Linked in, Netflix, Oracle, Sun Microsystems, Tesla Motors, Uber, and Yahoo.

The valley part of Silicon Valley refers to Santa Clara Valley where early high-tech firms first located (although increasingly high-tech start-ups are now based in San Francisco). The silicon part is a reference to the silicon chip or integrated circuit that revolutionized computer manufacture (and much else in electronic equipment production) by increasing processing speeds and reducing size, and first systematically developed and commercialized by Silicon Valley firms.

Key to the development of Silicon Valley was Stanford University, located in nearby Palo Alto, and its electrical engineering faculty (Saxenian 1996). Frederick Terman, a faculty member, later Dean, encouraged electrical engineering students to remain in the region. That is what William Hewlett and David Packard did in 1938 after inventing the audio oscillator in a Palo Alto garage. In 1951, Hewlett-Packard Corp., as it had become, along with further electrical engineering companies owned by other former students, such as the Varian brothers (Varian Associates), moved into Stanford Industrial Park. That cluster was further animated from 1956 by William Shockley's experiments in nearby Mountain View. They significantly improved silicon-based transistors (semiconductors), an essential component of computers, leading in 1959 to the patenting of the first integrated circuit. Nine years later, Intel, founded by two students who had worked with Shockley, began commercial production of the silicon chip.

It wasn't only silicon that was important. Also required were large gobs of money, and of a particular type, venture capital. Investing in hi-tech is perilous. There are no guarantees of any return, or even that the original investment will be repaid. Investors must be of a certain type, willing to accept high risks and to have deep pockets. They are called venture capitalists, and began to come to Silicon Valley in 1959. It wasn't only venture capitalists who gave money, though. So did the US military, making Silicon Valley early on part of what President Eisenhower called, in 1961, "the military-industrial complex." The semiconductor industry was integrated into weapons research, design, and manufacture, receiving unstinting flows of federal funds.

The internet, invented in the late 1960s (see above), further consolidated Silicon Valley's position, becoming central to the operations of some of the wealthiest high-tech companies now operating in the region such as Dropbox, Google, Facebook, and Uber. These companies and others like them employ around a quarter-million information technology workers, the greatest concentration anywhere in the globe. Moreover, those workers themselves are global. Around 70% are foreign-born, making them, according to AnnaLee Saxenian (2006), the "new Argonauts" of an international pool of highly trained tech workers.

external to the firm, but internal to the district or region in which the firm is located. As more firms locate within a region or district, cost advantages from external economies increasingly accrue to those firms located within them. External economies include: the advantages of tapping into a nearby pool of trained labor; the ease of communication and sharing of resources among firms; lower inter-firm transportation costs; and a fine-grained specialization among firms that maximizes the division of labor and associated productivity.

Within contemporary economic geography we can recognize three different theories of hi-tech location.

i. *Agglomeration*: The first is a modern version of Marshall's thesis. It is most closely associated with the work of Allen Scott, and to some extent Michael Storper. They are concerned to emphasize the forces of agglomeration in creating the great powerhouse metropolitan economic regions of the world, including hi-tech districts and regions. In a joint paper, Scott and Storper (2015, p.6) write, "agglomeration is the basic glue that holds the city together." Agglomeration allows sharing of economic services among firms, including infrastructure; matching the right jobs within a firm to the right, best-qualified workers who are in the city; and learning from the enormous urban flows of information, which in turn stimulates innovation. These agglomerative forces cause hi-tech and creative economy sectors to cluster in districts or regions, making economic production necessarily territorially based. Specifically, Scott (2008, p.554) recognizes contemporary clustering in two distinct urban spaces: "more service and design-oriented sectors [tend toward] … specialized quarters lying toward the city center, whereas more technology intensive sectors (e.g. electronics, biotechnology, aerospace and so on) … occur in agglomerations or technopoles located in more suburban areas."

ii. *Cultural institutions*: A different approach to geographical clustering stresses the role of culture and institutions (Asheim 2000, p.413). This is the upshot of a now classic study, *Regional Advantage: Culture and Competition in Silicon Valley and Route 128*, carried out by AnnaLee Saxenian (1996) that compared the US hi-tech regions of Silicon Valley and Route 128 (see also Chapters 5 and 10). Her research on Silicon Valley began during the late 1970s as a Berkeley graduate student. Initially she didn't think Silicon Valley would last, expecting it to return to Santa Clara Valley dust. She was so wrong, of course. But because it lasted, and because Saxenian was there from early on, she could carry out repeated interviews with Silicon Valley insiders over a long time period, constructing a fascinating longitudinal record. It showed that central to the Valley's success was a strong social network connecting firm owners and key innovators, facilitated by place-specific institutions located in the Valley, both formal and informal. That larger network, in turn, was the basis for shared work ideas and practices, business customs, aims and aspirations, attitudes to risk, expectations about trust, and norms around innovation. The network anchored Silicon Valley firms, holding them in place, giving them solidity, but also providing a competitive impetus to reproduce and flourish.

To make her point about the benefit for hi-tech of a close cultural network, Saxenian compared Silicon Valley with Route 128 on the other side of the country around

Boston in New England. Both sites emerged in embryonic form during the 1950s, associated with spin-off companies that derived from university-based research in electrical engineering. In Silicon Valley's case, it was Stanford University (see Box 12.3: Silicon Valley), and in Route 128's case it was Harvard and especially the Massachusetts Institute of Technology (MIT) in Cambridge, Massachusetts. Silicon Valley firms moved increasingly south from Stanford along the Santa Clara Valley toward San Jose (Figure 12.5), while in Boston they ringed the belt-road, Massachusetts Route 128 (the "Magic Semicircle"). As Saxenian argued, though, Route 128 continually played second fiddle to Silicon Valley in terms of generated income, value of firms, and employment. What made the difference, she contended, was a different place-specific firm culture. Silicon Valley was defined by dense, flat social networks in which hi-tech firms were embedded, promoting a degree of informality, porousness, horizontal communication, and encouraging collective learning and practice. It may not quite have been "'laid-back' California," but Saxenian (1996, p.2) argued that the openness and degree of casualness in Silicon Valley promoted the transfer of ideas, experimentation, and entrepreneurial risk-taking. In contrast, the culture of Route 128 hi-tech firms was much more "'buttoned-up' East Coast" (p.2). Rather than a flat, mobile, and diverse social network, with intense interaction and workers smoothly moving from one job to another, Route 128 was dominated by large corporate entities that enforced a vertical (top-down) decision-making model, a strict corporate hierarchy of authority, and well-defined and policed boundaries. This difference in West Coast/East Coast culture, suggested Saxenian, made a difference to each region's ability to prosper economically from the hi-tech economy (interestingly, since Saxenian completed her study, Route 128 has become more like Silicon Valley, suggesting evolution).

Saxenian's work rested on the idea that propelling innovation is learning through interaction (Lundvall 1988). Silicon Valley did better than Route 128 because its culture and institutional network facilitated more and qualitatively richer forms of interaction, enabling learning for innovation to occur. Furthermore, Saxenian and others assume that one learns hi-tech knowledge only by being there. Knowledge is "sticky," only emerging in context-specific sites like Silicon Valley. If you are not there, you will miss out. This kind of knowledge is also termed tacit knowledge (Gertler 2003). Tacit knowledge is embodied and difficult to articulate unless it is shown in practice ("this is how you do it, like this"). Its opposite is codified knowledge, which can be formulated in words, or numbers, or even a diagram, and then easily disseminated ("read the manual!"). Because hi-tech innovation occurs through interactive learning, and because what is learned can be shown only by being there (tacit knowledge), hi-tech necessarily geographically clumps in districts and regions.

iii. *Creative Cities*: A third approach is Richard Florida's (2002a). The concentration of economic creativity and innovation within only a few geographical sites is for him due less to agglomeration, learning through interaction, or geographical stickiness of knowledge, than to the peculiar qualities of places that attract (or not) large numbers of footloose creative (hi-tech) people. Creative people, fully formed in their creativity, come first in Florida's scheme. They choose cities, and within cities particular intra-metropolitan spaces where they want to live and work. The places they choose are like magnets, each with a different attractive force (Figure 12.6). Work then

Figure 12.6 Richard Florida's theory of hi-tech talent and the preference for "cool" places. In this example for Texas, Austin wins hands down. Source: Trevor J. Barnes.

follows wherever the talented individual is drawn. Further, because creative individuals share in common with other creative individuals similar place-based aesthetic, cultural, social, and political preferences – they are all part of the same "creative class" – they will locate in similar kinds of places (Florida 2002b). In Florida's account, those places are politically tolerant, socially and culturally diverse, materially varied, cosmopolitan, "cool," vibrant, and "cutting-edge." Florida's work has been especially popular because it seems to offer the possibility of a check-list of changes that a place could and should make to attract talented people, and with them the creative economy that follows.

There is one more theoretical issue. Given that hi-tech activities cluster at globally scattered self-contained sites, how do those sites connect one with another to participate in larger processes of globalization? This is a problem because if hi-tech knowledge is primarily tacit knowledge, geographically sticky, then different hi-tech regions would struggle to interact because each hi-tech site is stuck in its own bubble. Bathelt, Malmberg, and Maskell (2004), using the metaphor of global pipelines, offer a solution (seemingly widely taken up, given that their paper is cited over 3,500 times). They begin by saying that the empirical evidence suggests that hi-tech knowledge is not as sticky as it might appear. Rather, it is in constant motion, moving between global hi-tech centers. Communication can occur, they suggest, because of a complex network of "global pipelines" that ensure even tacit knowledge becomes fully understood even half a world away. If knowledge is tacit, the kind that can only be shown, then the solution is simply to send people down the pipeline who show it. That's what flying business class enables. Moreover, tacit knowledge is not tacit knowledge for ever. It is always in process. Over time it becomes codified, in which form it then travels down other global pipelines as email, a text message, an Instagram, a fax, a courier delivery, or even by snail mail. Further, global pipelines are themselves continually in construction. They facilitate action at a distance over an ever-changing array of global places. Not all hi-tech centers have as many global pipelines as others, but for Bathelt et al., the more they have the better, providing yet more opportunities for interaction, learning, and innovation.

12.3.2   Putting hi-tech into action: innovation systems

As a form of economic activity, hi-tech has much to recommend it. Workers are generally well paid, with good benefits and working conditions (although hours can be excessive). Unlike, say, heavy industry, hi-tech is generally not detrimental to the environment (it provides green jobs and investment). The value of hi-tech companies can be enormous (represented, for example, by the unicorns), and they are frequently growth-oriented, large investors, with strong backward and forward linkages to other industries and services. Further, their forward-looking, optimistic, and sometimes radical and youthful culture can spur and invigorate existing business. For all these reasons, jurisdictions at all levels, from metropolitan to regional to national, are keen to encourage hi-tech through a range of policies, strategies, and initiatives.

The result has been concerted attempts by, variously, the nation-state, regional jurisdictions (states and provinces), and particularly local governments (metropolitan and city) to establish appropriate conditions, material and institutional, for fostering and promoting an innovation, high-tech economy. To do so means bringing together, coordinating within a network, and reproducing the varied institutions that create and market innovations. This larger assemblage is called an "innovation system." Typically, it comprises three main players, forming a "triple helix": universities, the research and development (R&D) wings of corporations, and government agencies. If the bond can be made strong among this triumvirate, with the innovation system cohering, then the nation, or the region, or the locality is transformed, taking on life in hi-tech.

But even if innovation systems are successful on their own terms (and the record is at best mixed), they necessarily heighten uneven development. In this sense, public policy is used to create spatial inequity. "Innovation systems … play a key part in producing and reproducing … [an] uneven geography over time," write Asheim and Gertler (2006, p.291). Technological innovation has done this from the industrial revolution in Britain onward (and likely even before that). Innovation redraws social and geographical relations (as illustrated in the first half of this chapter). Further, this occurs even within the region where innovation occurs. For example, there is uneven development within Silicon Valley. Incomes of high-tech workers have driven up house prices and rents, displacing lower income residents, driving them from the city. Even city bus stops were hijacked by "Google buses" picking up hi-tech employees from San Francisco, ferrying them daily to the Valley (Solnit 2014). Of course, an even more glaring uneven development is the comparison between Silicon Valley and places that lie outside that region, such as former industrial powerhouse cities that are now shadows of their former selves, like Detroit, Cleveland, and Buffalo. No unicorns there. Schumpeter was never concerned with the destructive part of creative destruction that attended major innovations. If he had been more of an economic geographer than an economist, we hope he would have been more sensitive to the *geography* of uneven development, less callous and blasé.

12.4   Conclusion

It is easy to become besotted by the success of technology, to fetishize it. To fetishize means seeing only the object itself and not the underlying associated social and, we would add, economic geographic processes and entailments. One falls in love with technology's products – the iPhone, Dropbox, the Airbnb reservation, the Tesla Model S. It is as if these technological products descended fully formed from the heavens.

In this chapter, we showed they did not. Technology never arrives fully formed, but always emerges hesitantly from many grounded processes, material and immaterial, but which always include the economic geographic. That was true from the very first stirrings of industrial technology in eighteenth-century Britain. There was both a micro-economic geography, seen in Figure 12.2, the basis of Adam Smith's idea of division of labor, and a larger macro-economic geography centered on the urban. The sites both where industrial technology was deployed and where new technological innovations were instigated were overwhelmingly cities. There was a special economic geography in those urban spaces that fueled and ignited technological change, which we tried to uncover.

There was also another issue. Clearly technology does some wonderful things. But it is double-edged, as Schumpeter made clear, both creative and destructive. Much of the work on hi-tech in economic geography seems more interested in the creative than the destructive. At times, it seems to fetishize, too concerned with the object itself. However, it is time to rebalance, to be concerned with the full set of processes at work, both creative *and* destructive; that is, to be critical.

Notes

1 If we allow for price inflation in the UK economy between 1867 when Marx was writing and 2017, 17.5p for a 90 hour week becomes equivalent to £19.08, or 21p an hour in 2017 values.
2 https://www.youtube.com/watch?v = FGoQoCm0EWI (accessed on July 6, 2017).
3 NUMMI was the acronym for New United Motor Manufacturing, Inc., the jointly owned automobile company created by GM and Toyota.
4 Hear them speak for themselves at https://www.thisamericanlife.org/radio-archives/episode/561/nummi-2015 (accessed July 6, 2017).
5 The quote is from the transcript of the updated This American Life radio episode about NUMMI 2015: https://www.thisamericanlife.org/radio-archives/episode/561/transcript (accessed July 6, 2017).
6 See note 5.

References

Asheim, B.T. 2000. Industrial Districts: The Contributions of Marshall and Beyond. In G. Clark, M. Feldman, and M. Gertler (eds), *The Oxford Handbook of Economic Geography*. Oxford: Oxford University Press, pp.413–431.

Asheim, B.T., and Gertler, M. S. 2006. The Geography of Innovation: Regional Innovation Systems. In J. Fagerberg and D.C. Mowery (eds), *The Oxford Handbook of Innovation*. Oxford: Oxford University Press, pp.291–317.

Bathelt, H., Malmberg, A., and Maskell, P. 2004. Clusters and Knowledge: Local Buzz, Global Pipelines and the Process of Knowledge Creation. *Progress in Human Geography* 28: 31–56.

Benjamin, W. 1940. On the Concept of History, Thesis IX. https://www.marxists.org/reference/archive/benjamin/1940/history.htm (accessed June 29, 2017).

Florida, R. 2002a. The Economic Geography of Talent. *Annals of the Association of American Geographers* 92: 743–755.

Florida, R. 2002b. *The Rise of the Creative Class*. New York: Basic Books.

Gertler, M.S. 2003. Tacit Knowledge and the Economic Geography of Context, or the Undefinable Tacitness of Being (There). *Journal of Economic Geography* 3: 75–99.

Lundvall, B-Å. 1988. Innovation as an Interactive Process: From User–Producer Interaction to the National System of Innovation. In G. Dosi, C. Freeman, G. Silverberg, and L. Soete (eds), *Technical Change and Economic Theory*. London: Pinter, pp. 349–369.

Markusen, A., Hall, P., Campbell, S., and Deitrick, S. 1991. *The Rise of the Gunbelt: The Military Remapping of Industrial America*. Toronto: Oxford University Press.

Marx, K., and Engels, F. 1969 [1848]. Manifesto of the Communist Party. Translated into English by Samuel Moore in cooperation with Frederick Engels. https://www.marxists.org/archive/marx/works/download/pdf/Manifesto.pdf (accessed June 29, 2017).

Marx, K. 2010 [1867]. *Capital: A Critique of Political Economy, Volume 1*. Translated into English by Samuel Moore and Edward Aveling, edited by Frederick Engels. https://www.marxists.org/archive/marx/works/download/pdf/Capital Volume-I.pdf (accessed June 29, 2017).

Saxenian, A. 1996. *Regional Advantage: Culture and Competition in Silicon Valley and Route 128*. Cambridge, MA: Harvard University Press.

Saxenian, A. 2006. *The New Argonauts: Regional Advantage in a Global Economy*. Cambridge, MA: Harvard University Press.

Schoenberger, E. 2015. *Nature, Choice and Social Power*. New York: Routledge.

Schumpeter, J.A. 1942. *Capitalism, Socialism and Democracy*. London: Routledge, Kegan & Paul.

Scott, A.J. 2008. Resurgent Metropolis: Economy, Society and Urbanization in an Interconnected World. *International Journal of Urban and Regional Research* 32: 548–562.

Scott, A.J., and Storper, M. 2015. The Nature of Cities: The Scope and Limits of Urban Theory. *International Journal of Urban and Regional Research* 39: 1–15.

Smith, A. 1977 [1776]. *An Inquiry into the Nature and Causes of the Wealth of Nations*. Chicago: University of Chicago Press.

Solnit, R. 2014. Diary: Get Off the Bus. *London Review of Books* 36: 35–36.

Storper, M., and Walker, R. 1989. *The Capitalist Imperative: Territory, Technology and Growth*. Blackwell: Oxford.

Williams, R. 1976. *Keywords: A Vocabulary of Culture and Society*. Oxford: Oxford University Press.

Chapter 13

Conclusion

13.1   The Scope of Economic Geography

T.S. Eliot (1943) famously wrote in his poem *Burnt Norton*, one of *The Four Quartets*, "human kind/Cannot bear very much reality".[1] Not so for economic geography, however. One of the reasons we are such aficionados of the discipline is that we appreciate its ability to sop up, sponge-like, large but different discharges of reality: the environment, the economy, politics, society, culture, history, and of course geography. Nothing seems to faze it. Economic geography rolls up its sleeves and gets down to work, soaking up and taking in whatever is there. Unlike "human kind," economic geography *can* bear very much reality. We demonstrated that throughout this book. It is one of the reasons we want you to be disciplinary aficionados too.

That capaciousness was brought home to one of us, Barnes when he last taught his third-year undergraduate course in economic geography at the University of British Columbia (UBC) (fall, 2016). What a period. During that 13-week term, reality charged pell-mell at the class. It was unrelenting, a deluge. But whatever came at it, economic geography was ready to cope. There was always an economic geographic angle, an interpretation, a reading, to help make sense of it, even when sometimes reality itself appeared senseless if not hopeless.

Barnes began each of his twice-weekly, one-hour-and-twenty-minute lectures with stories from that day's financial press, principally from Canada's national newspapers, either the *Globe & Mail's Report on Business* or the *National Post's Financial Post*. The economist John Maynard Keynes once said that all he needed to practice economics was to have at

Economic Geography: A Critical Introduction, First Edition. Trevor J. Barnes and Brett Christophers.
© 2018 John Wiley & Sons Ltd. Published 2018 by John Wiley & Sons Ltd.

hand a copy of *The Times* and the economist Alfred Marshall's (2013 [1890]) economics textbook *Principles of Economics*. Barnes told his students that all they needed to do economic geography was to read the financial pages of the newspaper and a selection of papers by economic geographers (in Barnes's class, found in a course reading package available from UBC's bookstore – although in the future they will have this book!).

The reality that assailed each class began even before the course formally started, the day after Labor Day, the first week of September. By then there was already the shadow of the UK Brexit decision taken just two months before. That story was pregnant with economic geography. The European Union (EU), from which Britain now wanted to detach itself after 43 years, had been above all an economic geographic project. The original designers of what became the EU self-consciously attempted from the beginning to construct a new spatial economic configuration. It began in 1951 as the European Iron and Steel Community with six members, later became the European Economic Community (EEC: "The Common Market"), and was now the EU (with 28 member states). At each stage, because of new members and new initiatives, fresh economic geographic links were forged that never existed before. A brand new economic geography was continually in construction. Its logic required one to bring an economic geographic sensibility to its understanding. Likewise, to appreciate now the consequences of UK Prime Minister Theresa May's strategy of "hard exit" from the EU required the same economic geographic sensibility – except in reverse.[2] The questions now were about the effects on the United Kingdom of detaching and separating from Europe, however. What will be the consequences for the country of severing now longstanding European economic geographic links? Who will Britain trade with? What will happen to London as a financial center when Britain leaves Europe (see Chapter 9)? Will Europe's financial center now be Frankfurt or Paris or Amsterdam? Where will the 850,000 Poles go who migrated to Britain for work? How will patterns of uneven development in the United Kingdom be ameliorated without EU money to help "less developed regions" such as Cornwall or South Wales? The point: one cannot consider the effects of Brexit without thinking economic geographically. Economic geography was woven into the warp and weave of the agreement with Europe that the United Kingdom originally signed and practiced. Hence, economic geography must also enter the analysis of the consequences of withdrawal. Just as well Theresa May was trained as a geographer at Oxford University! We hope she did some economic geography, though.

Brexit was the story on only the first day of class. As the course continued there were other stories, albeit no less economic geographic. Continuing throughout the term were stories about the debilitating economic geographic effects of the depressed price of oil. It was disastrous for Canada's primary oil-producing region, the province of Alberta, which specialized in extracting high cost "dirty oil" from its tar sand. Because of low oil prices, Alberta's annual government deficit for 2016 was close to $11 billion, its unemployment rate at 8.5%. It was not good even for Saudi Arabia, which had partly engineered those low oil prices by increasing oil production. Like Alberta, Saudi accumulated massive public deficits. Strangely, though, it was not too bad for US oil shale frackers in places like North Dakota, Colorado, and Texas. They not only survived but thrived with help from US financiers who bet on their economic geographic success. Then there were some strange economic geographic goings-on in Silicon Valley (Chapter 12). Apple launched iPhone 7 in September 2016, which tech pundits said no one would buy, but which sold so quickly that its manufacturer, Foxconn in Shenzhen, China (Chapters 1 and 10), a third of the way around the world, could not keep up with deliveries. Another Silicon Valley firm, Uber – on paper worth more than General Motors – through its subsidiary Otto, ran an experiment involving hauling a full trailer of Budweiser beer from Fort Collins to Colorado Springs (about 200 km) using a driverless truck. The economic geographic consequences of roads full of driverless trucks, and the loss of employment for about 3.5 million truck drivers who now currently operate in the United States, are stark and scary. Also during the fall there were stories about the beginning of a retail war that could reshape the economic geography of shopping. Amazon extended to 18 cities its online same-day-delivery grocery service to gain a share of the almost $1 trillion US grocery business currently dominated by Walmart. Walmart responded by upping the ante on its own online business. The potential end result might be the disappearance of brick and mortar supermarkets as we've known them.[3] Finally, there was another story local to Vancouver, where Barnes teaches, about an ongoing economic geographic war. It concerned Canadian oil pipelines. To sell the increased quantity of oil extracted from Alberta's tar sands, more pipelines were needed, especially to the Canadian West Coast with its access to Asian markets and higher oil prices. But few on the West Coast wanted pipelines in their backyards, or the increased marine tanker traffic that would ensue. Many First Nations peoples whose territories the pipelines would cross did not want them either. Environmentalists predict an Exxon Valdez-sized oil spill on the British Columbian coast when 400 oil tankers a year come into Vancouver's harbor to fill up. And Pacific whales that migrate and feed along Canada's West Coast do not want extra tanker traffic either, given their (mortal) sensitivity to booming engines and propeller noise. There is an economic geography even of whales.

The larger point is that, if you know what you are looking for and have the right kinds of critical conceptual tools at hand, it doesn't take much to tease out and to analyze an economic geographic story. Often the economic geography is not even hidden but visible in plain sight, as it was in all the cases discussed and taken over only a 13-week period. In our book, we wanted both to show you what to look for, and to provide you with the right conceptual tools. With that, economic geography and its importance will stare you in the face.

There was also one more story that proves the case. It was the biggest story not only of those 13 weeks but of the year, maybe of several years. On November 8, 2016, Donald Trump was elected to the US Presidency, beating his Democratic Party challenger, Hilary Clinton. So much of that event was ripe for an economic geographic analysis. Trump's electoral platform rested in large part, like the Brexit decision, on undoing economic geographic relationships that the United States through various trade deals had previously actively sought or at least acceded to. The 1994 North American Free Trade Agreement (NAFTA) made with Canada and Mexico was the most prominent. Trump said that deal would be torn up because it produced a bad economic geography. In his view, many American manufacturing jobs had shifted to Mexico as a consequence. Trump, instead, promised to reverse the economic geography of globalization by bringing back to America jobs that had been outsourced, certainly to Mexico as a result of NAFTA, but also to Asia, and to China in particular. He pledged a 15% import tax on goods arriving from China to discourage outsourcing. Trump was promising a fundamentally different economic geography. It appealed to those who seemingly were most adversely affected by the decline of the old economic geography, and by the rise of the present one marked by global trade (as was also the case for voters for Brexit in the United Kingdom; Lanchester 2016; Sharma 2016; Chapters 1 and 8).

Hilary Clinton's America

Donald Trump's America

Figure 13.1 Maps of majority Clinton and Trump counties for the US lower 48 states for the November 2016 US Presidential election. Maps drawn by Dillon Mahmoudi and David O'Sullivan.

That's brought out in Figure 13.1, which maps both Trump-majority and Clinton-majority counties for the November 2016 presidential election. Clinton's voter map comprised a set of "islands" of variable size, large on the East and West coasts as well as in the South West, but almost disappearing in the West and Center (Mountain and Plains States), and even more importantly, only small and very scattered in the Mid-West and the western parts of East Coast states like New York and Pennsylvania. Trump's map, in contrast, was almost a solid land mass, with just a few "lake-sized holes" (the mirror of the islands in the Clinton map). Where Trump won (and Clinton lost) was in the economic geography associated with the Mid-West and the western sections of those East Coast states. The "lake-sized holes" in those regions representing support for Clinton were sufficiently small to allow Trump to win. Or, to put it the other way, Clinton's islands were not numerous or big enough to prevent her from losing. More specifically, the economic geography that supported Trump's victory was in those regions of industrial and urban decline found in the former industrial heartland of the United States, the manufacturing belt (Chapter 1).[4] Too much of that region became part of Trump's "land mass," and not enough of it formed those "lake-sized holes" that would have enabled Clinton to win. It was the economic geography of falling population in Detroit, Cleveland, and Buffalo. It was uneven development. It was Neil Smith's (1984) seesaw of capital, and David Harvey's (1982) crisis of accumulation (Chapters 3 and Chapter 8).

For sure, there were factors other than economic geography at work in causing both the Trump victory and the Brexit decision and the socioeconomic schisms and political alienations that appear to have seeded them. As the *Financial Times*' economics commentator Martin Sandbu (2016a) observed in relation specifically to angry US white working-class voters, "not only trade, but technology, distributive norms, and welfare policy together make a quadruple whammy that has badly knocked the economic security of the US white working class. That in turn has led to a crumbling of the physical health, mental well-being and social status of individuals and communities." Nor can Trump's and (some) Brexiteers' racism and other overt prejudices be ignored. But economic geographic causes had a substantial impact. The changes that have gone on over the last 30 years or so have deeply affected working-class industrial workers in the Global North, and germane here specifically are those in the US old manufacturing belt and the UK's former northern and central industrial areas (Chapters 1 and 8). In these changes globalization has been to the fore, provoking significant distress and anxiety among certain social classes and economic geographic regions. Compared to Republican candidates in recent US presidential elections, Trump did best in places that *have* lost out to Mexico and, especially, to China (Cerrato, Ruggieri, and Ferrara 2016). So as Sandbu (2016a) says: "Trade may only be one part of the cause, but it is the focal point of much of the understandable anger." The Trump victory and Brexit were not inevitable, though. Conjunctural circumstances had to be aligned. And they clearly were during the summer and autumn of 2016. We believe a key element of those conjunctural circumstances was economic geography. It might be uncomfortable to discuss, too much reality to bear, but economic geography was undoubtedly a significant presence. It made discussing topics in Barnes's class that on the face of it were not directly about economic geography, like Brexit and the US election, both germane and revealing. Economic geography has relevance. It is not only academic. That's also why we admire the discipline and why we hope you will as well.

13.2   The Hope of Economic Geography

There is another reason too, and going to one more purpose of the book, that runs throughout: the potential of economic geography to provide a critical perspective. We tried to address in this book not only its main title, *Economic Geography*, but also its subtitle, *A Critical Introduction*.

Our interest in being critical is primarily worked out in relation to the internal operations of economic geography as a discipline. In that context, we recognized two forms of being critical (Chapter 1). First, it means you should always realize that there is nothing natural or pre-given about the discipline. It was never, "In the beginning was economic geography." As we variously argued in the first section of the book, economic geography was and is a constructed entity, the product of a set of contingent social, political, cultural, historical, and geographical forces. The knowledge it produces is local knowledge, rather than the distillation of some pure placeless or timeless rationality. Consequently, it is always knowledge from somewhere, and from someone who has the power to assert those knowledge claims. As we contended, it takes a lot of material and immaterial resources and social backing to make economic geographic knowledge. To be critical in our first sense means you should always be aware of the constructed nature of economic geographic knowledge, never taking it at face value. You should always raise questions about it by probing, picking at, and disputing it. That will not make you popular. It is easier to fall in, to accept uncritically what is said. When you don't, you become known as a contrarian, a gadfly, an awkward cuss. Why can't you just accept what you are told? We hope that by showing you the material messiness and social unevenness that lies behind the front stage of economic geography's outward performance you will never simply accept what you are told (even by us, especially by us). You will be an awkward cuss. You will be critical.

Second, being critical also means taking a position, making a stand, not being afraid to criticize a body of work from your perspective. As we stressed, this does not mean laying waste to different perspectives wholesale. Critique is not pure negativity. Following Wendy Brown (2005), who we quoted in Chapter 1, critique is a form of recouping an object by putting it within your perspective, and seeing it in a different light. In the case of economic geography, we were most concerned to recoup the economy as an object (see especially Chapter 2). That was the focus of our critique. *From our perspective*, the economy is not, as it is portrayed in standard economics, some closed, hermetically sealed pure object, autonomous, regulated by a set of internal, law-like strictures derived from an invariable human rationality. Rather, in our recouped version, it is open-ended, muddied by the non-economic – the social, the political, the cultural, the historical, and especially the geographical – and shot through with various kinds of inequalities and differences around power, resources, and status. Once we recouped the economy in this form, it became the basis for us making evaluations, for carrying out critique within the discipline. Given this recouped conception of the economy, how do various works carried out by economic geographers stack up? Are they compatible with it? Do they provide a novel, creative interpretation and advance? Or do they go nowhere? Or worse, following David Harvey (1972), are they "counter-revolutionary" (Chapter 5)? Do they take us backward, to the un-recouped notion of the economy, such as in "Geographical Economics" (Chapter 2) that typically does exactly that?

This last is a particularly timely question. With economic geography's importance increasingly widely apparent in the wake of Brexit and Trump, even the most mainstream of economists have suddenly "discovered" geography. At the *Financial Times*, for example, (the generally excellent) Sandbu (2016b) has called for an "economics of place," citing approvingly his colleague, economics editor Chris Giles's (2016) declaration that "geography matters." It is easy for economic geographers such as us to be frustrated by this woefully tardy recognition. It is easy to scream, "We told you so." It is easy to be maddened by the fact that for many newspaper and magazine readers, not to mention politicians and policymakers, the idea that "geography matters" doubtless will only be accepted now that "they" (economists, with their model-based Truths) are saying something that "we" (geographers) have been saying for decades. "Geography matters!" was after all the titular rallying cry of a brilliant 1984 volume assembled by the British economic geographers Doreen Massey and John Allen (Massey and Allen 1984). It is not news that geography matters to the economy. Further, it is also easy to fret that in now "discovering" geography, economists will continue to ignore decades of relevant economic geographic work – as Paul Krugman and the other pioneering "Geographical Economists" did (Chapter 4), as Edward Glaeser did in his *Triumph of the City* (Chapter 10) – and will make out like the "discovery" really is theirs. But the right reaction is to be analytically, not viscerally, critical. It is to ask what *kind* of economy these accounts invoke – and, equally crucially, what kind of geography? It is to ask *why* until so recently "the economics profession and the political elites" showed, by Sandbu's (2016a) own admission, "blithe unconcern" not just about economic geography but more importantly about "whether trade was causing deindustrialisation, social exclusion and rising inequality." For the defense, Sandbu submits that "as the sun set on the 20th century, it was entirely natural to associate freer trade with greater fairness, prosperity and political harmony." But it wasn't "entirely natural" at all. The only thinking that pretends to be "entirely natural" is non-critical thinking. It is ideology. It is covert theory that masquerades as mere "common sense," and which theory as critique (Chapter 5) constantly interrogates.

Of course, all of this is *our* critique, from *our* perspective. It is not one we think you should necessarily share. The point is to be clear about your perspective (whatever it is), and then use it to make judgments, to assess, to gauge what you read in the name of economic geography. That is what we have tried to do in this book. We hope you do the same, and do it to this very book.

So, this is being critical within and around economic geography. But there are other ways to be critical that we certainly don't want to exclude. That we have not discussed them explicitly is only because they are not directly relevant to the purpose of our book. They remain very important, however.

An obvious way to be critical that we haven't discussed in depth is to use economic geography to expose and criticize in the world various injustices, unfairnesses, forms of oppression, misuses of power, and bigoted prejudices. Many of the works we reviewed, especially in the second section of the book (Chapters 8–12), have this as an end. Those works can be couched at all scales from exposing and criticizing the unfairness of global uneven development (Chapter 8), to revealing the deleterious effects of international cities like London in scooping up global talent and instituting a new urban imperialism (Chapter 10), to uncovering the oppression of specific individuals over the course of their work histories (Linda McDowell's project discussed in Chapter 4).

Note that being critical here means for the most part using words. Most obviously, that is so in academic publications like books and journal articles, but also in speaking in lectures and seminars to students, or writing blogs, or contributing to web-based forums, or forging alliances by speaking with social movements, unions, and community organizations, or even going out on marches and protests singing "chants for changing the world".[5] T.S. Eliot had another famous line in *The Four Quartets* (in *East Coker*): words "are a raid on the inarticulate." We believe words can also be a raid on injustice, oppression, power, and prejudice. That might not always seem so obvious. The sword seems mightier than the pen. What's important is action, not "mere" words. Even words by David Harvey, or Doreen Massey, or Linda McDowell may not be enough. We don't believe so. Words have a per-formative force. They in fact get things done. Indeed, they may sometimes be the only means by which to get things done. The late Christopher Hitchens (2000, p.xiv), possibly everyone's gadfly, an extremely awkward cuss, wrote, "there are things that pens can do, and swords cannot. And every tank, as Brecht said, has a crucial flaw. Its driver. Suppose that driver had read something good lately, or has a decent song or poem in his head …." It is in this sense that Hitchens believed, following Shelley, that "poets are the unacknowledged legislators of the world" (quoted in Hitchens 2000, p.xiii).

We don't go that far. We don't claim that economic geographers are the unacknowledged legislators of the world. But economic geographers have had their moments, and through their words they have produced material effects both good and bad. Their texts are never innocent. They don't just describe the world, but help re-make it. Consequently, as people who live in that world, we have a right and responsibility to be critical, to applaud and support when we think it is justified, or to challenge and criticize. Moreover, the texts, the words, of economic geographers that make a difference need not always be the big and obvious ones. One of our favorite quotes is from the American literary critic Frank Lentricchia (1983, p.10), who wrote, "struggles for hegemony are sometimes fought out in (certainly relayed through) colleges and universities; fought undramatically, yard for yard, and sometimes over minor texts of Balzac: no epic heroes, no epic acts." Similarly, the work of being critical in economic geography need not be epic for it to be important.

There is one more point about being critical. A critical economic geography should not only be in the business of exposing and explaining the inflictions of the present, but also of trying to improve the world for the future. It should pursue an anticipatory utopian impulse. To use the economic geographer August Lösch's phrase (1954 [1940], p.4), the task is "not to explain our sorry reality, but to improve it." For Lösch this meant mobilizing a series of complex equations and geometries (spatial science, Chapters 3 and 4). While this is unlikely the preferred mode among contemporary critical economic geographers, it might be one mode. The important point (at least for us) is that critique should be directed from some sense of what a better world should look like. This requires that critical theory contain an imaginative capacity to reconfigure the world and our place within it, that it foreshadow a different, better kind of economic geography. It is true, as David Harvey (2000, chapter 8) argues, that most utopian projects when realized on the ground turn into their antonym, dystopia. The twentieth-century French-Swiss architect Le Corbusier's "machines for living" became the Liverpool "piggeries"; the bucolic suburban ideal of the nineteenth-century US landscape architect Frederick Law Olmstead became gated communities; and the twentieth-century American urban planner cum activist Jane Jacobs' "community of eyes" became inner city neighborhoods of video-camera surveillance and steel-barred windows and doors. But for Harvey, this is not the result of utopian thought per se, but a consequence of the market capitalism in which it is materially embedded. Harvey (2000, p.195) writes:

> There is a time and place in the ceaseless human endeavor to change the world, where alternative visions, no matter how fantastic, provide the grist for shaping powerful political forces for change. I believe we are precisely at such a moment. Utopian dreams … never entirely fade away. They are omnipresent in the hidden signifiers of our desires.

In sum, there is only the common difficult and halting task of offering social critique, of making use of different vocabularies to see if they can produce a better world, a better economic geography. It is what Richard Rorty (1999) called "social hope," and what we call the hope of economic geography. This insistent and pressing task confronts both us and you.

Notes

1 An online version is available at http://www.coldbacon.com/poems/fq.html (accessed July 6, 2017).

2 Since we originally wrote that sentence in December 2016, the results of the British election of June 8, 2017, have cast doubt on whether Prime Minister May's hard-exit strategy remains politically feasible, although it still appears almost certain that the United Kingdom will leave Europe. It is another reason to continue closely reading the financial press and practicing economic geography.

3 Again, since we originally wrote that, in June 2017 Amazon bought Whole Foods, an upscale US grocery chain, for almost $14 billion, which may imply Amazon is going back to bricks and mortar retailing. However, there has also been speculation that Amazon took over Whole Foods primarily to obtain information about its customers which it will use to lure more people to become online Amazon Prime shoppers.

4 Tom Hazeldine (2017) has constructed a similar map but for Brexit leavers in the United Kingdom following the June 8, 2016, referendum. His argument, like ours, is that Brexit was a consequence in large part of the economic geography of deindustrialization that affected especially the old UK manufacturing regions of the North-East and North-West. Hazeldine (2017, p. 53) estimates "six million Leave votes" came from "England's historic industrial regions," and it was they who made the difference.

5 https://everychant.wordpress.com/the-chants/(accessed July 6, 2017).

References

Brown, W. 2005. *Edgework: Critical Essays on Knowledge and Politics.* Princeton, NJ: Princeton University Press.

Cerrato, A., Ruggieri, F., and Ferrara, F. 2016. Trump Won in Counties that Lost Jobs to China and Mexico. *The Washington Post*, December 2.

Giles, C. 2016. The Poor Suffer While Britain Avoids Straight-Talking. *Financial Times*, December 14.

Harvey, D. 1972. Revolutionary and Counter-Revolutionary Theory in Geography and the Problem of Ghetto Formation. *Antipode* 4: 1–13.

Harvey, D, 1982. *Limits to Capital.* Chicago: University of Chicago Press.

Harvey, D. 2000. *Spaces of Hope.* Berkeley: University of California Press.

Hazeldine, T. 2017. Revolt of the Rust Belt. *New Left Review* 105: 51–79.

Hitchens, C. 2000. *Unacknowledged Legislation: Writers in the Public Sphere.* London: Verso.

Lanchester, J. 2016. Brexit Blues. *The London Review of Books* 38: 3–6.

Lentricchia, F. 1983. *Criticism and Social Change.* Chicago: University of Chicago Press.

Lösch, A. 1954. *The Economics of Location*, 2nd edn, trans. W.H. Woglom with the assistance of W.F. Stolper. Originally published in German in 1940. New Haven, CT: Yale University Press.

Massey, D., and Allen, J., eds. 1984. *Geography Matters!: A Reader.* Cambridge: Cambridge University Press.

Marshall, A. 2013 [1890]. *Principles of Economics.* London: Palgrave Macmillan.

Rorty, R. 1999. *Philosophy and Social Hope*. London: Penguin.

Sandbu, M. 2016a. The Shock of Free Trade. *Prospect*, June 16.

Sandbu, M. 2016b. Place and Prosperity. *Financial Times*, December 16

Sharma, R. 2016. Globalisation as We Know It Is Over – and Brexit Is the Biggest Sign Yet. *The Guardian*, 28 July. https://www.theguardian.com/commentisfree/2016/jul/28/era-globalisation-brexit-eu-britain-economic-frustration (accessed June 29, 2017).

Smith, N. 1984. *Uneven Development: Nature, Capital, and the Production of Space*. Oxford: Blackwell.

Index

Page locators in *italics* indicate figures/tables/boxes.